Information Modeling for Interoperable Dimensional Metrology

Yaoyao (Fiona) Zhao · Robert J. Brown
Thomas R. Kramer · Xun Xu

Information Modeling for Interoperable Dimensional Metrology

Springer

Yaoyao (Fiona) Zhao
National Institute of Standards and Technology
100 Bureau Drive
Gaithersburg
MD 20899
USA
e-mail: fiona.zhao@nist.gov

Thomas R. Kramer
National Institute of Standards and Technology
100 Bureau Drive
Gaithersburg
MD 20899
USA
e-mail: thomas.kramer@nist.gov

Robert J. Brown
Mitutoyo America Corporation
965 Corporate Blvd
Aurora
IL 60502
USA
e-mail: Robert.Brown@mitutoyo.com

Xun Xu
Department of Mechanical Engineering
University of Auckland
PO Box 92019
Auckland
New Zealand
e-mail: xun.xu@auckland.ac.nz

ISBN 978-1-4471-2166-4 e-ISBN 978-1-4471-2167-1
DOI 10.1007/978-1-4471-2167-1
Springer London Dordrecht Heidelberg New York

British Library Cataloguing in Publication Data
A catalogue record for this book is available from the British Library

Cover design: eStudio Calamar S.L.

Printed on acid-free paper

Springer is part of Springer Science+Business Media (www.springer.com)

This book is dedicated to my parents. For their unconditional love, endless support and encouragement

Yaoyao (Fiona) Zhao

I would like to thank my colleagues and teachers for their guidance and wisdom and to thank my family for their support. Eternal gratitude to my wife, Jean and to my kids, Zachary and Anika for providing Dad with the time to do this important work

Robert J. Brown

Foreword

After its people the single most valuable asset of any organization is information, and the effective utilization of that information is critical to quality, innovation, competitiveness, and even corporate survival. Although manufacturers currently enjoy a panoply of innovative products from a variety of vendors, these benefits are accompanied by information exchange incompatibilities, which come with costs such as missed opportunities, product quality shortcomings, data translation costs, data quality problems, and unnecessary software license and training fees. Data exchange problems hamper effective information utilization. The technical term for the effective utilization of information is *interoperability*.[1] In summary, manufacturers would like to achieve interoperability, while avoiding unnecessary costs and while still exploiting the broad array of product options. But is this possible, and if so, how?

We will begin with how manufacturers currently pursue interoperability: the translation approach, the single vendor mandate approach, and the information exchange standards[2] mandate approach.

In the translation approach, a manufacturer chooses to use systems and components from multiple vendors to support the enterprise. To achieve interoperability, data translators (from one proprietary format to another) must be built and maintained.

In the single vendor mandate approach, a manufacturer mandates a single vendor's product line throughout the enterprise. As long as the mandate is fully implemented (not always possible), interoperability is achieved at the manufacturer's plants. However, since the manufacturer's suppliers also provide products and services to other manufacturers, who commonly mandate products from

---

[1] *Interoperability* is the successful performance of required tasks by two or more agents requiring the exchange of information.

[2] An information exchange standard is a common (non-proprietary) language constraining the information transferred between activities performed by devices, software, or humans, whose goal is to enable effective encoding and decoding of information to successfully perform required tasks.

competing vendors, the suppliers will suffer a variety of interoperability costs, such as data translation, the payment of unnecessary fees, training fees, and fixing data quality errors. These costs are passed on to the manufacturer in the form of increased fees for products and services from the supplier. The manufacturer also experiences loss of agility and higher product and component costs due to its commitment to the single vendor. Furthermore, when the manufacturer experiences a merger or acquisition (a common experience), massive and sudden retraining and translation are commonly required to achieve interoperability between the incompatible software systems of the newly merged organizations.

In the information exchange standards mandate approach, a manufacturer chooses to use systems and components from multiple vendors to support the enterprise, but only if vendor products demonstrably exchange information in specified standard (non-proprietary) formats. To achieve interoperability, standards for the information used by the manufacturer must exist and be implemented by the manufacturer, its suppliers, and its system vendors.

Manufacturers achieve interoperability through some blend of these three paths. The information exchange standards mandate option offers both freedom of choice and no translation requirement. It appears that the standards approach might be the path to interoperability that is the most cost-effective while still maintaining freedom of product choice. For this path to be available, however, manufacturers, suppliers, and system vendors must support standards development efforts.

Close examination of the costs suffered under a proprietary-based data exchange environment reveals that having a single standard for each data interface would eliminate a multitude of costs and risks unique to the other two options.[3] Furthermore, it is plainly cheaper to develop and maintain one standard versus developing and maintaining a large number of proprietary formats for the same underlying information,[4] as long as all product vendors worldwide adopt and comply with the standard. The latter is achieved when a critical mass of manufacturers mandates the standard, and the standard continues to meet the changing needs of all these manufacturers.

Having a language standard adopted by vendors worldwide does not ensure that the language will be encoded and decoded correctly, so the standards-based path must (and commonly does) include conformance and certification definitions and requirements.

Therefore, we conclude that the standards-based path to interoperability is the optimal one from a standpoint of information quality, cost, risk, and a host of other reasons. A common objection at this point is: Do not information exchange standards constrain product innovation? Constraining the language used between tools supporting dimensional metrology activities does not constrain innovation in

---

[3] Horst J, Hartman N, Wong G (2010) Metrics for the cost of proprietary information exchange languages in intelligent systems, PerMIS'10, September 28–30, Baltimore, MD, USA.

[4] Horst J (2009) Reduce costs and increase quality with information exchange standards for manufacturing quality. CMM Quarterly, Sept. 4, 2009, www.cmmquarterly.com, Special DMSC Edition.

those activities or innovation in the design of those tools, as long as the interface standard is expeditiously maintained and well-supported by a broad range of manufacturers. However, admittedly, this is a consummation devoutly to be wished, but alas, rarely evidenced in reality.

This line of reasoning establishes that information critical to accomplish dimensional metrology activities can and should be defined by the industry in standard formats (i.e., languages), as long as the standards keep up with information definition needs required by the steady stream of new product innovations requiring them. To the degree that these requirements are satisfied, the entire industry can provide products that are less expensive, more innovative, feature-rich, and of higher quality. Arguably, this will grow the market for all players (user, supplier, and vendor), benefitting all.

So, why are not information exchange standards more widely developed and mandated? The answer to this would require another essay, however, the two top reasons are (1) the dearth of support for information exchange standards by manufacturers, and (2) the natural resistance of solution providers, particularly the large ones, to information exchange standards, due in part to concern about loss of market share.

Happily, there have been significant information exchange standards successes in the dimensional metrology sector. Dimensional metrology information exchange standards have been developed and implemented since the early 1980s. These standards are described in some detail in this book, but here is a brief summary of key successes and new standards ventures, starting with the first standard dimensional metrology programming language, the Dimensional Measuring Interface Standard (DMIS). DMIS has had broad market penetration and, for those manufacturers who have mandated DMIS enterprise-wide, savings have been substantial.

Starting in the early 2000s, the Inspection Plus-Plus (I++) Group of European automotive manufacturers generated, and currently maintain, a widely adopted open specification called the I++ DME (Dimensional Measurement Equipment) Interface, which defines commands from a dimensional metrology program execution module to a coordinate measuring machine controller. As with DMIS, manufacturers who have mandated I++ DME enterprise-wide have enjoyed substantial savings.

Building on these successes and lessons learned, the Dimensional Metrology Standards Consortium (DMSC), which is currently the official DMIS development organization, has recently introduced a holistic approach to dimensional metrology information exchange standards development, called the Quality Information Framework (QIF).

The importance of improving and maintaining the quality of manufactured goods can hardly be overemphasized. Ours is a litigious age with an instantaneously worldwide news cycle, within which a company can be ruined in a moment over a single product defect. Dimensional metrology is essential for ensuring product quality, and improving product quality, in a way that is cost-effective, timely, and error-free, is therefore of great value.

Dimensional metrology solutions providers, manufacturers, and suppliers are encouraged to join in the exciting and rewarding work of information exchange standards research and development.

John A. Horst
Program Manager
Engineering Laboratory
National Institute of Standards and Technology (NIST)
Gaithersburg, MD USA

Preface

Dimensional metrology is the science of measurement and its corresponding accuracy, precision and uncertainty. To measure is to ascertain a numerical value, in terms of some physical unit, of a quantity, quality, or dimension. To inspect is to determine compliance to a specification by measuring, gauging, or other means of examination. Often, measurements are performed to verify and inspections are performed to accept. In its most basic form, dimensional metrology can be thought of as the determination of lengths, angles, and other geometric relationships. In the world of manufacturing, dimensional measurement and dimensional inspection are synonymous with dimensional metrology. Dimensional metrology is an important subject because it is essential for making parts correctly. It is based on complex 3D geometric entities and their relationships. These geometries are associated with a large, diverse knowledge base that has many interconnections with entities such as the measurement process, the language of measurement, devices, standards, traceability, and statistics.

However, there is more to the dimensional measurement process than just analyzing the dimensions and tolerances of manufactured components. The product design specifications must be taken into account in planning the measurement process; the measurement process must be executed to obtain appropriate measurement data; the data must be analyzed and the results reported in a way that accepts or rejects the component and provides feedback to the manufacturing process and the production management process. These processes are supported by many software applications, including those that are incorporated into machine tools. The entire dimensional measuring system is most effective if the software applications are seamlessly integrated together at the information interfaces. In manufacturing industry, dimensional metrology data is very important because it is intimately tied to a company's product quality and performance assessment efforts. Metrology data has to be shared easily with production scheduling, design, purchasing, and many other manufacturing company functions. Ideally, a manufacturer should be able to acquire and store any type of dimensional measurement information in the same format regardless of the type of equipment used to acquire it.

The concept of dimensional metrology interoperability can be defined as the ability of two system components to communicate correctly and completely with each other—with minimal effort to either the component user or the component vendor. Component-to-component interoperability using open standards reduces training costs, allows best-in-class component choices, and provides a more competitive technology provider environment—thus providing the promise of reduced cost for Original Equipment Manufacturers, technology providers, suppliers, and consumers. The main challenge to achieve dimensional metrology interoperability is to specify a minimum set of standards to provide coverage for the information exchanges required that will also enable integration for the full range of software applications presently available and likely to be available in the future.

The hardware and software of dimensional metrology systems have had significant development in the past few years. However, the translation of data between the different components of dimensional metrology systems remains a major non-value-added cost for manufacturing industries. Past quality standards and specifications have been developed in isolation, each targeting a single dimension of a dimensional metrology system. At present, no national or international standard exists to provide for the interoperable exchange of data between the various data producers and consumers within dimensional metrology systems in industry. This book focuses on investigating and illustrating the hardware and software elements of dimensional metrology systems, information processed in and generated from each of the elements, existing data models, and the interoperability situation in dimensional metrology systems.

The history of dimensional metrology and the nature of workpiece surfaces are introduced first in the book. As an integrated element of the complete manufacturing system, the importance and functions of dimensional metrology systems are exemplified. Information modeling theory and languages are then introduced to give readers basic knowledge for the appreciation of interoperability issues. In the second part of the book, the four main elements of a dimensional metrology system are described in detail, namely product definition, dimensional metrology planning, dimensional metrology plan execution, and quality data analysis and reporting. The activities in each of these four elements and their functions are introduced first; then the information modeling techniques and existing data models are analyzed and illustrated. The aim of achieving interoperable dimensional metrology is not only to save the data translation cost but also to provide sufficient and timely data for advanced quality control. Industries of different scales need different types of quality data for quality control. Therefore, quality control and information modeling for small to medium industry and global industry is also discussed.

This book has nine chapters. Chapter 1 (Introduction) provides basic knowledge of dimensional metrology—its history and its relationship with manufacturing processes.

Chapter 2 (Practices of Information Modeling) introduces the basic knowledge of information modeling. The most commonly used contemporary information modeling languages are discussed such as UML language, IDEF1X language, EXPRESS language, and XML language.

Chapter 3 (Product Definition and Dimensional Metrology Systems) first discusses the activities that generate product definition information. The most commonly employed produce design approaches are discussed in this section followed with a brief introduction of features and tolerances in product design. Then, an inclusive discussion of existing data models and standards representing product design information is presented. It is followed with the introduction of product lifecycle management information such as product data management information and key characteristic management information. Existing data models representing product lifecycle management information are discussed at the end of this chapter.

Chapter 4 (High-level Dimensional Metrology Process Planning) presents the state-of-the-art computer-aided inspection process planning (CAIPP) research works. It reviews the CAIPP research from its beginning in the mid 1980s. The research trend of CAIPP research is divided into two parts: conceptual development and system module development. Existing information data models of CAIPP systems are then discussed in the second half of this chapter including STEP AP 219, AP 238 and QMP data models.

Chapter 5 (Low-level Dimensional Metrology Process Planning and Execution) first introduces the hardware—different types of dimensional measurement sensors—in dimensional metrology execution systems. This section covers the most common dimensional measurement execution systems such as CMMs, portable devices, and on-machine measurement systems. Then, software systems of these measurement systems are discussed. Current data models for dimensional metrology plan execution are investigated and presented at the end of this chapter.

Chapter 6 (Quality Data Analysis and Reporting) introduces, at the beginning, the basic data fitting theories in modern computational metrology. The mathematical representations of geometric elements and geometry data fitting criteria are discussed in detail in this section. It is followed by the discussion of information modeling of measurement data analysis and reporting. Both proprietary data models and standard data models are introduced. At the end of this chapter, the commercial application of quality data analysis and reporting are presented including business intelligence, quality and production engineering such as First Article Inspection, PPAP, SPC, etc.

Chapter 7 (Dimensional Metrology Interoperability Issues) tackles the interoperability issues in modern dimensional metrology systems. The information exchange between product definition, process planning, execution, and data analysis and reporting are discussed in detail. Then, this chapter lays a road map for achieving interoperable dimensional metrology.

Chapter 8 (Dimensional Metrology for Manufacturing Quality Control) presents the current quality control technologies, such as six sigma, based on dimensional measurement data. Information modeling requirements for different manufacturing industries are discussed such as for small to medium industry and global manufacturing industries. The data models for different types of quality control are presented at the end of this chapter.

Chapter 9 (Outlook for the Future of Dimensional Metrology Systems Interoperability) first discusses the technology adoption lifecycle, which enlightens different paths to achieve interoperability in dimensional metrology systems. Then, research trends and emerging standard effort for dimensional metrology systems integration are presented.

This book has three groups of people as its potential audience, (1) senior undergraduate students and postgraduate students conducting research in the areas of dimensional metrology design, process planning, execution, data analysis and reporting, and their integration; (2) researchers at universities and other institutions working in these fields; and (3) practitioners in the R&D departments of an organization working in these fields. This book differs from other books that also have dimensional metrology as the focus in two aspects. First of all, integration is an essential theme of the book. Secondly, information modeling and interoperability are the focuses of this book.

The book can be used as an advanced reference for a course taught at the postgraduate level. It can also be used as a source of information about modern dimensional metrology technologies and contemporary applications. The basic theories and knowledge of dimensional metrology and information modeling are introduced at the beginning of the book. This is followed by the detailed explanations of each element of a typical modern dimensional metrology system.

Yaoyao (Fiona) Zhao
Robert J. Brown
Thomas R. Kramer
Xun Xu

Contents

Chapter 1
Introduction

Dimensional metrology is the measurement of the deviations of a workpiece from its intended size and shape. The aim of dimensional metrology is to ensure that the size and shape of the workpiece conform to the designer's wish. This in turn ensures that the workpiece will assemble into associated assembly workpieces and the static characteristics of the workpiece have therefore been satisfied. In engineering, there are three types of metrology: dimensional, surface, and physical. This chapter first gives an inductive discussion of these three types of metrology and surface irregularities that can be measured by them.

The different kinds of surface irregularities have different origins in the manufacturing process. Dimensional metrology is related to both manufacturing process and workpiece function. Measuring the geometry of a workpiece can be important in controlling the manufacturing process and optimizing the function of the workpiece. Dimensional measurements in manufacturing can be categorized into three types: in-process measurements, in situ measurements, and remote measurements. Each type of measurements has its advantages and disadvantages. They are employed throughout manufacturing processes to control different types of parameters. The detailed introduction of these dimensional measurement systems and their functionalities are given in the remainder of this chapter.

1.1 Dimensional Metrology Versus Surface Metrology and Physical Metrology

Dimensional metrology is the measurement of the deviations of a workpiece from its intended size and shape, which are from the size and shape specified on the drawing [1]. It is taken to include such features as deviations from roundness, straightness, flatness, cylindricity and so on. Dimensional metrology is one of the three types of engineering metrology. The other two types of metrology are surface metrology, which is the measurement of surface texture, and physical and

Y. Zhao et al., *Information Modeling for Interoperable Dimensional Metrology*,
DOI: 10.1007/978-1-4471-2167-1_1,

chemical metrology, which is the measurement of physical and chemical condition of the workpiece.

The best way to place the role of engineering metrology is to consider what needs to be measured in order to enable a workpiece to work according to the designer's aim—one has to measure in order to be able to control. Assuming that the material has been specified correctly and that the workpiece has been made from it, the first thing to be done is to measure the dimensions. These will have been specified on the drawing to a tolerance. Under this heading is included the measurement of length, area, position, radius and so on. Therefore, dimensional metrology is a first aim because it ensures that the size of the workpiece conforms to the designer's wish. This in turn ensures that the workpiece will assemble into an engine, gearbox, gyroscope and so on; the static characteristics have therefore been satisfied. This by itself is not sufficient to ensure that the workpiece will satisfy its function; it may not be able to turn or move, for example. Surface metrology, as the second group of measurements, ensures that all aspects of the surface geometry are known and preferably controlled. If the shape and texture of the workpiece are correct and the design is sound, then it will be able to move at the speeds, loads, and temperatures specified in the design; the dynamic characteristics have therefore been satisfied [2].

The final group of measurements concerns the physical and chemical condition of the workpiece. This will be called physical metrology. It includes the hardness of the materials, both in the bulk and in the surface layers, and the residual stress of the surface, both in compression or in tension, left in the material by the machining process or the heat treatment. It also includes measurement of the metallurgical structure of the material, and its chemical construction. All these and more contribute to the durability of the component, for example its resistance to corrosion or fatigue. Physical metrology therefore is the third major sector of engineering metrology: the long-term characteristics.

As a general rule, all three types of measurement must take place in order to ensure that the workpiece will do its assigned job for the time specified; to guarantee its quality. It is impossible to divorce any of these disciplines completely from each other. After all, there is only one component and these measurements are all taken on it. This book focuses primarily on the first and the most important type of metrology—dimensional metrology. Dimensional metrology, in common industrial usage, designates the measurement of the deviations of a workpiece from the specified dimensions and geometric characteristics.

1.1.1 Workpiece Surface Properties and Imperfections

Before introducing the details of dimensional metrology, it is important to understand the workpiece surface properties and what types of imperfections may appear on workpiece surfaces. Surface properties are taken to mean the breakdown of the surface geometry into basic components based usually on some functional requirement. These components can have various shapes, scales of size, distribution in space and can be constrained by a multiplicity of boundaries in height and position. The suitability of a workpiece for a given purpose depends on its internal properties and its surface condition. The workpiece properties include material properties, internal discontinuities such as shrink holes, and internal imperfections such as segregations. The surface condition comprises the properties of the surface border zone. These are chemical, mechanical and geometrical properties. The chemical and mechanical properties comprise chemical composition, grain, hardness, strength and inhomogeneities. The properties of the surface border zone may be different from those in the core zone. The geometrical properties are defined as deviations from geometrical ideal elements of the workpiece. Geometrical ideal elements (features) are parts of the entire workpiece surface that have unique nominal geometrical forms (e.g. planes, cylinders, spheres, cones and tori). They can also be derived for example as axes, section lines, generator lines, lines of highest points and edges [3, 4].

There are three widely recognized causes of geometrical deviation [2]:

1. The irregularities known as roughness that often result from the manufacturing processes. Examples are (a) the tool mark left on the surface as a result of turning and (b) the impression left by grinding or polishing.
2. Irregularities, called waviness, of a longer wavelength caused by improper manufacture. An example of this might be the effects caused by a vibration between the workpiece and a grinding wheel.
3. Very long waves referred to as errors of form caused by errors in slideways, in rotating members of the machine, or in thermal distortion.

Often, the first two are lumped together under the general expression of surface texture, and some definitions incorporate all three. Some surfaces have one, two or all of these irregularities. The different kinds of surface irregularities have different origins in the manufacturing process. In order to control the manufacturing process, these irregularities are recommended to be assessed separately. These surface irregularities often have different effects on the suitability of the surface for its purpose. In order to specify the permissible function-related deviations, the different surface irregularities should be specified separately. Furthermore, the depths of the irregularities vary over large ranges. For assessment of different kinds of irregularities, different kinds of measuring instruments with different magnifications and different profile diagrams with different ratios of horizontal to vertical magnifications are used.

It is, therefore, very important to know the common surface imperfections which constitute surface irregularities. Figure 1.1 shows a list of possible surface defects defined in ISO 8785:1998 [5]. It should be clear that the imperfections are not related to the surface roughness or waviness. A reference surface is usually specified onto which defect characteristics are projected. The reference surface is determined over a specified surface area, or over a limited part of the surface area related to the size/dimensions of a single imperfection, the size of the area being sufficient to assess the imperfection while suppressing the influence of form

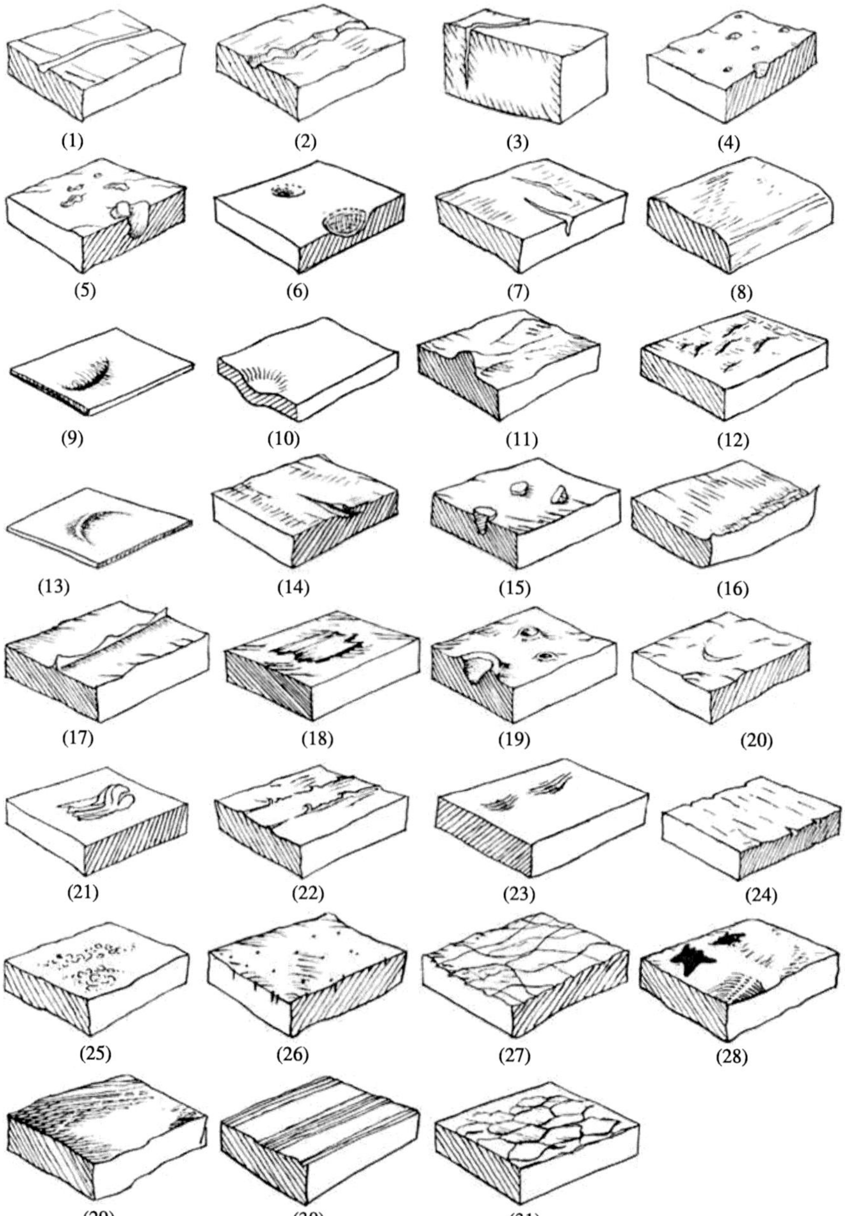

Fig. 1.1 Specific types of surface imperfections. *1* Groove, *2* scratch, *3* crack, *4* pore, *5* blowhole, *6* shrinkage hole, *7* fissure/chink/crevice, *8* wane, *9* concave buckle, *10* dent, *11* wart, *12* blister, *13* convex buckle, *14* scale, *15* inclusion, *16* burr, *17* flash, *18* deposits, *19* crater, *20* lap, *21* scoring, *22* chip rest, *23* skidding, *24* erosion, *25* corrosion, *26* pitting, *27* crazing, *28* spot/patch, *29* discoloration, *30* streak, *31* cleavage/flaking

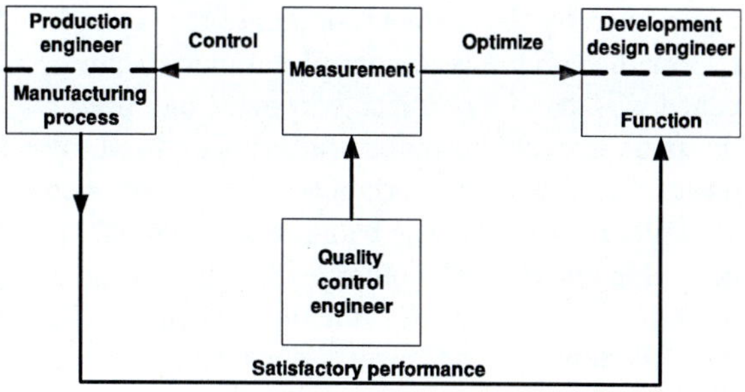

Fig. 1.2 Importance of dimensional metrology

deviation on the assessment. It usually coincides with the area adjacent to the defect. The dimensional characteristics of a surface imperfection include: imperfection length, width, depth, height, and surface imperfection area.

The definitions of the different kinds of irregularities (deviation) are rather uncertain. There are no distinct borderlines. Therefore it was discussed in ISO whether to define borderlines in terms of defined spacings of irregularities, in terms of defined ratios between spacings and depths of irregularities, in terms of defined ratios between spacings of irregularities and feature lengths. However, it was decided to retain the definitions according to the causes of the irregularities [6–9].

There is another distinction, namely that between micro- and macro-deviations. Macro-deviations are those that can be assessed with the usual measuring devices for the assessment of size, form, orientation and location. Micro-deviations are assessed with roughness- or waviness-measuring instruments. Macro-deviations are assessed over the entire feature length; while micro-deviations are assessed from a representative part of the surface. There is no distinct borderline because sometimes parts of the waviness will contribute to the result of the measured macro-deviations and sometimes parts of the form deviations will contribute to the result of the measured micro-deviations [3].

1.1.2 Importance of Dimensional Metrology

Dimensional metrology is related to both manufacture and function (Fig. 1.2). Manufacture includes all aspects of the manufacturing process such as machine performance, tool wear, and chatter, whereas function includes all functional properties of the surface of components such as the tribological regimes of friction, wear and lubrication. The geometry of a workpiece can be important in two quite different applications: one is concerned with controlling the manufacture, including the process and the machine tool; and the other is to help optimize the function of the workpiece. Many of these uses fall under the title of tribology. Manufacturing process and function are not completely independent of each other. Controlling the manufacturing process helps repeatability and hence quality of conformance. Functional optimization helps the designer and thereby assists in the quality of design.

Historically the correct function of the workpiece was guaranteed by one of the two methods. One is to control the manufacturing process. In practice what happened was that a workpiece was made and tried out. If it functioned satisfactorily the same manufacturing conditions were used to make the next workpiece and so on for all subsequent workpieces. It soon became apparent that the control of the workpiece geometry was being used as an effective go-gauge for the process and hence the function. The other method is to try to assemble the workpiece and see if it works for the designed functions. If the workpiece does not assemble, continuous modification or rework is carried out until the workpiece assembles and fulfills its designed functions. Obviously what is required is a much more flexible and less remote way of guaranteeing functional performance; it should be possible, by measuring parameters of the surface geometry itself, to predict the function. The conventional method is very much a balancing act and unfortunately the delicate equilibrium can be broken by changing the measurement parameters or the production process or even the function. This may seem an obvious statement but within it lays one of the causes of everyday problems in engineering.

Dimensional metrology cannot simply be regarded as an irritant to the manufacturing processes of the workpiece. The smallness of the magnitude of the measurements does not imply their importance is small. Dimensional metrology is actually absolutely critical in many applications, and it provides masses of information that can be invaluable to a manufacturer if extracted correctly from the surface geometry. In almost every example of its use, the criticism can be raised that it is not well understood and, even where it is; it is not properly specified especially on drawings. The importance of dimensional metrology and surface geometry should be recognized not just by the quality control engineer or inspector but also, more importantly, by the designer. It is he or she who should understand the influence that the surface has on behavior and specify it accordingly. In today's manufacturing industry, dimensional metrology can be found in nearly every shop floor. For example, Coordinate Measuring Systems (CMSs), which employ Coordinate Measuring Machines (CMMs) for dimensional measurements, are widely employed in manufacturing industry to generate surface measurements from a measured part.

1.2 Dimensional Metrology and Manufacturing Processes

In order to produce a workpiece, raw material is required, equipment such as a machine tool is needed to effect the process, the design drawing or data from the design is needed, and machining process data and toolpaths information is also required to describe how to remove the material from raw stock to produce designed shapes on the workpiece. In order to transform raw material into a workpiece having the desired shape, size and surface quality, it has to be processed by some means. There are many different ways in which this transformation can be achieved such as cutting with single or multiple tool tips, abrasive machining,

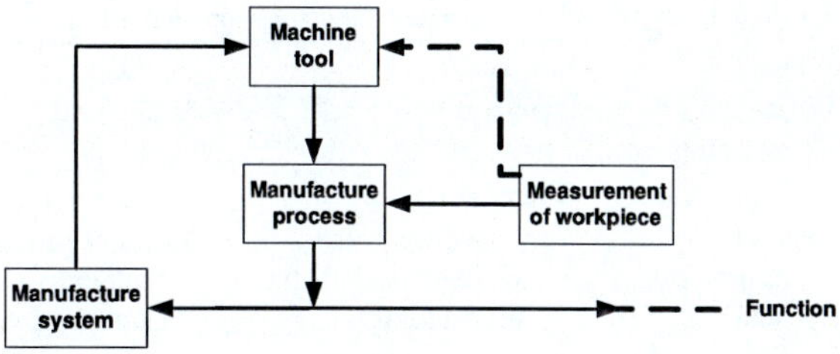

Fig. 1.3 Role of measurement and manufacturing processes

forming, casting, etc. Each has its own particular advantages and disadvantages. Some workpieces are produced by one process and others by many. The manufacturing process considered in this book is the cutting process, especially milling. This process is most common for generating the primary dimensions and geometry of the workpiece and involves an axis of rotation somewhere in the generation. The role of workpiece geometry in manufacturing processes is shown in Fig. 1.3. There is a two-way interaction: the first interaction is concerned with the nature of the geometric characteristics produced on the surface by the manufacturing process; and the other interaction concerns ways in which the surface roughness and form can be used to detect changes in the process and also, in some cases, the machine tool.

Typical variables in a milling or turning process are cutting speed—workpiece peripheral speed relative to the tool, axial feed—the advancement of the tool per revolution of the workpiece or the tool, the shape of the tool and the depth of cut of the tool into the workpiece material. There are other very important aspects that contribute a considerable difference to the form and surface finishing. These include the absence or presence of coolant and, if present, its constitution and the method of supply, whether fluid, mist, or drip, and so on. In addition to these effects is the effect of the machine tool itself.

Using the measurement of the workpiece as a check on the manufacturing process is well established but there are a number of issues as to where to measure, what to measure and when to measure. There are three types of measurement in manufacturing processes [10] listed as follows, all of which can be used to provide measurement data for monitoring and adjusting manufacturing processes. These types of measurement and their properties are also summarized in Table 1.1.

1. In-process measurement

    i. with On-Machine Measurement (OMM), which takes place as the workpiece is being made.
    ii. with portable measurement, where the workpiece surface is tested when the part has been made but not relocated. The surface instrument, which is hand-held, has somehow to be perched on the part when the machining has stopped and then the measurement recorded.

Table 1.1 Types of measurement and their properties

Where?	In-process	In situ	Remote
What type of measurement?	Dynamic	Static	Integrated
Type of control	Adaptive control	Statistical control	Long-term traceability
Type of information obtained	Very specific process or tolerance-satisfaction information	Medium process or tolerance-satisfaction information	Comprehensive workpiece geometry information
Purpose of information	Process control	Process monitor	Machine tool monitor
Speed	Very fast Little/No operator intervention	Time to record and judge	Functional judgment
Outcome	Working shift controlled	Quality of conformance assured	Quality of the workpiece assured

2. In situ measurement
   The workpiece is removed from the machine and measured with an instrument located near the machine tool.
3. Remote measurement
   The workpiece is taken to a properly equipped inspection room to be inspected on a CMM.

1.2.1 Dimensional Metrology and Closed-Loop Manufacturing

Closed-loop Manufacturing (CLM) is a method for optimizing the efficiency of a manufacturing process. It involves the use of measurement technology (metrology), most often touch sensor probes, to determine actual part dimensions as well as values of machine tool characteristics [11]. The elements of CLM are comprised of reliable machines, robust processes, automatic data collection, continuous improvement, and efficient and accurate analysis. Each element is supported by various methods which when combined deliver a complete closed-loop solution. The CLM cycles consist of measurement, data collection, data analysis, and process adjustment. Figure 1.4 illustrates these elements and the cycle of CLM.

The loop is closed when the measurements are controlled and when they are utilized to improve the manufacturing process. The data that is collected during the measurements is the basis for various kinds of optimization loops targeting different aspects of the manufacturing process. This data is not only used and applied within the machine's control when adjusting offsets, but also utilized by engineers who analyze the process data over time to evaluate retargeting of dimensions, to modify tolerance requirements or to use the gained knowledge to better design parts for their producibility. The benefits of utilizing CLM for manufacturing processes can be summarized as follows:

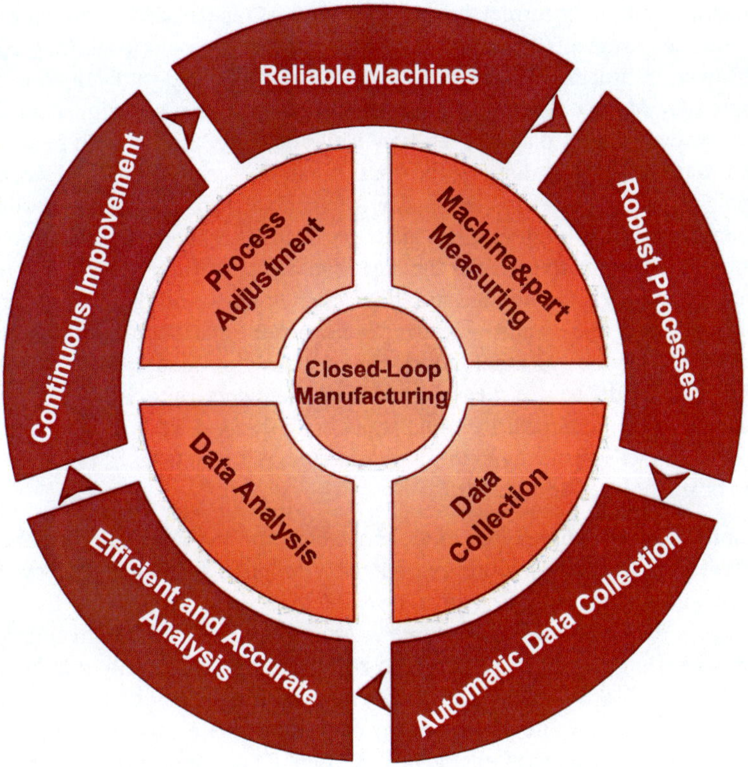

Fig. 1.4 Elements of CLM [12]

1. improving the reliability of the machine,
2. providing a more controlled environment,
3. lessening the human error involved with offset modification,
4. providing assurance of an acceptable accuracy level of the machinery, and
5. providing automated workpiece setup and tool setup.

Different types of CLM systems utilize different measurement operations to collect data. There are three types of CLM loops in a manufacturing system.

1. *Process Control Loop*. An immediate closed loop exists between the machine tool and the measuring system. The Computer Numerical Controlled (CNC) program takes measurements instantly after one machining operation, compares the results to tolerance limits and adjusts offsets to compensate for the deviation between the actual and the desired state. In-process measurement is commonly employed for this type of closed loop.
2. *Process Improvement Loop*. The process improvement loop is neither automated nor instant after machining but depends on the manufacturing engineer evaluating the process. Part dimensional data that was collected from the manufacturing process over a period of time provides an understanding of whether the

process is nominally correct, how well the process repeats, and how accurate it is. In situ measurements are commonly used for this type of closed-loop control. However, in-process measurement data is also considered for evaluation.

3. *Design-for-Manufacturability Loop*. If interpreted from a design engineering perspective, process data can tell how well and how easily certain features can be produced. A dissatisfying process capability causes scrap, rework, and repair. Some part features are easier to machine than others. The design-for-manufacturability loop suggests that the design engineer analyses manufacturing process data to consider the degree of difficulty in producing one feature over another. Providing that fit, form and function of the feature are not compromised, the feature offering the better manufacturability may be preferred. Obviously, other considerations such as assemblability and cost influence should be considered by design engineers as well. In this type of closed loop, remote measurements are commonly used. Data collected through in-process measurement and in situ measurement is also utilized for analysis for this type of closed-loop control.

Among the above three types of CLM systems, process control loop is the most important one with an obvious reason: only when a manufacturing system is adaptive and stable enough, can it provide high quality parts for the rapidly changing modern manufacturing industry. An important goal of the manufacturing system with a process control loop is to reduce process variability and bias to as small a level as is economically justifiable. Process bias is the difference between a parameter's average value and its designed value. Bias errors are a steady-state deviation from an intended target and while they do cause unacceptable products, they can be dealt with through calibration procedures. On the other hand, process variability is a continuously changing phenomenon that is caused by alterations in one or more manufacturing process parameters. It is inherently unpredictable and therefore more difficult to accommodate. However, the real-time process parameter measurements in CLM can provide the information needed to deal with unexpected excursions in manufacturing systems. The concept of conventional CLM is not a complex concept. However, the collection of the necessary process data can be a challenge. To decide when and where to conduct measurement for collecting necessary process data, it is important to understand common process variation and error sources in manufacturing systems.

1.2.2 Process Variations and Error Sources

Manufacturing operations are driven by cost requirements that relate to the value of a particular product to the marketplace. Given this selling price, the system works backwards to determine what resources can be allocated to the manufacturing portion of the cost equation. Then, production personnel set up the necessary resources and provide the workpieces that are consumed by the market.

Everyone is happy until something changes. Unfortunately, the time constant associated with change in the manufacturing world has become very short. Requirements often change even before a system begins producing parts, and even after production is underway there are typically many sources of variability that impact on the cost/quality of the operation. Variability associated with scheduling changes is to be accommodated by designing flexibility into the basic manufacturing systems. However, the variability that is related to changing process conditions is best handled by altering system performance at a more basic level.

Error conditions often occur where one or more process parameters deviate significantly from the expected value and the process quality degrades. The sensitivity of the process to these variations in operation conditions depends on the point in the overall manufacturing cycle at which they occur as well as the specific characteristics of a particular process disturbance. Amplitude, a frequency of occurrence, and a direction typically characterize these process errors [13, 14]. In a machining operation, the typical result is a lack of synchronization between the tool and part locations so that erroneous dimensions are produced.

Over time, the amplitude of process errors is typically limited to a specific range either by their inherent nature or by the operator's actions. For example, shop temperature profiles tend to follow a specific pattern related to cutting forces, and cutting tools are replaced as they wear out. As multiple process error sources interact, the result is typically a seemingly random distribution of performance characteristics with a given "normal range" that defines the routine tolerances achievable within a given set of operations. On the other hand, trends such as increased operating temperatures due to a heavy workload, coolant degradation, and machine tool component wear, have a non-random pattern that continue over time until an adjustment is made [14].

One solution to the problem of process variation is to build a system that is insensitive to all disturbances; unfortunately, this is rarely practical. A more realistic approach is to use a manufacturing model that defines the appropriate response to a particular process parameter change. This technique can be very successful if the necessary monitoring systems are in place to measure what is really happening within the various manufacturing operations. This approach works because manufacturing processes are deterministic in nature: a cause-and-effect relationship exists between the output of the process and the process parameters [14]. Events occur due to specific causes, not random chance, even though an observer may not recognize the driving force behind a particular action. If the key process characteristics are maintained at a steady-state level, then the process output will also remain relatively constant. Conversely, when the process parameters change significantly, the end product is also affected in a noticeable manner. By measuring the important process parameters in real-time and performing appropriate adjustments in the system commands, great improvements can be achieved in increasing product quality and lowering production costs [13]. Process variability hinders the efforts of system operators to control the quality and cost of manufacturing operations. This basic manufacturing characteristic is caused by the inability of a manufacturing system to do the same thing at all times and under all conditions. Machining operations typically exhibit a much higher degree of process control. However, variability is still present in relatively simple operations such as attempting to control a feature's diameter and surface finish without maintaining a constant depth of cut, coolant condition, temperature, tooling quality, etc. Inspecting parts and monitoring the value of various process parameters under different operating conditions helps collect process variability data. However, the following questions must be answered before qualifying the process variability: what parameters can and should be measured; when should measurement take place; how much variation is acceptable; is bias a problem (it is usually a calibration issue); what supporting inspection data is required; and does the process model accurately predict the system operation? In industry, the Error Budgets (EB) method [13] is commonly used to answer most of these questions. It categorizes system errors and understands the impact of altering the magnitudes of the various errors, and selects a viable approach for meeting the desired performance goals. After the EB procedure, a system error model is obtained by conducting a series of experiments through which a relationship is established between individual process parameters and the quality of the workpiece. Once the system error model has been validated, a reliable assessment can be made of the impact of reducing, eliminating, or applying a suitable compensation technique to the different error components.

Process parameter information can be used to monitor the condition of a manufacturing operation as well as provide a process control signal to a feedback algorithm. If any of the key process parameters deviates, an error is known to have occurred. The error can come from three sources: machine accuracy related, tooling accuracy related, and workpiece setup related. Machine accuracy can be described in a generic term that is how accurate the tool path can be. In general, this characteristic is influenced by two categories of error: quasi-static error and dynamic error. Quasi-static errors are process disturbances that change relatively slowly and have long time constants. The result of this type of system error is usually observed as the degradation of part form due to the inaccurate positioning of the tool with respect to the workpiece. Quasi-static errors are related to the machine structure and design (the geometry and kinematics of the machine) and normally caused by slowly varying forces that act on the machine and thermally induced strains in the machine tool. Dynamic errors are mostly related to a machine's servo system, such as vibration, spindle errors, axes motion errors, etc. Sometimes vibration also occurs on the workpiece itself, which causes errors. The workpiece vibration error is not related to the machine's servo system but relates to workpiece design and certain machining parameters, such as cutter spindle speed and feed rate. This type of error occurs in relatively high frequency and in short time constants. It is associated with the travel of a machine's moving elements and can be discussed in relation to typical machine axes. With the type of linear carriage shown in Fig. 1.5, it is possible to measure six individual error elements due to the six degrees of freedom. The result of this type of error can be observed in many forms, such as the dimensional or geometric errors on the workpiece. These errors are normally the combination of these six error elements.

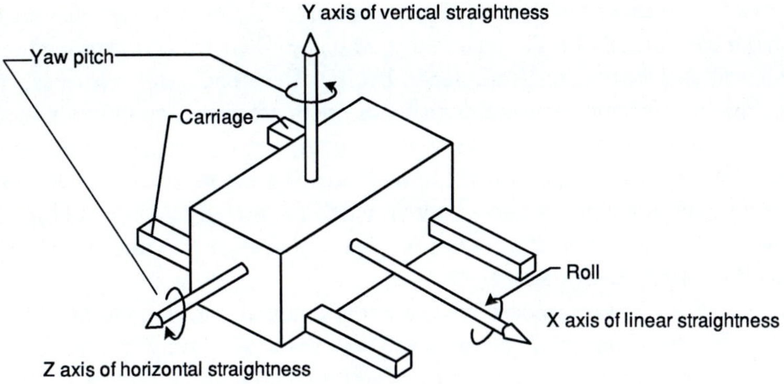

Fig. 1.5 Six error elements in linear axis motions

Tooling accuracy related errors can also be categorized in two groups: workpiece fixturing and cutting tool errors. Workpiece fixturing error is mostly caused by wrong clamping force, which makes the workpiece deform. Cutting tool errors are mostly caused by misalignment between cutter and machine axis and errors in the shape of the cutting tool, such as tool deflection. The third common source of error is from workpiece setup. The misalignment between the workpiece and its desired home location on the machining table often causes workpiece geometry to be dislocated from the designed location.

Once a correct manufacturing system error model is established, key process parameters can be identified. Constant monitoring of these key parameters and their error occurrences enables correct compensation or adjustments to the control of manufacturing processes. Therefore, a stable and accurate manufacturing system is achievable. For example, if errors due to component deflection, machine geometry, etc. are relatively constant, then tool offsets based on the condition of the cutting tool can improve the system performance. However, if adjustments are made based on historical data rather than immediate monitoring, then the system is vulnerable to unexpected changes in factors such as tool performance, material characteristics, operator induced changes in feeds and speeds. Offsets that are based on product certification results are a little better, since there is a closer tie with the "current process", but the delay between production and inspection can still cause difficulties. The three types of measurements introduced at the beginning of Sect. 1.2 can be categorized into two groups: in-process measurement and off-line measurement which are discussed in detail in the following sections.

1.2.3 In-Process Measurement

The concept of in-process measurement and control has to do with (a) measuring a process variable while that variable can still be influenced and (b) applying a corrective feedback to the machine that affects the process so as to encompass those sources of error that normally occur during the process and thus eliminate error from the variable on the resultant workpiece. Figure 1.6 illustrates the basic concept of in-process measurement. During the machining process, OMM instruments or portable measurement instruments provide a continuous measurement that can be in the form of an analog signal or a digital data word, which is compared with the required dimension derived from the part design. The result of the comparison is a compensatory signal which is applied to the machine control so as to restore the dimension within its allowable range on either the part being machined or subsequent parts.

The use of limited in-process measurement coupled with the monitoring of the key process parameters of manufacturing processes as a substitute for extensive post-process measurements is becoming more realistic and attractive in achieving fully automated manufacturing processes [13]. In-process measurements, compared with in situ measurements and remote measurements, offer the best alternative for real-time manufacturing process monitoring and control as long as the time required to collect data is not an unacceptable cost to production operations. In order to be useful, in-process measurement data must be easily obtained; the prediction of system performance must be accurate and useful to the process operator.

Apart from improving the accuracy and consistency of manufacturing operations, in-process measurements of critical parameters can be used to provide real-time assurance that the workpiece quality is being maintained at the desired level. Aside from the obvious step of measuring one or more critical dimensions on a finished workpiece, additional process data can be collected to qualify the process before the part is removed from the machine tool. The data collected through in-process measurement can also be used to analyze the process consistency, the deflection/size errors, and tool wear errors. Therefore, in-process measurement has the advantages of providing direct, continuous, real-time measurements of those part attributes that are defined in the acceptance-tolerance criteria for the workpiece. The advantages of employing in-process measurements are summarized as follows [15–21].

1. Cost and time saving through
   - i. reducing lead-time required for gages and fixtures,
   - ii. minimizing need for design, fabrication, maintenance of hard gages, fixtures and equipment,
   - iii. reducing inspection queue time and inspection time, and
   - iv. eliminating rework of nonconforming product.
2. Changing from "reactive" inspection to "proactive" control by
   - i. integrating quality control into product realization processes,
   - ii. using characterized and qualified processes to increase product reliability,
   - iii. focusing resources on prevention of defects instead of detection in the end (a post-mortem process),

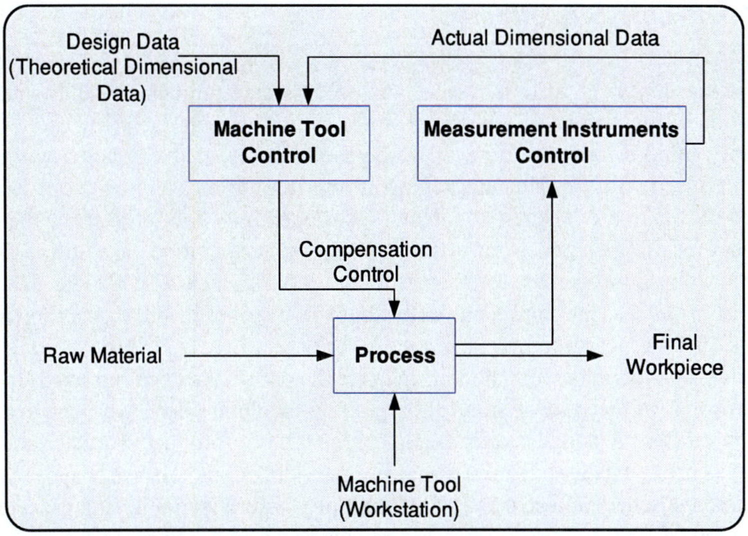

Fig. 1.6 Concept of in-process measurement

iv. utilizing real-time process knowledge and control, and part acceptance/disposition, and
v. enhancing small lot acceptance capability.

3. Elimination of non-value added operations such as lot inspection, sampling plans, receiving inspection, design, fabrication and maintenance of hard gages, and reworking nonconforming parts.
4. Agile machining.

However, doubts may be raised as to the validity of dimensional measurement on the same machine that makes the part and the accuracy that a portable measurement device can provide. While measurements performed by a cutting machine are subject to some of the same error producing factors as the cutting progress, the errors that are most difficult to eliminate through machine maintenance and certification can easily be detected and accounted for with in-process measurement. For example, machine flexing, tool wear, and vibration will all be absent during measurement. Additional error compensation techniques such as laser measurement, ball-screw compensation, and measuring pre-cut proofing parts for future reference can also be applied to compensate for other machine inaccuracies. The ability to rectify manufacturing errors caused by problems such as these has led to the acceptance of in-process measurement in manufacturing systems [22].

1.2.4 Off-Line Measurement

Off-line measurement, also known as post-process measurement, implies a product evaluation sequence that is performed at a point in the manufacturing cycle that is relatively remote from the time at which the features of the finished workpiece were produced. Both in situ measurement and remote measurement can be categorized as off-line measurement. However, depending how efficiently the measurement results gathered from in situ measurement operations are utilized in the entire machining processes, sometimes in situ measurement can also be treated as a type of in-process measurement. In this book, we regard in situ measurement as a type of off-line measurement.

At this location in the manufacturing cycle the parts have been removed from the machine and, in the case of small lot sizes, the machine may have been set up to produce a different workpiece. Examples of the types of product characteristics that might be checked include feature dimensions and locations, surface shapes, weight, hardness, resiliency, and so on [2]. One difficulty with this monitoring procedure is that rework of any features which are rejected in certification can be difficult to accomplish due to the problem of realigning the workpiece on the machine as well as the possible necessity of reconfiguring the machine for that particular operation. Also, an additional problem results from the evaluation of the quality of the product features at a relatively remote point in the manufacturing cycle. The difficulty can be that additional products are produced in the interim period between the time when a particular part is completed and when it is finally inspected. In the event of an excursion in the process, a number of deviant parts will be produced during the delay interval between the fabrication and inspection stages.

If the value of a manufactured product is relatively small then it may be acceptable to just scrap the defective units. However, this may create an undesirable atmosphere of apparent indifference to product quality. In any event, for expensive workpieces, it is necessary to have a high acceptance rate for the fabrication process. One way of accomplishing this objective is to employ a process which rarely results in a rejected feature. If this is not possible, then it is necessary to shorten the time interval between fabrication and inspection as much as possible. Then process excursions are detected more rapidly which effectively reduces the number of additional defective parts that are manufactured during the waiting period. An additional technique that is effective in this type of situation is the use of control charts. Control charts can enhance the early detection of process shifts so that the production of out of tolerance parts is avoided. However, a process for fabricating expensive parts which, under normal circumstances, utilizes most of the available tolerance band is like an accident waiting to happen. When this is coupled with a significant delay in the inspection cycle then the situation becomes quite precarious. Obviously, the smaller the normal process variation the better (within cost-effective limits) since this provides a buffer in the event of an unusual occurrence, but this is especially important in those situations in which the post-process activities are delayed significantly.

One example of a relatively simple post-process monitoring operation is a system to detect a faulty tapped hole in a workpiece that is produced on an automated manufacturing line. This inspection operation usually occurs within a relatively short time after the machining cycle since it is necessary to evaluate the part status before further operations are attempted. The types of problems that could occur in this drilling and tapping process are that the drill or tap could break and become imbedded in the material or that the hole or threads could be missing because of a damaged tool at a machining station. In the event that a defective part is detected, then it must be prohibited from continuing through the normal cycle. In addition, it is necessary to signal an operator that a malfunction has occurred so that the problem can be corrected. In a more sophisticated system, a tool change cycle can be utilized to correct the problem of a damaged tool without halting the process.

A more complex post-process monitoring system would be involved in a gear manufacturing operation. Determining the quality of a helical gear is significantly more complicated than detecting the presence of a tapped hole in a workpiece. Some of the part parameters that may need to be examined on a precision gear include eccentricity, pitch, profile, tooth spacing and helix angle. In addition, an attempt to rework this type of workpiece would be much more involved than just drilling or tapping a new hole. It is readily recognized that maintaining the normal process variations at a low level is extremely important since the process defects are only detected after it is too late to make an easy correction for a given workpiece.

Post-process measurement is the most common mode of inspection [23]. Regardless of the level of in-process gauging, post-process inspection is usually required to certify the part to all conditions specified by engineering. These conditions include dimensional errors, errors of form, surface roughness and material integrity. While all these measurements can be (and were) made with low cost equipment such as measuring gauges, the process is slow and requires special inspection jigs and fixtures as well as periodic calibration with "master" parts that are individually crafted to serve as a reference for production. Also, performing precise measurements on complex parts rapidly using manual inspection methods increases the chances of human error. Errors at this stage can be extremely expensive since post-process measurements are often used to verify process settings, and the endorsement of an incorrect setting may result in the scrapping of an entire batch of products. The availability of CMMs has provided a means of achieving improved measurement throughput with greater accuracy and precision than manual measurement.

1.3 Summary

Dimensional metrology is the measurement of the deviations of a workpiece from its intended size and shape. The aim of dimensional metrology is to ensure that the size and shape of the workpiece conform to the designer's wish. This in turn ensures that the workpiece will assemble into associated assembly workpieces and the static characteristics of the workpiece have therefore been satisfied. Roughness, waviness, and errors from form are the three common causes of geometrical deviations. They often combine and exhibit as all different kinds of surface geometrical irregularities. The ISO standard defines 31 types of surface irregularities. The different kinds of surface irregularities have different origins in the manufacturing process.

Dimensional metrology is related to both manufacturing process and workpiece function. Measuring the geometry of a workpiece can be important in two quite different applications: one is concerned with controlling the manufacturing process, including the process and the machine tool; and the other is to help optimize the function of the workpiece. Dimensional measurements in manufacturing can be categorized into three types: in-process measurements, in situ measurements, and remote measurements. Each type of measurements has its advantages and disadvantages.

When dimensional measurement are controlled and used to improve the manufacturing process, the manufacturing loop is closed. This type of manufacturing is thus called closed-loop manufacturing. CLM is a method for optimizing the efficiency of a manufacturing process in which dimensional metrology is a crucial element. To decide where and when to conduct measurements in a manufacturing process, it is important to know the process variations and error sources. The key process characteristics of a manufacturing process must be identified and process parameters are then monitored. When a process parameter deviates, an error occurs. There are dynamic errors and tooling errors. The former is related to the machine's servo system. The latter can be divided into two groups: workpiece fixturing errors and cutting tool errors.

The use of limited in-process measurement coupled with the monitoring of key process parameters of manufacturing processes is more suitable than extensive post-process measurements in achieving a fully automated manufacturing process. In-process measurement and control of a manufacturing process involves measuring workpiece dimensional and geometrical variables together with process variables and applying a corrective feedback to the machining process.

Off-line measurement evaluates a workpiece after the manufacturing processes are carried out and the workpiece is removed from the machine. Thus, the rework of any features on the workpiece can be difficult due to the problem of realigning the workpiece on the machine as well as reconfiguration of the machine tool. However, because off-line measurements are able to provide high accuracy and comprehensive workpiece geometry measurement, off-line measurements are usually required to certify the workpiece to all conditions specified by design.

References

1. Dotson C (2006) Fundamentals of dimensional metrology, 5th edn. Thomson Delmar, New York
2. Whitehouse DJ (1994) Handbook of surface metrology. IOP Pub, Bristol

3. Henzold G (2006) Geometrical dimensioning and tolerancing for design, manufacturing and inspection: a handbook for geometrical product specification using ISO and ASME standards, 2nd edn. Butterworth-Heinemann, Oxford
4. Henzold G (1995) Handbook of geometrical tolerancing: design manufacturing, and inspection. Wiley, West Sussex
5. ISO (1998) ISO 8785:1998: geometrical product specification (GPS)—surface imperfections—terms, definitions and parameters
6. ISO(1997) ISO 4287:1997: geometrical product specifications (GPS)—surface texture: profile method—terms, definitions and surface texture parameters
7. ANSI (2009) ASME B46.1–2009 surface texture, surface roughness, waviness and lay
8. BS (1990) BS 1134-2:1990: assessment of surface texture. Guidance and general information
9. DIN (1982) DIN 4760: form deviations; concepts; classification system
10. Whitehouse D (2002) Surfaces and their measurement. Hermes Penton Ltd, London
11. Jesse C (2001) Process controlled manufacturing at Pratt & Whitney. Soc Autom Eng trans 110(01):227–232
12. Zhao YF (2009) An integrated process planning system for machining and inspection. Department of Mechanical Engineering, Ph.D. thesis, University of Auckland
13. Barkman WE (1989) In-process quality control for manufacturing. M. Dekker, New York
14. Barkman WE (2000) Handbook of industrial automation. In: Richard ELH, Shell L (eds) In-process measurement. Marcel Dekker, New York
15. Kamath R (2000) On-machine inspection and acceptance (OMIA). University of Missouri-Rolla, USA
16. Cho MW et al (2004) A computer-aided inspection planning system for on-machine measurement—Part II: local inspection planning. KSME Int J 18(8):1358–1367
17. Lee H, Cho MW, Yoon GS, Choi JH (2004) A Computer-aided inspection planning system for on-machine measurement—Part I: global inspection planning. KSME Int J 18(8): 1349–1357
18. Chung SC (1999) CAD/CAM integration of on-the-machine measuring and inspection system for free-formed surfaces. In: Proceedings of American society for precision engineering, vol 20. American Society for Precision Engineering, Raleigh, pp 267–270
19. Renishaw (2000) Renishaw MTP products technical specification H-2000-3020-09-A. Probing systems for CNC machine tools. http://resources.renishaw.com/details/Probing+systems+for+CNC+machine+tools+technical+specifications%2821408%29. Accessed 8 Apr 2011
20. Zhao YF, Xu X, Xie S (2008) STEP-NC enabled on-line inspection in support of closed-loop machining. Robotics Comput Integr Manuf 24(2):200–216
21. Kim KD, Chung SC (2001) Synthesis of the measurement system on the machine tool. Int J Prod Res 39(11):2475–2497
22. Davis TA et al (2006) Flexible in-process inspection through direct control. Meas J Int Meas Confed 39(1):57–72
23. Wang B (1997) Integrated product, process and enterprise design. Springer, Berlin

Chapter 2
Practices of Information Modeling

In recent years, information technology has become increasingly important in the manufacturing enterprise. Effective information sharing and exchange among computer systems throughout a product's life cycle has been a critical issue [1]. Dimensional metrology systems, as an essential part of manufacturing enterprise, face the same information interoperability issue. The concept of dimensional metrology interoperability is defined as "the ability of two system components to communicate correctly and completely with each other—with minimal cost to either component user or component vendor, where the components can come from any vendor worldwide" [2]. This concept is used to address the issues that complicate the measurement process. Formal information modeling languages that describe information requirements unambiguously together with unambiguous specifications for methods of reading and writing (or storing and retrieving) modeled data contribute an enabling technology that facilitates the development and integration of a networked computer environment in dimensional metrology that behaves consistently and correctly.

This chapter first describes the basics of information modeling and the typical information modeling process such as how information models are used to define data requirements and how information models enable information sharing and exchange. Several commonly used information modeling methodologies, modeling languages, and implementation methods are discussed here. Then, the four elements of a typical dimensional metrology system are described: product definition, measurement process planning, measurement plan execution, and analysis and reporting. The functionalities of these elements and their sub-systems are also discussed in detail.

2.1 Basics of Information Modeling

Information modeling is a technique for specifying the data requirements that are needed within the application domain [3]. An information model is a representation of concepts, relationships, constraints, rules, and operations to specify

Y. Zhao et al., *Information Modeling for Interoperable Dimensional Metrology*,
DOI: 10.1007/978-1-4471-2167-1_2, © Springer-Verlag London Limited 2011

data semantics for a chosen domain of discourse [4]. The advantage of using an information model is that it can provide a sharable, stable, and organized structure of information requirements for the domain.

In the 1970s, the Standard Planning and Requirements Committee (SPARC) of the American National Standards Institute (ANSI) published a three-schema architecture for database management systems [5]. The three schemas include an external schema—the user view of the information, an internal schema—the computer view of the information, and a conceptual schema—a logical, neutral view of the information. The conceptual schema is a single, integrated definition of the data within an enterprise that is unbiased toward any single application of data and independent of how the data is physically stored or accessed. It provides a consistent definition of the meanings and interrelationship of the data in order to share, integrate, and manage the data. The need to define conceptual schemas has led to the development of semantic modeling techniques. An important benefit of having a fully developed, semantic information model is that the model can be used to define various applications. During the 1970s, the relational data model was introduced to represent the conceptual schema level [3, 4]. As the relational database management system (DBMS) design techniques grew, the need to design shared databases was recognized. Information modeling techniques provide a way to develop specifications for sharing and exchanging data.

2.1.1 Information Modeling Methodologies

There are different practices in developing an information model. The underlying methodologies for the recent modeling practices are based on three approaches:

- The entity-relationship (ER) approach,
- The functional modeling approach, and
- The object-oriented (O-O) approach.

The ER approach focuses on how the concepts of entities and relationship might be applied to describing information requirements. The emphasis of the functional modeling approach is placed on specifying and decomposing system functionality. The O-O approach focuses on identifying objects from the application domain first then operations and functions.

The ER approach is based on a graphical notation technique [4]. Various ER extensions have been introduced since then. The basic constructs in an ER model are the entity type, the relationship type, and the attribute type. The notation is easy to understand and the technique has been useful in modeling real problems [6]. The functional approach addresses the system's processes and the flow of information from one process to another. It uses objects and functions over objects as the basis. The approach often uses data-flow diagrams. A data-flow diagram shows the transformation of data as it flows through a system. The diagram consists of processes, data flows, actors, and data stores. This approach has been in wide use. In the objected-oriented approach, the fundamental construct is the object, which incorporates both data structures and functions. The building blocks in the O-O model are object classes, attributes, operations, and associations (relationships). The objected-oriented approach has the following advantages: easier modeling of complex objects, better extensibility, and easier integration with O-O database models and O-O programming code.

Choosing an appropriate modeling methodology is a judgment that must be made at the beginning of the modeling work. In general, an information model, developed in any methodology, is a representation of entities, attributes, and relationships among entities. However, each information model has a different emphasis; the emphasis often depends on the viewpoint that represents a specific person or organization associated with the model. Occasionally there are multiple viewpoints for the model. The viewpoints of the model help to decide the type of information modeling methodology to be used. For example, the ER approach is a better selection if data requirements are at the higher levels of detail. In the case where functions are more important and more complex than data, the functional approach is recommended. The O-O approach, however, may provide better extensibility and may be more compatible with the intended implementation environment. The disadvantage of the ER model is its lack of preciseness in supporting the detailed levels. Very often the data requirements of the application may need to be changed and most changes are function related; if the information model was developed using the functional approach, these changes may lead to a major modification to the model. Finally, the major obstacle for using the O-O approach is that the approach requires a critical paradigm shift in thinking compared with other data modeling approaches—from considering only the data to considering both the data and the functions. A hybrid approach combining ER, functional modeling approach, and/or object-oriented (O-O) approach has also been adopted and used by many.

2.1.2 Information Model Development Process

A good-quality information model should have the following characteristics: complete, sharable, stable, extensible, well-structured, precise, and unambiguous. In general, the contents of an information model include a scope, information requirements, and a specification.

The initial phase for developing an information model starts with the definition of the scope of the model's applicability. The scope specifies the domain of discourse and the processes that are to be supported by the information model. It is a bounded collection of processes, information, and constraints that satisfy some industry need. For manufacturing, the scope statements include the purpose as well as viewpoints of the model, the type of product, the type of data requirements, the supporting manufacturing scenario, the supporting manufacturing activities, and the supporting stage in the product life cycle. The scope definition may be supported by an activity model and/or a data planning model. An activity model is a representation of the application context, data flows, and the processes of the application. It is a mechanism for gathering high level information requirements. A data planning model provides a high level description of the data requirements for the information model, as well as the relationships among the basic data components. It is used as a roadmap to establish interfaces across a wide range of data. A well-defined scope should be accurate, unambiguous, and meet the industrial need. During the course of the modeling, the scope should be revisited and may be refined. Since the scope provides the boundaries of the application domain, it also serves as a guideline for evaluating the "completeness" of the information model.

After the scope is defined, the next phase is to conduct a requirements analysis. There is no standard method for collecting information requirements. However, requirements analysis may be accomplished by: literature surveys, standards surveys, domain experts' interviews, industrial data reviews, and state-of-the-art assessments. Depending on the scope, the analysis may include today's manufacturing practices, traditional practices, and near-future needs. It is important to capture data requirements accurately for the application scope while performing the requirements analysis. Industry reviews of the result of the analysis will help to ensure the completeness and correctness of the information requirements. As the result of the requirements analysis, information requirements should be documented. The definition of each identified information item should be included in the document. This document becomes the strawman for developing the information model.

After the detailed scope and information requirements are defined, the next phase is to develop the model. This phase transforms information requirements into a conceptual model. The information model is independent of any physical implementation, and it should be developed using a formal modeling language. Each information requirement should be expressed in the model. The model should be sufficiently detailed to describe the data needs of the application fully. To actually develop the information model, three types of design approaches can be taken: a top-down design, a bottom-up design, and a mixed or inside-out design. While the most effective way is to take the top-down design approach for modeling, it may not be possible or appropriate in all cases. An optimal design approach may depend on the individual application environment. Conceptualizing information requirements starts with grouping concepts and identifying the model's units of functionality. After that, an abstraction process will be performed to establish the model's structure for each functionality. This abstraction process, which structures information requirements into entities, objects, or classes, may include generalization, specialization, aggregation, classification, and association. Classification is the grouping of objects with the same data structure and operation. Generalization, specialization, aggregation and association are used for establishing relationships among the model's elements. Generalization and specialization identify the "inheritance-from" and "inheritance-to" relationships, respectively. Aggregation identifies "subset-of" relationships. Association identifies "dependency" relationships. If the structure of the model is established graphically, it must then be laid out according to the syntax of the selected modeling language.

Take the STandard for the Exchange of Product model data, also known as STEP as an example, STEP standards consist of numerous Application Protocols (APs). Each AP is focused on defining information for a particular application domain. APs are the Parts of STEP standards intended to be implemented for industrial use. When the AP concept was first introduced in STEP, an AP had three parts:

- Application activity model (AAM)—a model of the activities and data flows of the application
- Application reference model (ARM)—a model of the data needed for a particular application
- Application interpreted model (AIM)—an encoding of the ARM in terms of the STEP integrated resources. This is the model that is intended for implementation in systems that use STEP.

The AAM of a to-be-developed application protocol is a model of the activities and data flows of the application. AAMs are built using IDEF0, which is a graphical method of modeling activities and data flows. Activities are represented as boxes, while data, actors, and constraints are represented by arrows. In the IDEF0 approach, an aggregated model is built first to show the big picture with three to six activities. Then one or two rounds of refinement are performed, with each activity at an upper level being expanded into an entire page at the next level down. Once the AAM stage is completed and an ARM has been built, the AAM plays no further role. The ARM of an application protocol is a model of the data needed for a particular application. The model is given using the terminology of the application so that the model can be understood by practitioners of the application (who are involved in the development of the model). The process of building an ARM usually includes workshops at which domain experts decide what entities should be defined and what their attributes should be. ARMs may be written in EXPRESS, EXPRESS-G, or IDEF1X. The information modeling language is less important than the content at this stage. The AIM of an application protocol is an EXPRESS model of (exactly) the information in an ARM but encoded in terms of the STEP integrated resources. The encoding is done using mapping tables, the format of which is formally defined and is uniform across STEP. Because extensive knowledge of the integrated resources, the format of mapping tables, and strategies for mapping are required to do the mapping, the encoding can only be done by a STEP expert.

It is necessary to point out that the previously introduced data modeling process is almost always repeated multiple times during the development of a data model in practice. During each repetition of the data modeling process, the data model is refined and improved.

2.2 Information Modeling Languages

An information modeling language is a formal syntax that allows users to capture data semantics and constraints. In 1976, an Entity Relationship graphic notation was introduced to develop relational data models [4]. Since then, languages for information modeling have continued to evolve: the Unified Modeling Language (UML) [7], the Integrated Computer Aided Manufacturing (ICAM) Definition Language 1 Extended (IDEF1X) [8], the EXPRESS language [9, 10], and eXtensible Markup Language (XML) Schema [11] are some of the most commonly used information modeling languages. In this section, these languages will be described in certain detail.

The Integration DEFinition for information modeling (IDEF) language was developed in the U.S. Air Force ICAM Program between 1976 and 1982 [3]. The objective of the ICAM Program was to increase manufacturing productivity through the systematic application of computer technology. IDEF includes three different modeling methods: IDEF0, IDEF1, and IDEF2 for producing a functional model, an information model, and a dynamic model respectively. IDEF1X is an extended version of IDEF. Improvements included enhanced graphical representation, enhanced semantic richness, and simplified development procedures. The language is in the public domain. It is a graphical representation and is designed using the ER approach and the relational theory. It is used to represent the "real world" in terms of entities, attributes, and relationships between entities. Normalization, that eliminates redundancy and arranges a collection of data according to its inherent logical structure, is enforced by KEY Structures and KEY Migration. The language identifies property groupings to form complete entity definitions.

EXPRESS was created as ISO 10303-11 for formally specifying the information requirements of a product data model. The language is part of a suite of standards informally known as the STEP standards and was first introduced in the early 1990s [12]. EXPRESS is a textual representation. In addition, a graphical representation of a subset of EXPRESS called EXPRESS-G is available. EXPRESS is based on programming languages and the O-O paradigm. A number of languages have contributed to EXPRESS in particular: Ada, Algol, C, C++, Euler, Modula-2, Pascal, PL/1, and SQL. EXPRESS consists of language elements that allow an unambiguous object definition and specification of constraints on the objects defined. It uses the SCHEMA declaration to partition models, and it supports specification of data properties, and constraints. It does not support the concept of having an object own functions, and it supports functions only for stating constraints.

UML is a modeling language for specifying, visualizing, constructing, and documenting the artifacts, rather than processes, of software systems. It was conceived originally by Grady Booch, James Rumbaugh, and Ivar Jacobson. UML was approved by the Object Management Group (OMG) as a standard in 1997 and was quickly adopted by the software engineering disciplines as a convergence of several different modeling techniques that had existed in the past. The language is non-proprietary and is available to the public. It is a graphical representation. The language is based on the objected-oriented paradigm. UML contains notations and rules and is designed to represent data requirements in terms of O-O diagrams. UML organizes a model in a number of views that present different aspects of a system. The contents of a view are described in diagrams that are graphs with model elements. A diagram contains model elements that represent common O-O concepts such as classes, objects, messages, and relationships among these concepts.

The XML schema, as a structural definition, was published as a World Wide Web Consortium (W3C) Recommendation in May 2001. It was the first separate schema language for XML to achieve Recommendation status by the W3C. Because of the confusion between XML Schema as a specific W3C specification, and the use of the same term to describe schema languages in general, some parts of the user community referred to this language as WXS, which stands for W3C XML Schema, while others referred to it as XSD that stands for XML Schema Document—a document written in the XML Schema language. XML schemas serve as design tools establishing a framework on which implementations can be built.

A summary of language features is presented in Table 2.1, where the capabilities of UML, IDEF1x, EXPRESS, and XML schema are shown. Features presented in the table include the language representation form, the underlying methodology of the language, the source of the language, the availability of an accompanying exchange structure format, and language constructs describing object, attribute, constraint, algorithm, relationship, and abstraction. UML, IDEF1x, EXPRESS, and XML schema all can be used to create a conceptual model, and each has its own characteristics. Although some may lead to a natural usage (e.g. implementation), one is not necessarily better than another. In fact, the modeling practice is often more important than the language chosen. In the following sections, each of these four languages will be discussed in detail.

2.2.1 UML Language

The UML is the successor to the wave of Object-Oriented Analysis and Design (OOA&D) methods that appeared in the late 1980 and 1990s. The UML went through a standardization process with the Object Management Group (OMG) and is now an OMG standard. The UML is called a modeling language, not a method. Most methods consist, at least in principle, of both a modeling language and a process. The modeling language is the mainly graphical notation that methods use to express designs. The process is their advice on what steps to take in doing a design [13]. In this book, our focus is to describe the UML data modeling language. UML, as a data modeling language, is used to specify, visualize, modify, construct and document the artifacts of an object-oriented software-intensive

Table 2.1 Comparison of information modeling language features

	UML	IDEF1x	EXPRESS	XML Schema
Representation form	Graphics	Graphics	Text	Text
Methodology	Object-orientation	Entity-relationship	Object-oriented flavor	Specify document content, object-oriented flavor
Source	Object management group standard	In public domain	ISO standard	W3C group recommendation
Standard mapping exchange format	None	None	ISO 10303 Part 21	XML Data file
Object construct	Class, use class, interface, collaboration, active class, component, node, package, behavioral things (interaction, state machine)	Entity, view	Type, entity, schema	Schema, element, simpletype, complextype,
Attribute construct	Explicit attribute, derived attribute	Explicit attribute (primary key, alternate key, foreign key)	Explicit attribute, derived attribute, inverse attribute	Explicit attribute, sequence of elements
Constraint construct	Using "note"	Using "note"	Where-rule, unique, optional	MaxOccurs, minOccurs, enumeration, unique, maxInclusive, minInclusive, regular expression
Algorithm construct	Method, operation	Using "note"	Function, procedure (but only for constraints)	None
Relationship construct	Dependency, association, aggregation, generalization, realization	Cardinality, categorization	Supertype, subtype, aggregation type (array, bag, set, list), constructed type (enumeration, select), defined type	Implicit subtype implicit supertype, explicit list (only explicit aggregate), implicit list (by maxOccurs), choice
Abstraction construct	Abstract class, abstract operation	None	Abstract supertype	Explicit and implicit abstract supertypes

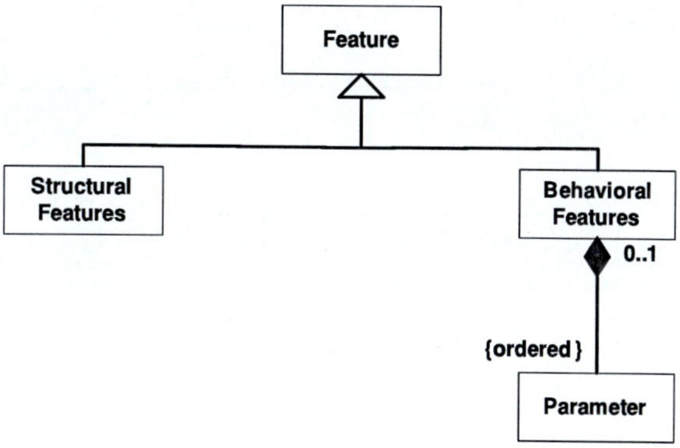

Fig. 2.1 UML meta-model extract

system under development. UML offers a standard way to visualize a system's architectural blueprints, including elements such as:

- Activities;
- Actors;
- Business processes;
- Database schemas;
- Logical components;
- Programming language statements;
- Reusable software components.

The UML, in its current state, defines a notation and a meta-model. The notation is the graphical content you see in models; it is the syntax of the modeling language. For instance, class diagram notation defines how items and concepts such as class, association, and multiplicity are represented. A meta-model is a diagram, usually a class diagram, which defines the notation. Figure 2.1 shows a small piece of the UML meta-model that shows the relationship among associations and generalization.

A UML meta-model contains three major categories of elements: classifiers, events, and behaviors. Each major category models individuals in an incarnation of the system being modeled. A classifier describes a set of objects; an object is an individual thing with a state and relationships to other objects. An event describes a set of possible occurrences; an occurrence is something that happens that has some consequence within the system. A behavior describes a set of possible executions; an execution is the performance of an algorithm according to a set of rules. Models do not contain objects, occurrences and executions, because they are the subject of models, not their content. Classes, events, and behaviors model sets of objects, occurrences and executions [14]. UML 2.2 has 14 types of diagrams divided into two categories shown in Fig. 2.2. Seven diagram types represent structural information and the other seven represent general types of behavior including four that represent different aspects of interactions.

The structural diagrams show the static structure of the objects in a system. That is, they depict those elements in a specification that are irrespective of time.

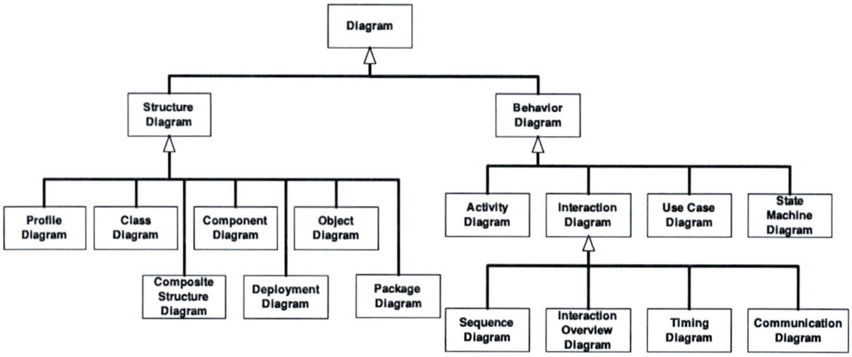

Fig. 2.2 The taxonomy of structure and behavior diagram of UML

The elements in a structure diagram represent the meaningful concepts of an application, and may include abstract, real-world and implementation concepts. For example, a structure diagram for an airline reservation system might include classifiers that represent seat assignment algorithms, tickets, and a credit authorization service. Structure diagrams do not show the details of dynamic behavior, which are illustrated by behavioral diagrams. However, they may show relationships to the behaviors of the classifiers exhibited in the structure diagrams.

Behavior diagrams show the dynamic behavior of the objects in a system including their methods, collaborations, activities and state histories. The dynamic behavior of a system can be described as a series of changes to the system over time. Behavior diagrams can be further classified into several other kinds as illustrated in Fig. 2.2 [14, 15]. The constructs contained in each of the UML diagrams can be found in UML specifications and many UML-related textbooks. Interested readers can refer to those books for more detailed descriptions.

2.2.2 IDEF1X Language

IDEF1X is a data modeling language for the developing of semantic data modes. It was developed for designing relational databases with a syntax designed to support the semantic constructs necessary in developing a conceptual schema. A conceptual schema is a single integrated definition of the enterprise data that is unbiased toward any single application and independent of its access and physical storage. A conceptual schema must have three important characteristics:

- It must be consistent with the infrastructure of the business and be true across all application areas.
- It must be extendible, such that, new data can be defined without altering previously defined data.
- It must be transformable to both the required user views and to a variety of data storage and access structures.

Because it is a design method, IDEF1X is not particularly suited to serve as an AS-IS analysis tool, although it is often used in that capacity as an alternative to IDEF1. IDEF1X is most useful for logical database design after the information requirements are known and the decision to implement a relational database has been made. Hence, the IDEF1X system perspective is focused on the actual data elements in a relational database. If the target system is not a relational system, for example, an object-oriented system, IDEF1X is not the best method. There are several reasons why IDEF1X is not well-suited for non-relational system implementations. IDEF1X requires, for example, that the modeler designate a key class to distinguish one entity from another, whereas object-oriented systems do not require keys to individuate one object from another. Further, in those situations where more than one attribute or set of attributes will serve equally well for individuating IDEF1X entities, the modeler must designate one as the primary key and list all others as alternate keys. Explicit foreign key labeling is also required. The resulting logical design IDEF1X models are intended to be used by the programmers who take the blueprint for the logical database design and implement that design. However, the IDEF1X modeling language is sufficiently similar to IDEF1 in that models generated from the IDEF1 information requirements can be reviewed and understood by the ultimate users of the proposed system [16–18].

The IDEF1X modeling technique was developed to meet the following requirements:

1. Be a coherent language.
2. Be teachable
3. Support the development of the conceptual schemas.
4. Be well-tested and proven
5. Be automatable

An IDEF1X model is comprised of one or more views (often presented in view diagrams representing the underlying semantics of the views), and definitions of the entities and domains (attributes) used in the views. Each IDEF1X model must be accompanied by a statement of purpose (describing why the model was produced), a statement of scope (describing the general area covered by the model), and a description of any conventions the authors have used during its construction. Author conventions must not violate any of the rules governing model syntax or semantics. The components of an IDEF1X view are:

1. Entities

    a. Identifier-Independent Entities
    b. Identifier-Dependent Entities

2. Relationships

    a. Identifying Connection Relationships
    b. Non-Identifying Connection Relationships
    c. Categorization Relationships
    d. Non-Specific Relationships

3. Attributes/Keys
   a. Attributes
   b. Primary Keys
   c. Alternate Keys
   d. Foreign Keys
4. Notes

Entities represent the things of interest in an IDEF1X view. They are displayed in view diagrams, and defined in the glossary. An entity represents a set of real or abstract things (people, objects, places, events, ideas, combinations of things, etc.) which have common attributes or characteristics. An individual member of the set is referred to as an "entity instance." A real world object or thing may be represented by more than one entity within a view. For example, John Doe may be an instance of both the entity EMPLOYEE and BUYER. Furthermore, an entity instance may represent a combination of real world objects. For example, John and Mary could be an instance of the entity MARRIED-COUPLE. An entity is "identifier-independent" or simply "independent" if each instance of the entity can be uniquely identified without determining its relationship to another entity. An entity is "identifier-dependent" or simply "dependent" if the unique identification of an instance of the entity depends upon its relationship to another entity. The difference between these two concepts is shown in Fig. 2.3a. If the entity is identifier-dependent, the corners of the box are rounded. The numbers are separated by a slash ("/").

Domain is another building block of IDEF1X. A "Domain" represents a named and defined set of values that one or more attributes draw their values from. In IDEF1X, domains are defined separately from entities and views in order to permit their reuse and standardization throughout the enterprise. A domain is considered a class for which there is a fixed, and possibly infinite, set of instances. For example, State-Code would be considered a domain, where the set of allowable values for the domain would satisfy the definition of a state-code (e.g. the unique identifier of a state) and might consist of the two-letter abbreviations of the states. Domains are considered immutable classes whose values do not change over time. In contrast, entities are time-varying classes; their instance data varies over time as the data is modified and maintained. As immutable classes, domain instances always exist in principle. Take, for example, the domain Date, each instance of date did or will exist, however all instances of date might not be used as instances in an entity containing a date domain. An illustration of domain hierarchy is shown in Fig. 2.3b.

An IDEF1X "View" is a collection of entities and assigned domains (attributes) assembled for some purpose. A view may cover the entire area being modeled, or a part of that area. An IDEF1X model is comprised of one or more views (often presented in view diagrams representing the underlying semantics of the views), and definitions of the entities and domains (attributes) used in the views. In IDEF1X, entities and domains are defined in a common glossary and mapped to

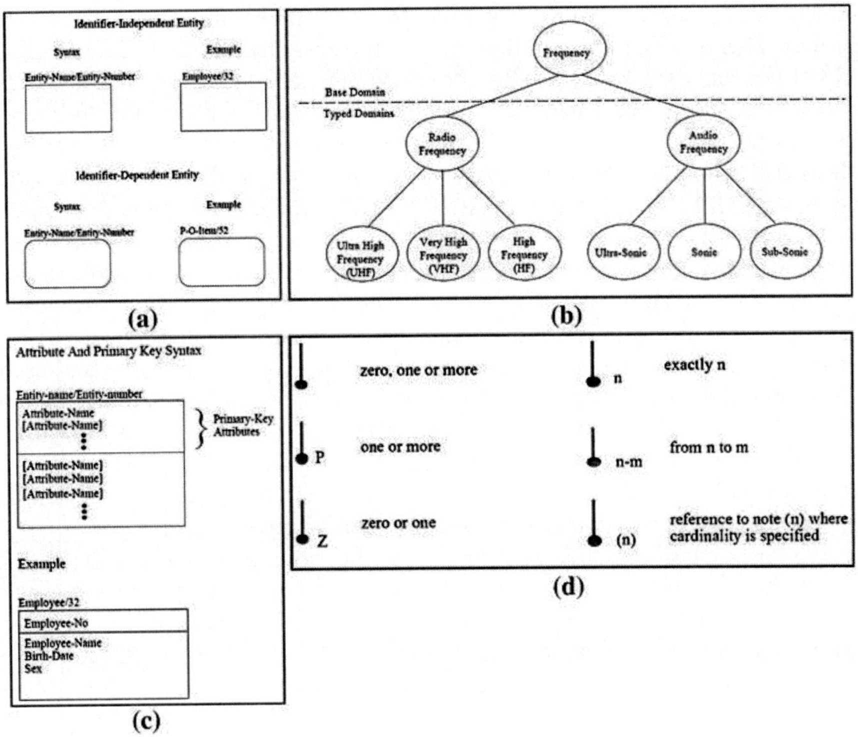

Fig. 2.3 IDEF1X building blocks. **a** Entity syntax. **b** Domain hierarchy. **c** Attribute and primary key syntax. **d** Relationship cardinality syntax

one another in views. In this way an entity such as EMPLOYEE may appear in multiple views, in multiple models, and have a somewhat different set of attributes in each. In each view, it is required that the entity EMPLOYEE mean the same thing. The intent is that EMPLOYEE be the class of all employees. That is, individual things are classified as belonging to the class EMPLOYEE on the basis of some similarity. It is that sense of what it means to be an employee that is defined in the glossary. Similarly, the domain EMPLOYEE-NAME is defined once, and used as an attribute in appropriate views.

A domain associated with an entity in a view is referred to as an "Attribute" of the entity. In an IDEF1X view, an "attribute" represents a type of characteristic or property associated with a set of real or abstract things (people, objects, places, events, ideas, combinations of things, etc.). Each attribute is identified by the unique name of its underlying domain. An "attribute instance" is a specific characteristic of an individual member of the set. An attribute instance is defined by both the type of characteristic and its value, referred to as an "attribute value." An instance of an entity, then, will usually have a single specific value for each associated attribute. For example, EMPLOYEE-NAME and BIRTH-DATE may be attributes associated with the entity EMPLOYEE. An instance of the entity EMPLOYEE could have the attribute values of "Jenny Lynne" and "February 27, 1953." An entity must have an attribute or combination of attributes whose values uniquely identify every instance of the entity. These attributes form the "primary-key" of the entity shown in Fig. 2.3c. For example, the attribute EMPLOYEE-NUMBER might serve as the primary key for the entity EMPLOYEE, while the attributes EMPLOYEE-NAME and BIRTH-DATE would be non-key attributes.

In an IDEF1X view, connection relationships are used to represent associations between entities. A "connection relationship" (also referred to as a "parent–child relationship") is an association or connection between entities in which each instance of one entity, referred to as the parent entity, is associated with zero, one, or more instances of the second entity, referred to as the child entity, and each instance of the child entity is associated with zero or one instance of the parent entity. For example, a specific connection relationship would exist between the entities BUYER and PURCHASEORDER, if a buyer issues zero, one, or more purchase orders and each purchase order must be issued by a single buyer. Figure 2.3d illustrates the relationship cardinality syntax.

If an instance of the child entity is identified by its association with the parent entity, then the relationship is referred to as an "identifying relationship", and each instance of the child entity must be associated with exactly one instance of the parent entity. For example, if one or more tasks are associated with each project and tasks are only uniquely identified within a project, then an identifying relationship would exist between the entities PROJECT and TASK. That is, the associated project must be known in order to uniquely identify one task from all other tasks. If every instance of the child entity can be uniquely identified without knowing the associated instance of the parent entity, then the relationship is referred to as a "non-identifying relationship." For example, although an existence-dependency relationship may exist between the entities BUYER and PURCHASEORDER, purchase orders may be uniquely identified by a purchase order number without identifying the associated buyer.

Categorization relationships are another major group of relationship in IDEF1X. They are used to represent structures in which an entity is a "type" (category) of another entity. A "categorization relationship" is a relationship between one entity, referred to as the "generic entity", and another entity, referred to as a "category entity". A "category cluster" is a set of one or more categorization relationships. An instance of the generic entity can be associated with an instance of only one of the category entities in the cluster, and each instance of a category entity is associated with exactly one instance of the generic entity. Each instance of the category entity represents the same real-world thing as its associated instance in the generic entity. For example, EMPLOYEE is the generic entity and SALARIED-EMPLOYEE and HOURLY-EMPLOYEE are the category entities. There are two categorization relationships in this cluster, one between EMPLOYEE and SALARIED-EMPLOYEE and one between EMPLOYEE and HOURLY-EMPLOYEE.

2.2.3 EXPRESS Language

EXPRESS is designed as a language for communicating information concerning data. It has much in common with some database definition languages and some programming languages, all of which can be used to define the structure of data. Unlike a database language, such as SQL [19], or a programming language, such as C [20], EXPRESS does not confuse the information modeling task with programming or database design tasks, and it is not specific to a particular programming or database system.

EXPRESS is similar to the data description half of object-oriented programming languages such as C++ [21]. That is, EXPRESS supports describing the data structure of an object, but objects do not have any executable functions or methods. In EXPRESS, the definition of a type of object is called an entity, rather than a class (the term used in C++). A property of an entity is called an attribute of the entity. For example, an attribute of a circle is its diameter. Like other object-oriented languages, EXPRESS supports parent/child relationships among entities. The parent is called a supertype of the child, which is called a subtype of the parent. A subtype entity inherits all the attributes of its supertypes [18, 22].

In order to support defining attributes, EXPRESS has built-in:

- simple data types (e.g. STRING and INTEGER)
- aggregates (e.g. ARRAY, LIST, and SET)

A logically complete set of entity definitions is called a schema. In addition to entity definitions, a schema can contain data type definitions and various kinds of constraints on instances of entities. EXPRESS includes a rich set of methods for describing constraints. In addition to making it possible to state rules about the attributes of a single entity, EXPRESS supports stating rules that apply to entire populations of instances of one or more data types.

EXPRESS is a completely generic modeling language and can, therefore, be used to model data objects of any type. It is a formal language for the definition of entity-attribute data models. Its original use was for the definition of standard data models describing 3D graphical representations of physical objects, i.e., Computer-aided Design (CAD) drawings. The EXPRESS language is completely declarative and implementation independent, making it well suited for the definition of standardized data models. On the other hand, EXPRESS is a data modeling language, which means it only defines entities and their properties, and does not define methods that might be applied to those entities in an application context [23].

The EXPRESS information model is organized into schemas (Fig. 2.4). These schemas contain the model definitions and serve as a scoping mechanism for subdividing large information models. In order to support stating complex rules, EXPRESS supports writing functions and has built-in:

- arithmetic operators and expressions (e.g. A+2)
- logical operators and expressions (e.g. A .AND. B)

```
SCHEMA example;

TYPE date = ARRAY [1:3] OF INTEGER;
END_TYPE;

TYPE materialType = ENUMERATION OF (steel,
                 stainless,bronze, Al, copper);
END_TYPE;

ENTITY thread
   ABSTRACT SUPERTYPE OF (ONEOF(male,female));
   diameter : REAL;
   pitch : REAL;
   starts : REAL;
   StartDate : date;
   material : materialType;
DERIVE
   worktime: INTEGER := days(StartDate);
END_ENTITY;

ENTITY female SUBTYPE OF (thread);
INVERSE
   bolt : SET [0:1] OF male FOR nut;
END_ENTITY;

ENTITY male SUBTYPE OF (thread);
   nut : OPTIONAL female;
END_ENTITY;

FUNCTION days(past : date) : INTEGER;
   (* This function calculates the number of dayss
     between the past date and the current date *)
END_FUNCTION;

END_SCHEMA;
```

Fig. 2.4 An example of EXPRESS schema

- numerical functions (e.g. cos(x))
- operators on aggregates (e.g. sizeof)
- methods of describing a set of objects (e.g. all circles with radius less than 1)
- entity equality test operators

EXPRESS functions may also be used for computing values of derived attributes. Figure 2.4 shows an example schema taken from the EXPRESS manual, Part 11 of STEP standard [9]. The schema says that a thread must be a male or a female thread. Each kind of thread has diameter and pitch, number of starts, type of material, manufacturing date (start date). Work time of a thread is calculated using the days function. A male bolt has to be used with a nut, in which case the female nut has an inverse relationship to the male bolt.

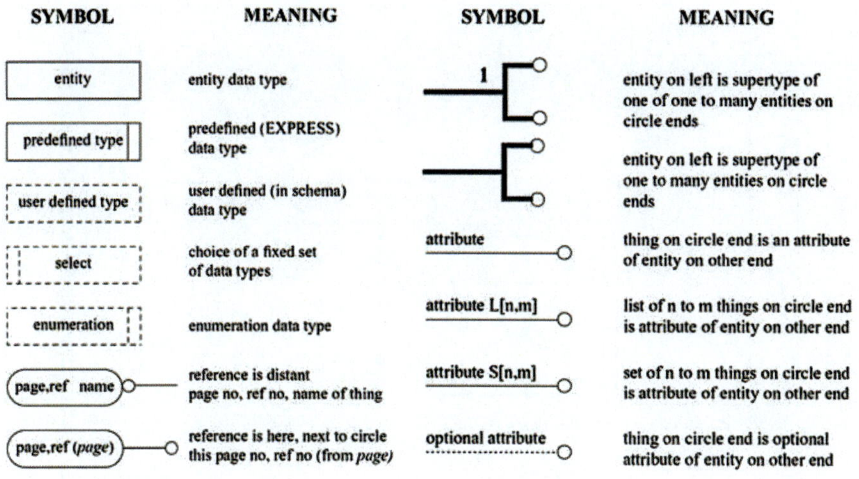

Fig. 2.5 EXPRESS-G notations

Another description method (also given in Part 11) is a graphical form of EXPRESS called EXPRESS-G. An EXPRESS-G diagram shows the following functions and their notations are shown in Fig. 2.5.

- entities in solid boxes
- simple data types in solid boxes with a double line on the right end
- defined data types in boxes with dashed borders
- enumeration data types in boxes with dashed borders and a double line on the right end
- subtypes as a thick solid line connecting a supertype entity to a subtype entity with a circle at the subtype end
- required attributes as a thin solid line connecting an entity to an attribute of the entity, with a circle at the attribute end and the name of the attribute (and any aggregate description) in text next to the line
- optional attributes as a thin dashed line connecting an entity to an attribute of the entity, with a circle at the attribute end and the name of the attribute (and any aggregate description) in text next to the line and more

The EXPRESS-G diagram for the EXPRESS schema shown in Fig. 2.4 is shown in Fig. 2.6.

2.2.4 XML Schema Definition Language

XML is a simple and flexible text format derived from SGML (ISO 8879) [24]. Originally designed to meet the challenges of large-scale electronic publishing, XML is now playing an increasingly important role in the exchange of a wide variety of data on the Web and elsewhere. The XML schema language is a

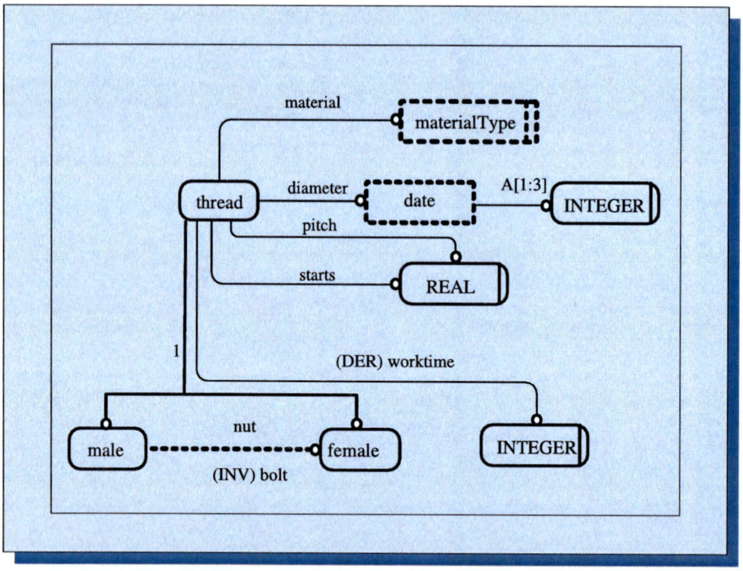

Fig. 2.6 An example of EXPRESS-G diagram

formalization of the constraints, expressed as rules or a model of structure, that apply to a class of XML documents [25]. XML schema was published as a W3C Recommendation in May 2001 [26]. It was the first separate schema language for XML to achieve Recommendation status by the W3C. Because of the confusion between XML Schema as a specific W3C specification, and the use of the same term to describe schema languages in general, some parts of the user community referred to this language as WXS, which stands for W3C XML Schema, while others referred to it as XSD that stands for XML Schema Document—a document written in the XML Schema language, typically containing the "xsd" XML namespace prefix and stored with the "xsd" filename extension. In the draft of the next version—XML Schema 1.1, the W3C has chosen to adopt XSD as the preferred name.

In many ways, XML schemas serve as design tools establishing a framework on which implementations can be built. Since formalization is a necessary ground for software designers, formalizing the constraints and structures of XML instance documents can lead to very diverse applications. An XML schema can be used to express a set of rules to which an XML document must conform in order to be considered 'valid' according to that schema. However, XML schema was also designed with the intent that a processor that makes a determination of a document's validity would also produce a collection of information adhering to specific data types. Such a post-validation information set can be useful in the development of XML document processing software. To produce such a set of information, data structures to hold the information must be defined. Automated tools have been built that will produce, for example, a set of C++ classes from an XML schema.

Validation is the most common use of schemas in the XML world. There are many reasons and opportunities to validate an XML document. For example, when an XML file is received, before importing data into a legacy system, a validation of the XML file to the schema is normally carried out. Another example is when an XML file is generated or hand-edited, before sending it out for an application; such validation is needed to check the possible errors in the XML file. In all the above cases, a schema helps to accomplish a substantial part of the job. Different kinds of schemas perform different kinds of validation, and some especially complex rules may be better expressed in procedural code rather than in a descriptive schema, but validation is generally the initial purpose of a schema, and often the primary purpose as well.

Validation can take place at several levels. Structural validation makes certain that XML element and attribute structures meet specified requirements, but does not clarify much about the textual content of those structures. Data validation looks more closely at the contents of those structures, ensuring that they conform to rules about what type of information should be present. Other kinds of validation, often called business rules, may check relationships between information and a higher level of sanity-checking, but this is usually the domain of procedural code, not schema-based validation.

XML schemas are also frequently used to document XML vocabularies, even when validation is not a requirement. Schemas provide a formal description of the vocabulary with a precision and conciseness that can be difficult to achieve in prose. It is very unusual to publish the specification of a new XML vocabulary without attaching some form of XML schema. The machine-readability of schemas gives them several advantages as documentation. Human-readable documentation can be generated from the schema's formal description.

Technically, a schema is an abstract collection of metadata, consisting of a set of schema components including element and attribute declarations, complex and simple type definitions [27]. These components are usually created by processing a collection of schema documents, which contain the source language definitions of these components. In popular usage, however, a schema document is often referred to as a schema. Schema documents are organized by namespace, which means that all the named schema components belong to a target namespace, and the target namespace is a property of the schema document as a whole. A schema document may include other schema documents for the same namespace, and may import schema documents for a different namespace. When an instance document is validated against a schema (a process known as assessment), the schema to be used for validation can either be supplied as a parameter to the validation engine, or it can be referenced directly from the instance document using two special attributes, xsi:schemaLocation and xsi:noNamespaceSchemaLocation. (The latter mechanism requires the client invoking validation to trust the document sufficiently to know that it is being validated against the correct schema.) XML Schema Documents usually have the filename extension ".xsd".

Figure 2.7 shows an example XML file. It is named example.xml. The file consists of a main element, "threads", which includes descriptions of two mating

Example
Thread data file: example.xml

```
<?xml version="1.0" encoding="UTF-8"?>
<threads xmlns="urn:Threads"
 xmlns:thr="http://www.w3.org/2001/XMLSchema-instance"
 thr:schemaLocation="urn:Threads example.xsd">
 <male startDate="2010-12-17">
  <id>1</id>
  <diameter>0.5</diameter>
  <pitch>0.1</pitch>
  <starts>2</starts>
  <startDate>2010-12-17</startDate>
  <material>steel</material>
  <nut>2</nut>
 </male>
 <female startDate="2010-12-17">
  <id>2</id>
  <diameter>0.5</diameter>
  <pitch>0.1</pitch>
  <starts>2</starts>
  <startDate>2010-12-17</startDate>
  <material>steel</material>
  <bolt>1</bolt>
 </female>
</threads>
```

Fig. 2.7 An XML example file

threads, each of which has several subelements. The lowest level subelements contain primitive data such as numbers or strings. Elements that contain subelements or carry attributes are said to have complex types, whereas elements that contain numbers (and strings, and dates, etc.) but do not contain any subelements are said to have simple types. Some elements have attributes; attributes always have simple types.

Figure 2.8 shows the XML schema corresponding to the example XML file in Fig. 2.7. It is named example.xsd and models the same sort of thread data as does the EXPRESS schema shown in Fig. 2.4. The threads schema consists of a "schema" element and a variety of subelements, most notably "element", "complexType", "simpleType", "sequence", and "choice", which determine the appearance and content of elements of instance documents. Each of the elements in the schema has a prefix xsd: which is associated with the XML schema namespace through the declaration xmlns:xsd="http://www.w3.org/2001/XMLSchema", that appears in the schema element. The prefix xsd: is used by convention to denote the XML Schema namespace, although any prefix can be used. The same prefix, and hence the same association, also appears on the names of built-in simple types, e.g. xsd:string. The purpose of the association is to identify the elements and simple types as belonging to the vocabulary of the XML Schema language rather than the vocabulary of the schema author.

A comparison of Fig. 2.4 with 2.8 shows interesting differences and similarities between EXPRESS and XML schema.

- In XML schema, there must be a single top-level element ("threads" in Fig. 2.8), and an instance document must start with that element. There is no comparable requirement for an EXPRESS schema, although each instance document corresponding to an EXPRESS schema must have an element to which everything else is connected. Under the XML schema, an instance document consists of instances one or more male or female threads.
- In EXPRESS, there is no built-in date type, so it is necessary to define one. XML schema has the date type built in, so an explicit definition of the type is not needed. If the type were not built into XML schema, the closest one could come to the EXPRESS definition of date would be to make a list containing exactly three positiveInts, since XML schema has no arrays.
- Both EXPRESS and XML schema support enumerations, as shown in the definitions of "materialType".
- Both EXPRESS and XML schema support subtyping. In both schemas, thread has male and female subtypes.
- XML schema itself does not provide for defining functions that may be used for expressing constraints or finding values of derived attributes, so Fig. 2.8 has no counterpart to the DERIVE and FUNCTION items of Fig. 2.4. It may be possible, however, to represent functions by using MathML http://www.w3.org/TR/2010/REC-MathML3-20101021.
- The idea of an "abstract" type is that it can be used as the parent of another type but cannot be instantiated in an instance document. In EXPRESS, it is necessary to make an explicit ABSTRACT declaration. In XML schema there are both explicit and implicit ways to make a type abstract. The implicit method is shown in Fig. 2.8. The abstractThreadType in that figure cannot appear in an instance document simply because there is nowhere abstractThreadType appears in the statements that define what can appear in an instance document. The definition of abstractThreadType could be modified by adding 'abstract="true"', but that would add no functionality (except that a schema processor would signal an error if some part of the schema tried to enable abstractThreadType to appear in an instance document).
- XML schema has no straightforward equivalent of the EXPRESS INVERSE statement. In the EXPRESS of Fig. 2.4, if a male thread has a nut, then the female thread that is the nut automatically has a bolt that is the correct male thread. In XML schema, the male thread may have a nut and the female thread may have a bolt, but the schema does not specify that they correspond correctly.
- XML schema provides two alternative methods (attribute or element) of assigning what is called an attribute in EXPRESS. To make this point, startDate has been included both as an attribute and as an element in Fig. 2.8. It is not clear why XML schema provides both methods. A common explanation is that metadata

Fig. 2.8 The XML schema of the example XML file

Example
Thread schema file: example.xsd

```
<?xml version="1.0"?>
<xsd:schema
 xmlns:xsd="http://www.w3.org/2001/XMLSchema"
 xmlns:thr="urn:Threads"
 targetNamespace="urn:Threads"
 attributeFormDefault="unqualified"
 elementFormDefault="qualified">

 <xsd:element name="threads"
  type="thr:threadType"/>

 <xsd:simpleType name="materialType">
  <xsd:restriction base="xsd:string">
   <xsd:enumeration value="steel"/>
   <xsd:enumeration value="stainless"/>
   <xsd:enumeration value="bronze"/>
   <xsd:enumeration value="Al"/>
   <xsd:enumeration value="copper"/>
  </xsd:restriction>
 </xsd:simpleType>

 <xsd:complexType name="threadType">
  <xsd:choice
   maxOccurs="unbounded">
   <xsd:element name="male"
    type="thr:maleThreadType"/>
   <xsd:element name="female"
    type="thr:femaleThreadType"/>
   </xsd:choice>
 </xsd:complexType>

 <xsd:complexType name="abstractThreadType">
  <xsd:sequence>
   <xsd:element name="id"
    type="xsd:unsignedInt"/>
   <xsd:element name="diameter"
    type="xsd:double"/>
   <xsd:element name="pitch"
    type="xsd:double"/>
   <xsd:element name="starts"
    type="xsd:positiveInteger"/>
   <xsd:element name="startDate"
    type="xsd:date"/>
   <xsd:element name="material"
    type="thr:materialType"/>
  </xsd:sequence>
  <xsd:attribute name="startDate"
    type="xsd:date"/>
 </xsd:complexType>

 <xsd:complexType name="femaleThreadType">
  <xsd:complexContent>
   <xsd:extension base="thr:abstractThreadType">
    <xsd:sequence>
     <xsd:element name="bolt"
      type="xsd:unsignedInt"
      minOccurs="0"/>
    </xsd:sequence>
   </xsd:extension>
  </xsd:complexContent>
 </xsd:complexType>

 <xsd:complexType name="maleThreadType">
  <xsd:complexContent>
   <xsd:extension base="thr:abstractThreadType">
    <xsd:sequence>
     <xsd:element name="nut"
      type="xsd:unsignedInt"
      minOccurs="0"/>
     </xsd:sequence>
    </xsd:extension>
   </xsd:complexContent>
  </xsd:complexType>
</xsd:schema>
```

(data about data) should be stored as attributes, and that data itself should be stored as elements, but in practice, the distinction is often obscure. See, for example, http://www.ibm.com/developerworks/xml/library/x-eleatt.html for a discussion of attributes versus elements in XML. At least one popular commercial XML manipulation system will signal an error if an element is omitted but not if an attribute is omitted.

Part 28 [28] of STEP ISO 10303 provides a mapping of EXPRESS to XML schema so that data governed by an EXPRESS schema may be represented in an XML instance document. The schema shown in Fig. 2.8 was not constructed using the Part 28 mapping rules.

2.2.5 Implementation of Information Models

An information model provides a sharable, stable, and organized structure of information requirements. It is developed to preserve independence from both usage and implementation. Implementation independence allows users to select their implementation methods. Three types of implementation methods are currently used by the manufacturing community [3]:

- data transfer via a working form, which is a structured, in-memory representation of data,
- data transfer via an exchange file, which is a file with a predefined structure or format, and
- data transfer using a database management system.

These implementation methods can be accomplished through programming languages. The first type of implementation method uses the mechanism that accesses and changes data without actually moving the data around. All shared data are stored in memory. The second type of implementation method requires a neutral file format for storing the data. Normally a standard or specification defines the neutral data format. The application systems read and write from the standardized data files. The third type of implementation method uses a DBMS where information is mapped onto and retrieved from databases. The selection of an implementation method is heavily dependent on the target environment where the application system resides.

Dimensional metrology systems consist of many software modules. These software modules not only interact with each other but also exchange information or commands with hardware systems. The information exchange between these software and hardware models and systems is where interoperability issues exist. The above sections introduced the most commonly used information modeling languages. All of the introduced information modeling languages have been chosen as the data modeling language for some of the standards and specifications for dimensional metrology systems. In the next section, the elements or subsystems that exist in a typical dimensional metrology system will be discussed.

2.3 Communications of Dimensional Metrology Systems

Information technology has played an important role in the continuing efforts of industries worldwide to make their manufacturing systems, processes, and overall enterprises more agile and productive [29]. Enterprise information technology is having a major impact on all manufacturing organizations, large or small. It especially has had an impact on the system applications for coordinating the complex array of manufacturing data, business decisions, work flows and processes, which have become increasingly important for today's enterprises to maintain their competitive edge. From that standpoint, information technology has played an important role in the effort to make organizations more integrated and productive. Dimensional metrology is a critical part of any manufacturing system. As it is introduced in Chap. 1, dimensional metrology is related to both manufacture and function. Whether the designed workpiece geometry and shapes are successfully manufactured or not depends on the verification of dimensional measurements.

However, it is important for the readers to understand that there is more to the dimensional measurement process than just analyzing the dimensions and tolerances of manufactured components. The product design specifications must be taken into account in planning the measurement process; the measurement process must be executed to obtain appropriate measurement data; the data must be analyzed and the results reported in a way that accepts/rejects the component and provides feedback to the manufacturing processes that produced the component. These processes are supported by many software applications, including those that are incorporated into machine tools. The entire dimensional measuring system is most effective if the software applications are seamlessly integrated together at the information interfaces. In industry, dimensional metrology data is very important because it is intimately tied to a company's product quality and performance assessment efforts. Dimensional metrology data has to be shared easily with production scheduling, design, purchasing, and many other manufacturing company functions [30]. Ideally, a manufacturer should be able to acquire and store any type of dimensional measurement information in the same format regardless of the type of equipment used to acquire it.

The concept of dimensional metrology interoperability is defined as "the ability of two system components to communicate correctly and completely with each other—with minimal cost to either component user or component vendor, where the components can come from any vendor worldwide" [2]. This concept is used to address the issues that complicate the measurement process. Component-to-component interoperability using open standards reduces training costs, allows best-in-class component choices, and provides a more innovative and competitive technology provider environment—thus providing the promise of reduced cost for Original Equipment Manufacturers (OEMs), technology providers, suppliers, and consumers. The main challenge to achieve dimensional metrology interoperability is to specify a minimum set of information exchange standards to provide coverage

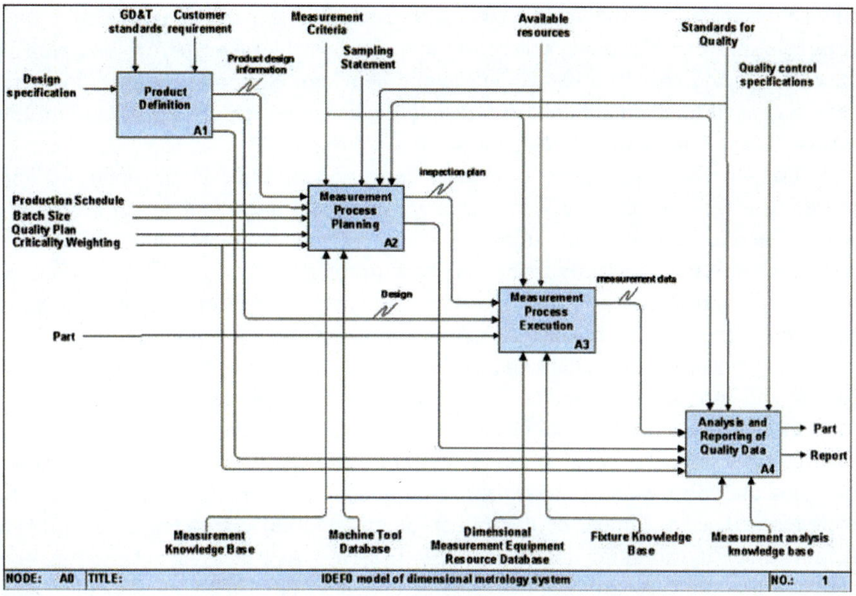

Fig. 2.9 The IDEF0 activity model of a typical dimensional metrology system

for the information exchanges required that will also enable integration for the full range of software applications presently available and provide a framework for those that are likely to be available in the future.

It is important for the readers to comprehend the main elements of a typical dimensional metrology system. The total dimensional metrology process can be divided into four major interacting elements: product definition, measurement process planning, measurement process execution, and analysis and reporting of quality data [31]. Figure 2.9 shows the relationship between these four elements in an IDEF0 activity model.

Product definition (activity A1 in Fig. 2.9) is the process in which a part is designed using CAD software based on customer requirements. From the perspective of dimensional metrology, the most important function of the product definition activity is to provide sufficient information to permit the generation of a downstream measurement process definition activity (activity A2 in Fig. 2.9). Such information must include things like part geometry, features, tolerances, and relevant part characteristics such as surface finish, reflectance, and material properties. The measurement process planning activity produces the process plan to measure the part so that the functionality of the part is ensured during/after the manufacturing process.

Then, the measurement process execution is carried out to execute measurement plans (activity A3 in Fig. 2.9). This activity is not as simple as it sounds. It needs to support not only the huge number of different types of measurement equipment, but also an almost limitless number of ways in which a complex part can be inspected. If the measurement process plan generated upstream is not complete and unambiguous, corrective actions must be taken before this plan is executed on the chosen measurement equipment. For example, a translation process may need to be carried out to translate the measurement process plan into the format that is compatible with the available equipment.

Following the measurement process execution activity is the analysis and reporting activity (activity A4 in Fig. 2.9). The most important functions of this activity are to receive input from measurement process execution and product definition activities, to analyze the part measurement data in terms of production definition requirements, to perform a statistical analysis of the measurement results and present them in the form of a statistical process control report, and to archive whatever measurement values and derived statistics are necessary for legal protection and support of other data retention compliance policies.

Each of these activities can be broken down into sub-activities. Some of these sub-activities involve only software modules, and some involve both software modules and dimensional measuring equipment hardware. A software module includes at least a library of compiled (or compilable) computer code with an Application Programming Interface (API) [32]. An API is a set of functions the software module can carry out that may be called by other software modules. The information communicated between these software modules is where the interoperability achieved or not. Interoperability is not a matter of how the information is generated within each sub-activity, but rather a matter of the assumed syntax and semantics of the information passed from one sub-activity to another.

John Evans, et al. from NIST [32] carried out a thorough analysis of the total metrology system and identified fifteen activities. This effort was an attempt to identify logical activities, whether or not these activities actually correspond to separable software modules in actual vendor products. These activities are summarized as following. This analysis produced a flow chart (Fig. 2.10) to display these fifteen activities and the producer and user interfaces between them.

1. Activity Coordination—the process of planning what other activities will take place and when they will take place, assigning resources to planned activities, and giving the orders necessary to carry out the plans.
2. CAD—the process of producing a design using a computer.
3. Hand-held Device Measuring—the process of measuring things with a hand-held device.
4. High-level Measurement Instruction Execution—the process of executing high-level inspection instructions, such as statements from a Dimensional Measuring Interface Standard (DMIS) program [33].
5. Inspection Planning—the process of deciding what to inspect and how to inspect it.
6. Inspection Programming—the process of producing a high-level inspection program, for example a DMIS program.

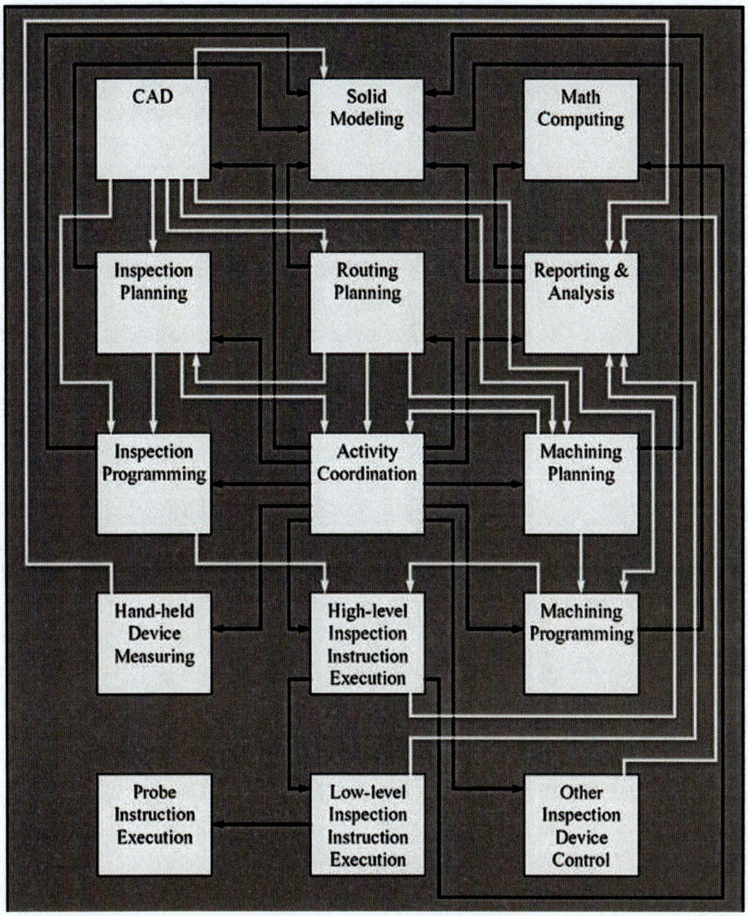

Fig. 2.10 Interfaces in a dimensional metrology system [32]

7. Low-level Inspection Instruction Execution—the process of executing low-level inspection instructions, such as set or get search distance, or probe a point.
8. Machining Planning—the process of deciding what operations should be performed on a machine tool to produce the part.
9. Machining programming—the process of generating a program that may be run on a machine tool controller to cut the part. This activity together with the previous activity deal with inspection functions that may be programmed on a machine tool for on-machine measurement.
10. Math Computing—the process of performing mathematical calculations. It is expected that only relatively sophisticated, difficult, or time-consuming calculations will be performed in Math Computing, such as fitting features to sets of points.

11. Other Inspection Device Control—the process of controlling other dimensional metrology equipment, such as theodolites and photogrammetry equipment.
12. Probe Instruction Execution—the process of executing instructions sent to a probe sensor.
13. Reporting and Analysis—the process of collecting inspection reports, analyzing data returned from inspection activities, and generating files and graphical representations of analyzed or unanalyzed data.
14. Routing Planning—the process of deciding which fabrication and inspection activities will take place at which workstations.
15. Solid Modeling—the process of building a representation of solid objects and performing calculations done on solid objects, such as determining the mass of an object, determining if a given point lies on the surface of an object, or Boolean subtracting one object from another.

Activity coordination is at the center of the chart as a user of ten other metrology activities. CAD for example is a producer of dimensional metrology data that interfaces with such activities as machine programming, routing planning, and inspection programming, among others. The white arrows in the figure show the interface from producer to consumer for data interfaces. The black arrows show the interface from user to used for active interface. Here, an active interface is one in which a command is given. The relationship between the two parties attached by an active interface may be either supervisor-subordinate or client–server. In both cases there is a user module and a used module. In a data interface, a producer puts data into a file system or data base system, and a consumer retrieves data from that file system or database system. Although there is no direct contact between producer and consumer, both must have the same understanding of the format and meaning of the data.

In today's manufacturing systems, it is common that several of these software modules are combined in commercial software systems. For example, the production definition software includes a CAD software module, allowing definition of part geometry and associated GD&T. The measurement process definition software includes solid modeling, inspection planning, and inspection programming modules. The measurement process execution software includes math computing, high-level inspection instruction execution, low-level inspection instruction execution, and probe instruction execution modules. The report and analysis software includes solid modeling, math computing, and reporting and analysis modules. The interfaces between these software modules inside a commercial system become invisible to users. However, the interfaces between these commercialized software systems are exposed. These commercial software systems can be categorized into four groups corresponding to the four main elements of dimensional metrology systems as it is shown in Fig. 2.11. Barriers (represented by the dotted lines in the figure) exist not only between these four elements but also within some of the elements themselves.

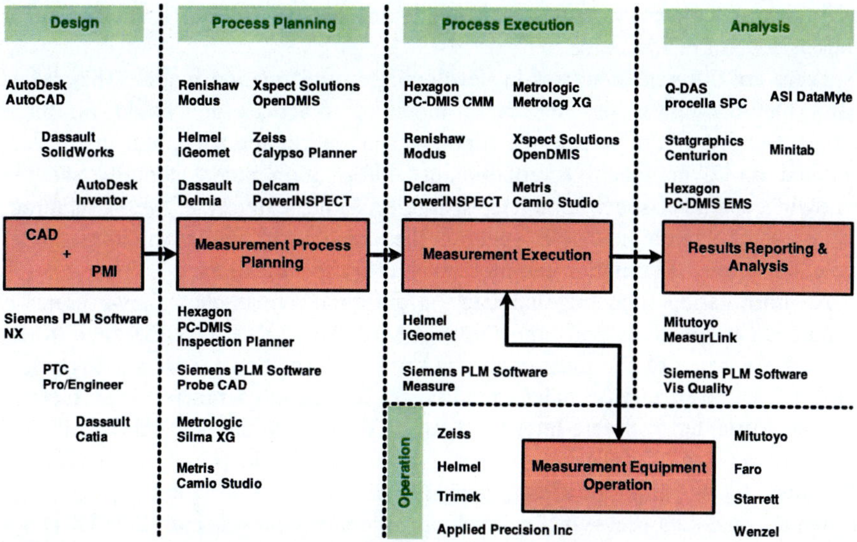

Fig. 2.11 Interoperability barriers between dimensional metrology commercial software systems [34]

For interoperable dimensional metrology, clear and unambiguous metrology information is needed to flow across each of these interfaces. This is accomplished best through the definition and worldwide implementation of information interface standards. However, there are multiple standards and specifications for each element of dimensional metrology systems. The information defined in these standards is sometimes overlapping. For example, the common data exchange standards for CAD software include Initial Graphics Exchange Specification (IGES) [35], STEP [36], Portable Document Format (PDF) [12], etc. These standards use different information modeling languages. They cover different aspects of design information in the data model; therefore they have advantages and disadvantages. In the next four chapters, the functionalities of each of the four dimensional metrology elements will be described in detail. The current standards and specifications that exist for the information exchange in each element, their information modeling techniques, and their data models are also discussed.

2.4 Summary

Information technology has become increasingly important in the manufacturing enterprise. Effective information sharing and exchange among computer systems throughout a product's life cycle has been a critical issue. Information modeling is a technique for specifying the data requirements that are needed within the application domain. An information model is a representation of concepts, relationships, constraints, rules, and operations to specify data semantics for a chosen domain of discourse.

There are different practices in developing an information model. The underlying methodologies for the recent modeling practices are based on three approaches: the ER approach, the functional modeling approach, and the object-oriented (O-O) approach. A good-quality information model should have the following characteristics: complete, sharable, stable, extensible, well-structured, precise, and unambiguous. In general, the contents of an information model include a scope, information requirements, and a specification.

An information modeling language is a formal syntax that allows users to capture data semantics and constraints. UML, IDEF1X, EXPRESS, and XML Schema are some of the most commonly used information modeling languages. UML is used to specify, visualize, modify, construct and document the artifacts of an object-oriented software-intensive system under development. UML offers a standard way to visualize a system's architectural blueprints. IDEF1X was developed for designing relational databases with a syntax designed to support the semantic constructs necessary in developing a conceptual schema. IDEF1X is not particularly suited to serve as an AS-IS analysis tool, although it is often used in that capacity as an alternative to IDEF1. IDEF1X is most useful for logical database design after the information requirements are known and the decision to implement a relational database has been made. EXPRESS is designed as a language for communicating information concerning data. It has much in common with some database definition languages and some programming languages, all of which can be used to define the structure of data. EXPRESS does not confuse the information modeling task with programming or database design tasks, and it is not specific to a particular programming or database system. XML schemas serve as design tools establishing a framework on which implementations can be built. XML schema can be used to express a set of rules to which an XML document must conform in order to be considered 'valid' according to that schema.

The concept of dimensional metrology interoperability is defined as "the ability of two system components to communicate correctly and completely with each other—with minimal cost to either component user or component vendor, where the components can come from any vendor worldwide". The total dimensional metrology process can be divided into four major interacting elements: product definition, measurement process planning, measurement process execution, and analysis and reporting of quality data.

Several software and hardware sub-systems exist in each of the dimensional metrology element systems. The interfaces between these software systems and also between software and hardware systems become exposed. They are where the interoperability issues occur. For interoperable dimensional metrology, clear and unambiguous metrology information is needed to flow across each of these interfaces. This is accomplished best through the definition and worldwide implementation of information interface standards.

There are multiple standards and specifications for each element of dimensional metrology systems. The information defined in these standards is sometimes overlapping. Different information modeling languages are also chosen for different standards, which may also cause interoperability issues.

References

1. National Research Council (1995) Information technology for manufacturing. National research council report
2. IMTI (2006) A roadmap for metrology interoperability. Integrated Manufacturing Technology Initiative (IMTI Inc), Oak Ridge
3. Lee YT (1999) An overview of information modeling for manufacturing systems integration. NISTIR 6382. National institute of standards and technology
4. Chen PP (2004) The entity relationship model towards a unified view of data. ACM Trans Database Syst (TODS) 1(1):1976
5. Tsichritzis D, Klug A (1978) The ANSI/X3/SPARC DBMS framework report of the study group on database management systems. Inform Syst 3(3):173–191
6. Badia A (2004) Entity-relationship modeling revisited. SIGMOD Rec 33(1):77–82
7. UML Unified Modeling Language (2010). http://www.uml.org/
8. I D Appleton Company (1985) Integrated information support system: information modeling manual, IDEF1—Extended (IDEF1X)
9. ISO (2004) ISO 10303-11: industrial automation systems and integration—product data representation and exchange—Part 11:description methods: the EXPRESS language reference manual
10. Schenck D, Wilson P (1994) Information modeling the EXPRESS way. Oxford University Press, New York
11. XML Extensible Markup Language (2010) http://www.w3.org/XML/
12. ISO (2008) ISO 32000-1:2008: document management—portable document format—Part 1: PDF 1.7
13. Fowler M, Scott K (2004) UML distilled: a brief guide to the standard object modeling language, 3rd edn. Addison-Wesley, Reading
14. OMG (2009) Object management group unified modeling language (OMG UML) superstructure version 2.2
15. Holt J (2004) UML for systems engineering: watching the wheels, 2nd edn. The Institution of Engineering and Technology, UK
16. IDEF Integrated DEFinition Methods (2010) http://www.idef.com/idef1X.htm(cited 19th of October 2010)
17. N.I.o.S.a.T. (NIST) (1993) Integration definition for information modeling (IDEF1X)
18. Kemmerer SJ (1999) STEP: the grand experience. National Institute of Standards and Technology, Gaithersburg
19. ISO (1992) ISO/IEC 9075:1992, Information technology—database languages—SQL
20. ISO (1990) ISO/IEC 9899:1990, Programming languages—C
21. C++ (2010) http://en.wikipedia.org/wiki/Microsoft_Visual_Studio (cited 22 October 2010)
22. Kramer TR, Xu X (2009) STEP in a Nutshell. In: Xu X, Nee AYC (eds) Advanced design and manufacturing based on STEP. Springer, UK, pp 1–19
23. Zhao YF (2009) An integrated process planning system for machining and inspection. Department of mechanical engineering. Ph.D thesis, University of Auckland
24. ISO (1986) ISO 8879:1986: information processing—text and office systems—standard generalized markup language (SGML)
25. E.V.d. Vlist (2002) XML Schema. O'Reilly Series: O'Reilly Media, Inc
26. W3C. XML Schema Definition Language (2001) http://xml.coverpages.org/schemas.html#W3CWorkingGroup (cited 22 October 2010)

27. W3C (2004) XML Schema Part 0: Primer 2nd edn
28. ISO (2002) ISO 10303-28:Industrial automation systems and integration—product data representation and exchange—Part 28: implementation methods: XML representations of EXPRESS schema and data
29. Shaw MJ, Seidmann A, Whinston AB (1997) Information technology for automated manufacturing enterprises: recent developments and current research issues. Int J Flex Manuf Syst 9:115–120 (Compendex)
30. Chalmers RE (2002) Metrology for manufacturing means business. Manuf Eng. 128(4)
31. Proctor F et al (2007) Interoperability testing for shop floor measurement. Performance Metrics for intelligent systems (PerMIS) Workshop
32. Evans J et al (December, 2001) Analysis of dimensional metrology standards. NISTIR 6847, National institute of standards and technology
33. ANSI (2004) Dimensional measuring interface standard, DMIS 5.0 Standard, Part 1, ANSI/CAM-I 105.0-2004, Part 1
34. Zhao YF et al (2009) Dimensional metrology interoperability and standardization in manufacturing systems. Comput Stand Interfaces 33:541–555
35. N.I.o.S.a.T. NIST (1988) Initial graphics exchange specification (IGES) Version 4.0, NBSIR pp 88–3813
36. ISO (1994) ISO 10303-1: industrial automation systems and integration—product data representation and exchange—Part 1: industrial automation system and integration—product data representation and exchange Part 1: overview and fundamental principles

Chapter 3
Product Definition and Dimensional Metrology Systems

Product definition is the process in which a part is designed using CAD design software based on customer requirements. One of the key activities in any product design process is to develop a geometric model of the product from the conceptual ideas, which can then be augmented with further engineering information pertaining to the application area. For example, the geometric model of a design may be developed to include material and manufacturing information so that it can later be used in Computer-Aided Process Planning and Manufacturing (CAPP/CAM) and quality control activities. A geometric model is also a must for any engineering analysis such as Finite Element Analysis (FEA). In mathematic terms, geometric modeling is concerned with defining geometric objects using computational geometry, which is often represented through computer software or rather a geometric modeling kernel. Geometry may be defined with the help of a wireframe model, surface model or solid model. Geometric modeling has now become an integral part of any CAD system.

From the perspective of dimensional metrology, the most important function of the product definition activity is to provide sufficient information to permit the automatic generation of a downstream measurement process activity. Such information must include things like part geometry, features, tolerances, measurement resource (e.g. CMM and sensor) specifications, and relevant part characteristics such as surface finish, reflectance, and material properties.

To support automatic dimensional metrology plan generation, in the simplest case, a product that consists of a single monolithic part provides a good example. The part must be decomposed into geometric features. Dimensions and tolerances must then be assigned to a geometric feature, or set of features. Datum features must be defined in such a way that they are appropriate both for manufacturing the part and for inspecting it. It is not uncommon that datum features are not the same for manufacturing and for inspecting purposes. Surface texture information must be included in the model, along with relevant information about the orientation or lay of the surface texture to be measured. Accurately extracting this type of information would require interaction with the manufacturing process plan, which defines the process used to create the surface that is to be measured. Therefore;

Y. Zhao et al., *Information Modeling for Interoperable Dimensional Metrology*,
DOI: 10.1007/978-1-4471-2167-1_3,

a process definition that defines the manufacturing and measuring process must be interconnected with elements within the product definition. Furthermore, the process requires resources (sensors, fixtures, machines), and therefore a resource definition that supports the process definition must be represented [1]. However, this does not exist in the current manufacturing world due to many technical and business reasons. This chapter will firstly introduce the product definition activity; then the concepts of design geometry, feature, tolerances and their information modeling techniques are described. It is followed by a detailed account of current information models and standards for the representation of product definition information. The remainder of the chapter provides a discussion of product life-cycle and traceability information management in the product design phase, which presents a very important aspect of the automation of dimensional metrology systems.

3.1 Product Definition

Although development of computer-aided design systems started as early as the 1960s, its progress was severely hampered by the capability of the computers at that time. A decade later, CAD development and implementations began to enter the commercial market. Initially, with 2D in the 1970s, it was typically limited to producing drawings similar to hand-drafted drawings. Advances in programming and computer hardware, notably solid modeling in the 1980s, allowed more versatile applications of computers in design activities. Key products were the solid modeling packages. Among them are Romulus™ (ShapeData) and Uni-Solid (Unigraphics®) based on PADL-2 and the release of the surface modeler Catia® (Dassault Systems); all were released in 1981. Autodesk® was founded 1982 and its product, AutoCAD® soon became one of the most successful 2D CAD systems. The next milestone was the release of Pro/Engineer® (Pro/E® for short) in 1988, which heralded greater usage of feature-based modeling methods and parametric linking of the parameters of features. Also of importance to the development of CAD was the development of B-rep solid modeling kernels (engines for manipulating geometrically and topologically consistent 3D objects) such as Parasolid® (ShapeData) and ACIS® (Spatial Technology Inc.) at the end of the 1980s and beginning of the 1990s. This led to the release of many affordable, mid-range packages such as SolidWorks® in 1995, SolidEdge® (Intergraph™) in 1996, and IronCAD® in 1998. Today, CAD has become one of the main tools for product design and development [2].

The bulk of the development in commercial CAD systems has been in modeling the form of products, for example in providing techniques to assist in the representation of form using conventional drawings or new modeling techniques. The driving force behind CAD has been the desire to improve the productivity of the designer by automating the more repetitive and tedious aspects of design, and also to improve the precision of the design models. New techniques have been

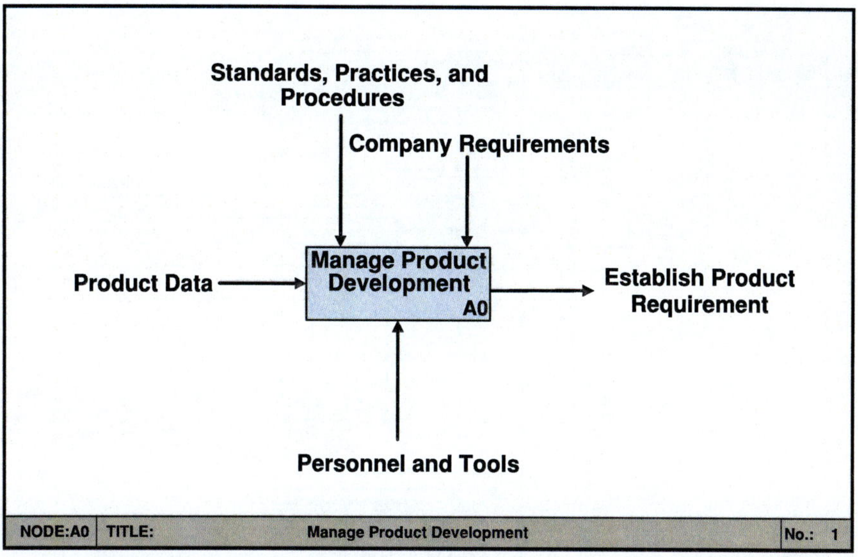

Fig. 3.1 IDEF0 diagram of management product development activity

developed in an attempt to overcome perceived limitations in conventional practice—particularly in dealing with complexity—for example designs as complex as automobile bodies, or as intricate as integrated circuits. Computer-aided design therefore enables the designer to tackle a task more quickly and accurately, or in a way that could not be achieved by other means.

In principle, CAD could be applied throughout the design process, but in practice its impact on the early stages, where very imprecise representations such as sketches are used extensively, has been limited. It must also be stressed that at present CAD does little in helping a designer in a more creative and intuitive way such as generation of possible design solutions, or in those aspects that involve complex reasoning about the design—for example in assessing, by visual examination of drawings, whether a component may be (easily) made, or whether it matches the specifications. These aspects are, however, the subjects of considerable current research. In practicing concurrent engineering, there is a pressing need for CAD systems to interface or integrate design with all the downstream activities, e.g. manufacturing, quality control, and marketing.

3.1.1 Product Design Activity

In a product design process, there are a number of activities. Figure 3.1 shows the highest level activity—the management of product development, in which designers use tools (i.e., CAD software) to establish product requirements based on company requirements and standards, practices and procedures.

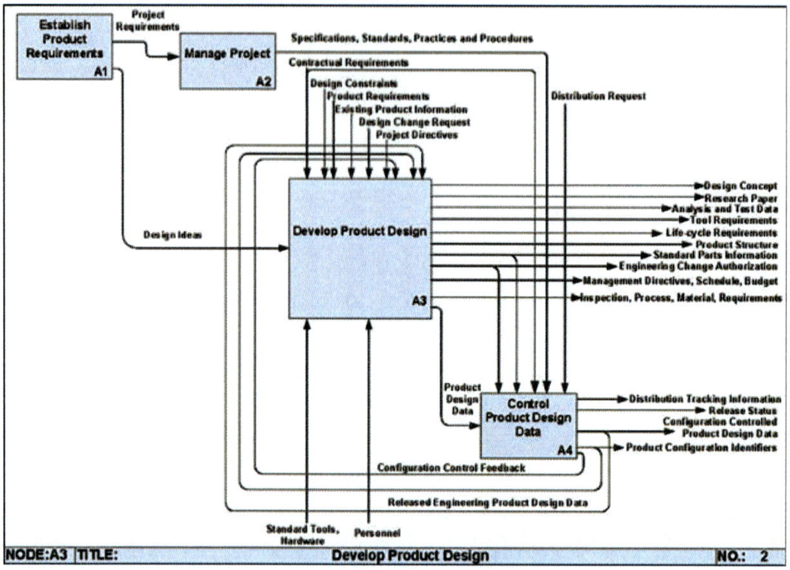

Fig. 3.2 IDEF0 diagram of develop product design activity

Activity A0 can be decomposed into four sub-activities shown in Fig. 3.2, within which activity A3 is the core of product design process. Activity A3 produces a representation of the form, fit and function of a product. This representation is then tested and released for the downstream processes. The product design representation must comprise the following information [3]:

- identification of a product to an organization's customers and of the components which comprise the product;
- description of the shape of the components of a product;
- assemblies of components;
- assemblies as components of higher assemblies;
- documentation of formal change and release of designs for the product;
- tracking of the history of the product as it goes through the formal initiation, change and release process;
- identification of qualified suppliers for the product or the design of the product.

In today's manufacturing industry, the product design activities are mostly carried out using a CAD system. So far, CAD systems have been described in very general terms. More specifically, they can be thought of as comprising (Fig. 3.3):

- hardware: the computer and associated peripheral equipment;
- software: the computer program(s) running on the hardware;
- data: the data structure created and manipulated by the software; and
- human knowledge and activities.

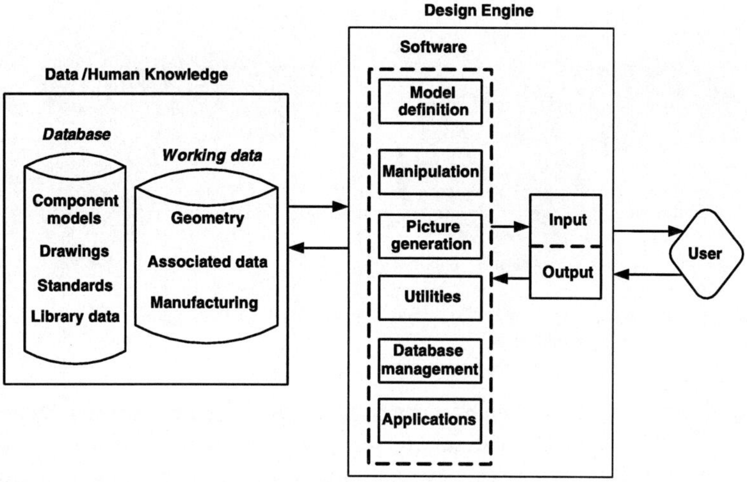

Fig. 3.3 The architecture of a computer-aided design system

3.1.2 Major Product Design Approaches

The early CAD systems were primarily based on building geometry with specific dimensions and creating geometry with specific initial relationships to existing geometry. When a line was drawn for example, it could not be changed except by redrawing it. That is, neither its position nor its length could be changed by changing the values associated with it. At the preliminary design stage, design engineers are often not sure what configurations will satisfy the design requirements. This leads to various modifications in product configurations and inevitably leads to changes in the geometric models and dimensions. It is therefore important for any CAD systems to have the functionalities to support such modifications.

To overcome this inflexibility of the early generation of CAD systems, many new approaches have been developed since. Three of the popular ones are feature-based design, parametric and variational design.

3.1.2.1 Feature-Based Design

Most of the contemporary CAD/CAM systems, such as Pro/Engineer®, Catia® and NX™, utilize a feature-based design approach. This is an approach by which both B-rep and Constructive Solid Geometry (CSG) methods are used for model construction. While B-rep is usually the underlying geometric representation scheme, CSG is used as the front-end of the software. Instead of simple solid primitives, form features are used for modeling purposes. By definition, features are viewed as information sets that refer to aspects of form or other attributes of a part, in such a way that these sets can be used in reasoning about design of the part or the assemblies they constitute.

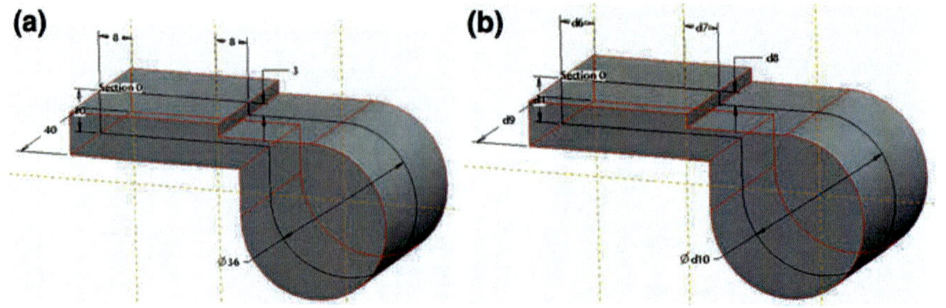

Fig. 3.4 Dimensions shown as values and parameters; **a** Dimensions shown as values; **b** Dimensions shown as parameters

A product model can be built by using features; this is known as design by features or feature-based design. One can start either with a more or less complete geometric model and define form features on it, or start from scratch by combining form features from a standard library. Designing with pre-defined form features can reduce the number of input commands substantially. This is especially advantageous in re-design. The parametric representation of features provides a powerful way to change features with respect to their dimensions. Features can serve as functional elements for designers. However, it is worth noting that design features often differ from the features used in "downstream" application features, e.g. measurement features and manufacturing features. Further discussions of feature-based design and feature definitions are in Sect. 3.2.

3.1.2.2 Parametric Design

Parametric design is a method of linking dimensions and variables to geometry in such a way that when the values change, the part changes as well—hence the dimension-driven capability. Take a part shown in Fig. 3.4 as an example. The dimensions are given in two forms for the 2D sketch based on which the solid part is created; true value form (Fig. 3.4a) and parameter form (Fig. 3.4b). This implies that the CAD system treats all dimensions as variables that can be changed any time and almost anywhere, be it in the modeling mode or drawing mode. The geometry is of course governed by these dimensions in the parameter form.

Being variables, dimensions can be obtained by means of parametric relations and equations. Take the same part shown in Fig. 3.4 as an example. One can establish a relationship between dimensions "d6" and "d7" as "d6 = d7". This way, three pieces of design intents are assumed,

- If d7 changes, d6 changes to the same value;
- d7 is a "strong" dimension in the sense that it can be changed any time and also governs d6;
- d6 is a "derived" dimension in that direct modification to it is not possible.

A parametrically defined model can also perform design modifications and creation of a family of parts in remarkably quick time compared with the redrawing required by a traditional CAD. In recent years, almost all CAD systems have adopted this approach. More conveniently, parametric modification can be accomplished with a spreadsheet or script, as well as by manually changing dimension text in the digital model and/or its associated drawings.

3.1.2.3 Variational Design

Variational design is a design methodology that utilizes fundamental graph theory and robust constraint-solving techniques to provide constraint-driven capability. As this definition indicates, parametric design and variational design have much in common. In practice, terms "parametric" and "variational" have been used almost interchangeably in technical and particularly commercial contexts. From the viewpoint of the end user, the two types of systems are similar to the extent that it is not always straightforward to determine from the outside which type of system one is using [4]. In fact, variational design may be considered as a superset of parametric design. Therefore, it is more general than parametric design. This book does not wish to make clear distinctions between these two types of design schemes. This said, in variational design constraints are typical types of modeling means and they are often modeled as relations between various geometric entities and dimensions.

Depending on different CAD systems, different types of relations can be defined such as equality, constraint, conditional and simultaneous equations. Equality relations set a parameter on the left side of an equation equal to an expression on the right. A relation that limits the permissible values for a dimension is a constraint relation. A conditional relation is used to assign values to variables only when specific criteria are satisfied. Simultaneous equations use the value from one relation to obtain the results for another relation. Figure 3.5 shows some of the constraints in a sketch. In Fig. 3.5a, two sets of collinear constraints (shown as symbols) applied to the center lines; two horizontal lines are constrained; and the two arcs join the two horizontal lines through the four tangent points. Also note that the sketch symmetry is implied through the two dimensions (18.00 and 12.50). Alternatively, symmetry can be defined explicitly as shown in Fig. 3.5b. In essence, relations use operators and functions in equations to control dimensions or parameters.

3.2 Features and Tolerances in Product Design

Features can be thought of as 'engineering primitives' suited for some engineering tasks. They originate in the reasoning processes used in various design, analysis and manufacturing activities, and are therefore often strongly associated with

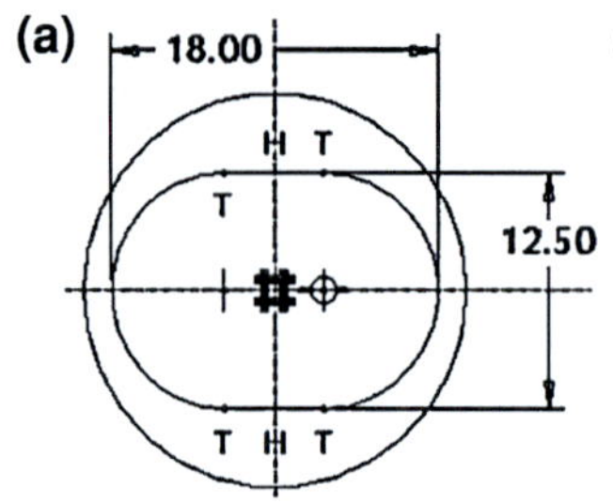

Fig. 3.5 Constraints in sketches

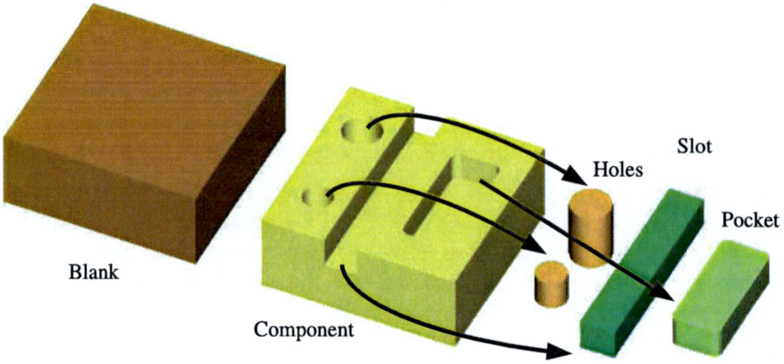

Fig. 3.6 A component with four machining features

particular application domains. This explains why there are many different definitions for features. According to Shah and Mäntylä [4], a feature should have,

- a physical constituent of a part;
- a generic shape that can be mapped to;
- engineering significance; and
- predictable properties.

In the context of computer-aided manufacturing systems and dimensional metrology, several more specific definitions have been suggested. One of such examples is, "*A feature is referred to as a distinctive or characteristic part of a workpiece, defining a geometrical shape, which is either specific for a machining process or can be used for fixturing and/or measuring purposes*" [5]. Another more generic feature definition example is, "*A feature is a generic shape that carries some engineering meanings*" [6]. A component with four machining features is shown in Fig. 3.6, where the pocket may require an end mill, the holes may require drilling operations and the slot may require a slotting cutter or a slotting cutter and an end mill.

Feature technology has the advantage of storing relevant information for applications during the design process, as well as offering the possibility for considering manufacturing, assembly, and measurement processes. However, when dealing with features, there is a further complication: viewpoint dependence. That is, depending on the application domain, one could have different views towards the same, or combination of, feature(s) on a part.

When a part is designed by features, the resulting model is not usually in a form convenient for other applications such as manufacturing process planning. Indeed, design features are stereotypical shapes related to a part's function, its design intent, or the model construction methodology, whereas manufacturing features are stereotypical shapes that can be made by typical manufacturing operations [7]. In this context, design features may also be called function features.

Table 3.1 illustrates a part whose different geometrical entities may be of interest for different applications. Of particular relevance in this book are the design features, tolerance features and measurement features.

Various types of feature by using a sub-classification of features were distinguished by Shah and Mäntylä [4]. These sub-classifications include form features, tolerance features, assembly features, functional features, and material features. Form features, tolerance features and assembly features are all closely related to the geometry of a part, and are therefore called geometric features. The same part may also be viewed differently by different applications. Each application may have its own "way of looking at an object" or definition of the object, with features relevant for that application. There can be, for example, a design, finite-element, machining, molding and assembly view of a part.

Apart from features, another type of essential information generated through the product definition process is tolerances. Tolerances treat the uncertainty with which the realized shape or measurements of a real manufactured object compare to their design ideals. If all parts could be manufactured perfectly as designed, there would be no need for tolerancing practices. However, it is certain that this cannot be done for finite cost in any but the most trivial cases. In the product design world, tolerances are noted on the CAD design or design drawing per standard notations such as ANSI Y14.5 [8] or ISO 1101 [9] and ISO 14405 [10]. There are two main classes of tolerances, dimensional and geometric. Geometric tolerances are the more complex of these two types. Geometric tolerances provide more flexible means for controlling shape than do dimensional tolerances. They achieve this by enabling tolerances to be defined independently of explicit dimensions. This enables tolerances to be specified that are more closely related to the functional requirements of the design, such as strength and fit. Dimensional tolerances are the less complex of the two classes of tolerances. They address the acceptable deviation of individual dimensions on a manufactured object. An ideal or nominal dimension is always expressible as a linear—one-dimensional—quantity. It may be an angle or a distance, and may refer to a size or a location. A tolerance on a dimension is therefore expressible as a range of values within which the manufactured dimension may fall in order to be acceptably close to the ideal value.

Table 3.1 Domain specific features

Engineering significance	Interested geometric entities
Rib design feature (the middle section) used for structural support	
Profile group tolerance feature (all side surfaces of the bigger pocket) used to apply a profile tolerance all around the pocket	
Pocket machining feature (all surfaces inside the bigger pocket) required to define the surfaces which in turn defines the volume to be removed	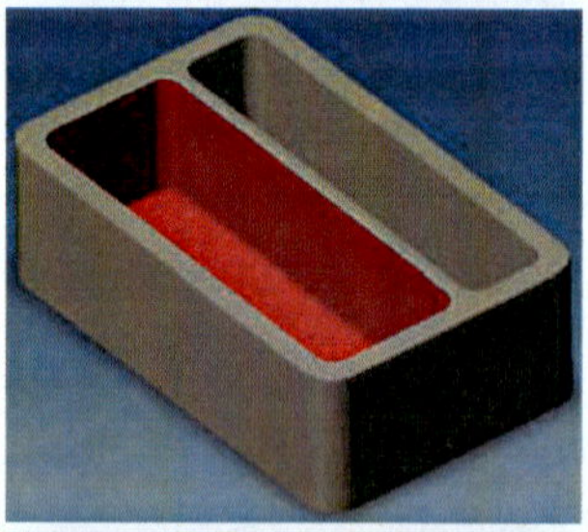
Multiple measuring features (planes and fillets inside the bigger pocket) necessary to perform inspections using a CMM for example	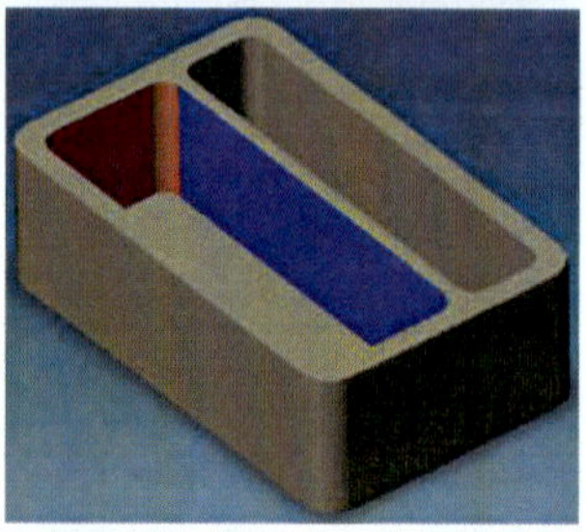

The reader needs to understand different types of information generated through the product design activity. To the concern of dimensional metrology, the design information can be categorized into the following categories:

- Design geometry and feature representation information,
- GD&T information, and
- Measurement feature information, which is deduced from the association between GD&T and design geometry/feature.

This book provides detailed discussions of the above information in the following sections.

3.2.1 Design Geometry and Feature Representation

Features are usually classified into different categories to enable designers to access the feature data and manufacturing engineers to generate process plans for a group of features which have some common geometric, topological or other properties. Such categories/classes are normally further divided into sub-classes such that classes and sub-classes form a hierarchy. This classification structure is known as feature taxonomy. Since the taxonomy of features is often of a hierarchical nature, the attributes of a class are inherited by its sub-classes. The method of classifying features is largely dependent on the feature representation schemes and the application domains of the feature data. Several design-oriented feature taxonomies have been reported. Dixon et al. [11] proposed a design-with-features taxonomy based on static and kinetic features. Static features are primarily structural in their functional intent, whereas kinetic features entail motion or energy transfer to meet their functional intent. The static features have been sub-classified into primitives, intersections, add-ons, macros and whole forms. Wilson and Pratt distinguished explicit feature and implicit feature taxonomies [12, 13]. Explicit features consist of four categories:

(1) through hole
   (a) face: complete or partial
   (b) edge
   (c) vertex
(2) depression
   (a) rotational: complete or partial
   (b) prismatic
(3) protrusion
   (a) rotational: complete or partial
   (b) prismatic
(4) area
   (a) with attributes
   (b) without attributes

Implicit features are further grouped into two groups:

(1) modifier
   (a) face
   (b) edge
   (c) vertex

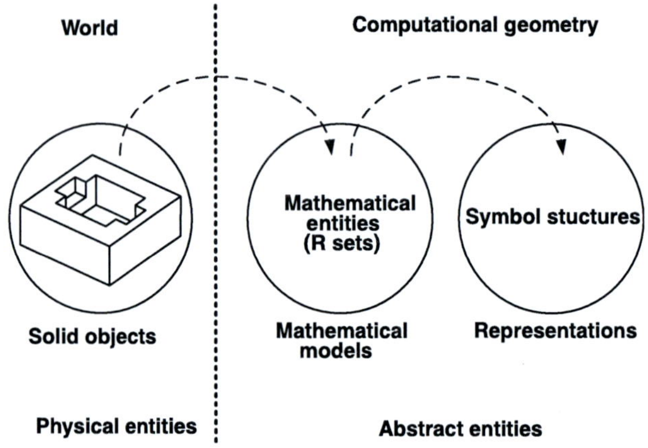

Fig. 3.7 Two-stage approach to geometric modeling CAD systems

(2) generic

 (a) prismatic
 (b) rotational
 (c) sweep

A hierarchical design feature taxonomy was proposed by Giacometti and Chang [14]. Features are categorized into composite and atomic features. Composite features can be subdivided into kinematics features and fixture features. Atomic features can be note features, volumic features, or library features. Feature taxonomy can also be application dependent. Wingard proposed a feature taxonomy based on application, shape, family of parts, etc. [6]. The two main groups of features are: compound features and atomic features. Compound features can be further subdivided into pattern features, complex features, and assembly features. Atomic features can be further grouped into part features, modifier features, and grouping features. The process of mapping an existing solid object on a computer-internal representation is shown in Fig. 3.7.

A representation scheme is defined as a relation s: (M → R), where M is the mathematical modeling space and R is the representation space [15]. There are six different representation schemes for solid modeling listed in the following, among which CSG and B-rep are the most common ones.

- CSG
- Sweep representation
- B-rep
- A combination of the above-mentioned types of representation schemes

In a CSG representation scheme, a solid object consists of a finite number of primitives (i.e., box, cylinder, cone, etc.). In this way, a solid object, for example, can be represented by a binary tree. The nodes of this tree represent the operations

Table 3.2 CSG representation scheme

Primitives	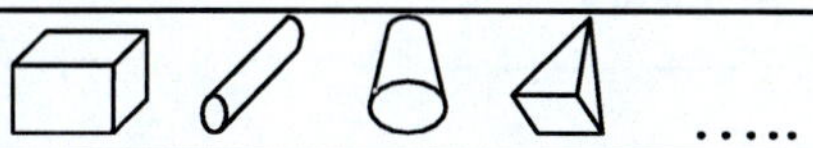		
Boolean operators	**Union** ∪	**Intersection** ∩	**Difference** ∖
Tree structure ○ Boolean operator □ Primitive	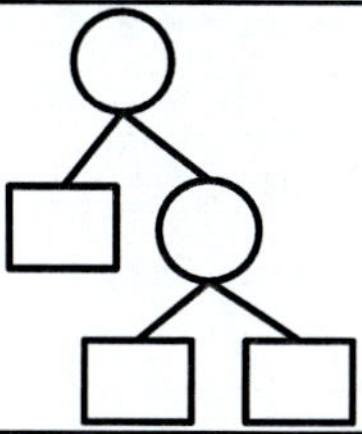		

that are carried out (i.e., union, intersection, difference) and the leaves represent the primitives as shown in Table 3.2.

In a sweeping representation scheme, a solid object is generated from a given geometry (i.e., contour, surface, body) by translation, rotation and displacement along any trajectory. In a B-rep representation scheme, a solid object is represented by its topological boundary (i.e., vertex, edge, loop, face, shell). The corresponding geometry is given by points, curves and surfaces as shown in Fig. 3.8.

Regardless of the type of features, it is fair to state that there are only two feature representation schemes, *surface* and *volume* representation scheme. They represent surface features and volume features, respectively. Surface features are form features represented by a number of faces (and possibly edges and vertices) that characterize a feature. They do not necessarily form a closed volume. In a B-rep model, features can be represented in either of these two schemes, whereas in a CSG model, features are always represented as volumes. Figure 3.9 shows a component together with the two types of feature representations, surface scheme (Fig. 3.9b) and volume scheme (Fig. 3.9c).

The representation of a part in terms of features is known as the feature model of the part, and the associated database is known as the feature data model. Features are represented in a feature model so that the relations between/among features can be kept. When using the B-Rep scheme in a feature model, features can be connected through individual faces, edges or vertices, and interference between features (i.e., feature interaction, which is an important operation in some applications) becomes relatively easy to detect. A comparison of CSG and B-rep solid modeling representation schemes is shown in Fig. 3.10.

3D CAD systems today represent parts using any of the aforementioned six representation schemes with history and parametric information. Two levels of parametric design are supported [16]:

(1) parameters associated with 2D sketches (geometric constraints between lines, arcs, etc.; algebraic relations between dimensional parameters);

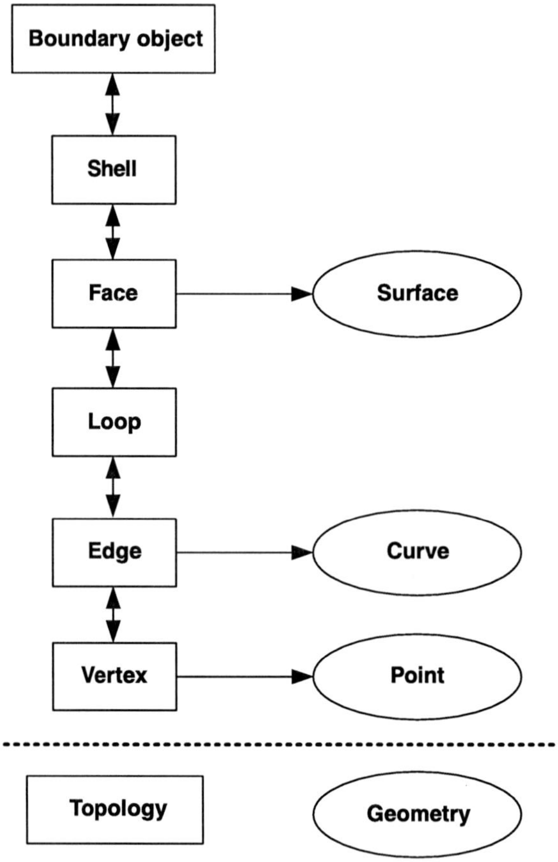

Fig. 3.8 B-rep representation scheme

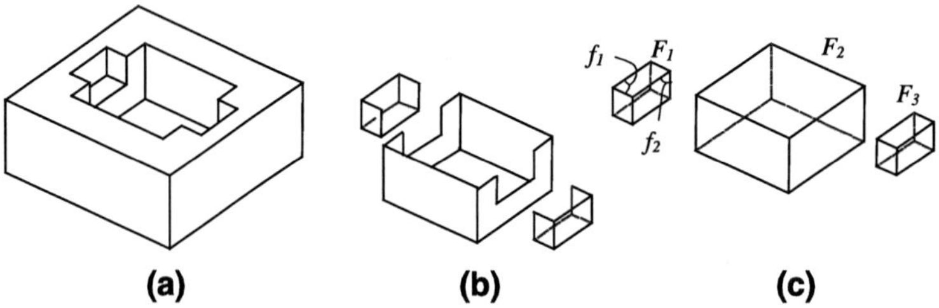

Fig. 3.9 Surface and volumetric feature representations

(2) parameters associated with 3D construction operations (sweeps, lofts, shells, fillets, blends, etc.).

The history tree records the procedure used for creating an object. Editing a model involves rolling back (retrieve back to a certain point in design history) to the

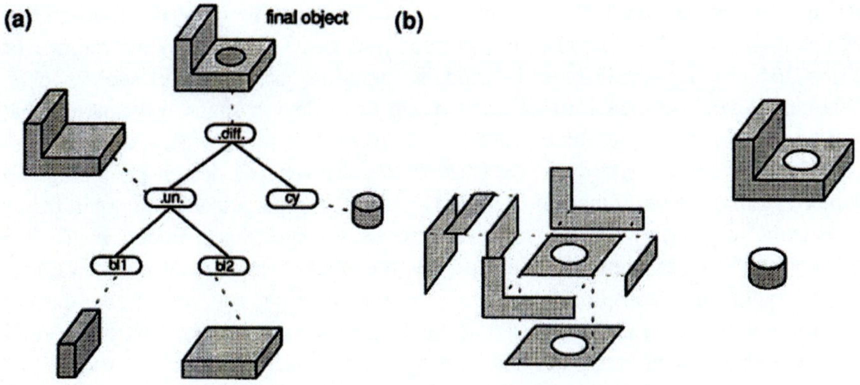

Fig. 3.10 A comparison of CSG and B-rep solid modeling representation schemes. **a** A solid object in CSG representation scheme. **b** A solid object in B-rep representation scheme

point of change in the history, making the change and rolling forward (return to the current design point). Changes to the sketch composition of parameters require a constraint solver[1] to recalculate all the dimensions. Changes to 3D parameters are usually made by direct assignment or sequential parameter calculation without the need for a solver. Thus, 3D CAD systems of today are hybrid in the sense that the representation is partly explicit (i.e., B-rep and parametric) and partly procedural (history). The representation data (i.e., CSG or B-rep) contains topology and geometry; parametric data contains constraints and explicit dimensions; history contains the sequence of construction operations in the form of a binary tree. CAD software also supports assemblies—collections of functionally related parts that need to be put together in a particular way to realize the function of the device. An assembly model contains: hierarchical relationships (components, sub-assemblies, assemblies), assembly relations (mating conditions, assembly level constraints), and components/sub-assembly positions (global or relative).

Every commercial CAD software system has its proprietary geometric modeling kernel. Given that the focus of this book is to discuss the information modeling techniques to facilitate interoperable dimensional metrology systems, the proprietary CAD data models are not the focus of this book. A brief discussion of current proprietary CAD data models is in Sect. 3.3.2.

Interoperability can only be achieved by efficient information exchange between different software systems. To achieve interoperability between CAD systems, a neutral data model for design geometry and feature information is needed. Most early work on data exchange of 3D CAD models, whether using formal IGES, STEP or de facto Drawing eXchange Format (DXF), Simulated

---

[1] A constraint solver is a computing program or module wherein relations between geometric variables are stated in the form of constraints. The constraints differ from the common primitives of imperative programming languages in that they do not specify a step or sequence of steps to execute, but rather the properties of a solution to be found.

Annealing Technology (SAT), focused on the final geometry of the model [17]. For example, the STEP Application Protocol (AP) 203 [18] allows the transfer of B-rep and closely related types of models, including assemblies of such models. Whereas, STEP AP 224 [19] carries the top level design shape information that can be identified as machining features. AP 203 is the most widely accepted and used international standardized data format for transferring design geometry and topology information. It may be regarded as an IGES replacement, though in fact it goes some way beyond the capabilities of IGES. Product shape models in AP 203 are explicit nonparametric models of the boundary representation and closely related types. For example, if a chamfer is created on an edge of a cube, the AP 203 file will contain information about the chamfer face, but details of the original edge will have been lost. Work is in progress on extensions for the transfer of parametric shape models and of procedural models defined in terms of their constructional history. The assembly models that can be exchanged using AP 203 are collections of positioned and oriented part models, suitable for the generation of parts lists and bills of materials. Methods for capturing feature-based inter-part relationships in assemblies are currently under development [20]. The detailed design geometry information modeling in IGES and STEP standards can be found in Sects. 3.3.3 and 3.3.4, respectively.

3.2.2 GD&T Information

The information required for GD&T and a symbology to communicate it on a part drawing have been standardized by the ISO committee as a set of standards [9, 10]. A similar system for GD&T has been developed into various national standards such as the National Standard of the United States of America ANSI Y14.5 [8], the German Standard DIN 7176, the British Standard BS 308 [21]. Some specifications in these national standards deviate from those defined in the ISO standards. However, they are not considered to be major interoperability issues. Table 3.3 presents a harmonized summary of the symbols of geometric characteristics from ISO 1101 and ANSI Y 14.5.

The geometric tolerances can be categorized into form, orientation, location, and run-out tolerances. Form tolerances are normally applied on individual features and datums are not needed; whereas the other three types of geometric tolerances are used to control related features and datums are required for most of them.

Apart from geometric tolerances that are used to control geometric variations, the international and national standards have also classified dimensional tolerances to control the size variations. This is because the types of variation that need to be controlled depend on functional and assembly requirements. For example, form needs to be controlled for smooth motion, perpendicularity is important for insertion of long features, and feature size and location must be controlled for proper assembly. Some important notes were drawn by Shah et al. [22] on an overview of the ANSI Y 14.5 standard:

Table 3.3 Symbols for geometric tolerances [23]

	Type of tolerance	Characteristic	Symbol	Datum needed
For individual features	*Form*	Straightness	⏤	No
		Flatness	⏥	No
		Circularity (roundness)	○	No
		Cylindricity	⌭	No
		Profile of a line	⌒	No
		Profile of a surface	⌓	No
For related features	*Orientation*	Angularity	∠	Yes
		Perpendicularity	⊥	Yes
		Parallelism	//	Yes
		Profile of a line	⌒	Yes
		Profile of a surface	⌓	Yes
	Location	Position	⌖	Yes or No
		Concentricity (for center points)	◎	Yes
		Coaxiality (for axes)	◎	Yes
		Symmetry	⌯	Yes
		Profile of a line	⌒	Yes
		Profile of a surface	⌓	Yes
	Run-out	Circular run-out	↗	Yes
		Total run-out	⌰	Yes

- Each geometric tolerance class is represented by a region (zone); the shape of the zone depends on the tolerance type and the feature being toleranced; the size depends on the tolerance value, material condition modifiers, and certain rules; the position/orientation of the zone depends on the tolerance type and datums.
- Datums are references for measurements; they are neither on the part nor on the gage, but simulated by the contact between the two; all tolerance relations (except size) are one-directional, i.e., datum-to-target.
- The number of datums used in a tolerance specification depends on the number of degrees of freedom that need to be controlled. When multiple datums are

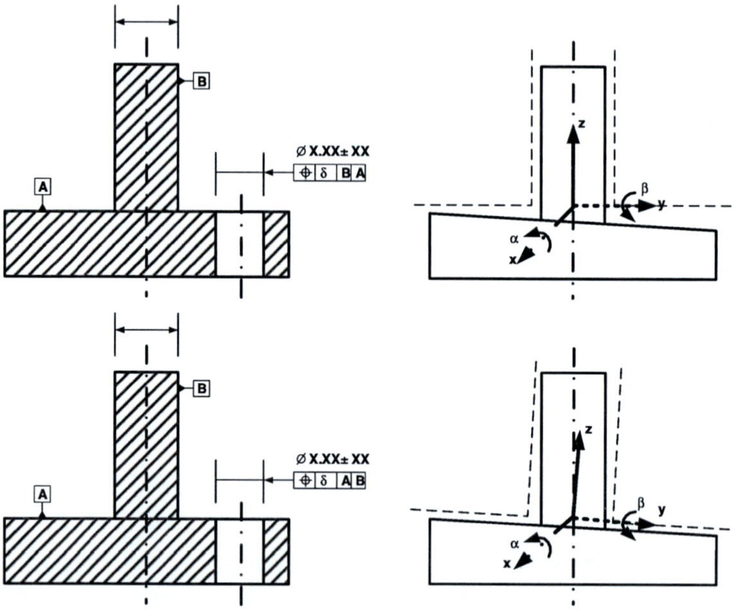

Fig. 3.11 Comparison of different datum precedence on measurement direction

used, the order in which they are specified creates a precedence order used to determine the coordinates and directions of control. As it is shown in Fig. 3.11, when the sequence of datums changes, the datum reference frame (shown in dotted lines) changes.

- Tolerances can be applied to resolved entities (axes, mid-planes), not just to boundary elements (faces, edges);
- In order to allow for trade-offs between feature size and certain types of geometric tolerances, such as position, the standards use material modifiers (MMC, LMC, RFS) to indicate what the geometric tolerance is when the size is at its largest or smallest value. When the feature size deviates from that value, a "bonus tolerance" is added to the geometric tolerance, i.e., trading position variation for size variation; when a modifier is applied to a datum feature of size, the geometric tolerance zones "shift" which is equivalent to a larger zone.

This standardized GD&T symbology communication provides a means for specifying the shape requirements of, and the interrelationships between, part features. Because no manufacturing process can make dimensionally perfect parts, designers must specify a region to allow dimensional variations in actual parts. This region is called the tolerance zone. The traditional view of tolerancing is that when the dimensional variation is within the allowable region, the part meets shape requirements; that is, the actual part is functionally acceptable.

Major geometric tolerancing theories and methods for mechanical design are usually categorized as the traditional plus/minus tolerancing theory and the modern tolerance zone theory. The traditional plus/minus tolerancing method is used for specifying allowable size variation around the nominal size. The modern tolerance zone method is used not only to specify allowable size variation but allowable variations of feature form and feature interrelationships. Traditional plus/minus tolerancing provides a basis for defining the limit of size used in dimensioning mechanical parts. The size tolerance indicates the quantity of the allowable variation of a dimension, either linear or angular.

The advantage of this traditional tolerancing method is that it is simple for designers to use. It is also simple for inspectors to verify using a micrometer, a caliper, or a protractor. However, there are several shortcomings in this approach. Only size tolerances and simple forms of positional tolerances are supported. There is no specification for form tolerances or complex feature interrelationships (including true position). As a result, assembly and alignment requirements cannot be represented or verified. Plus/minus tolerancing also lacks abstraction power in representing the tolerance of a mechanical part in CAD/CAM systems.

Modern tolerancing theory was developed to overcome shortcomings in traditional tolerancing theory. Modern geometric tolerancing methods are based on two major principles: the Maximum Material Condition (MMC) principle, also called Taylor's principle; and the Independence principle. The MMC requires an envelope which is the boundary surface of a similar perfect form of the nominal feature in the design. The envelope must totally contain the feature and must meet the shape requirements. The similar perfect feature is the feature at the maximum material size limit (the worst case). The Independence principle makes a clear distinction between size tolerance and form tolerance. It requires tolerancing for size without any reference to form or location tolerances. The latter must be defined separately, when necessary. ISO 1101 and ASME Y14.5 are based on the MMC principle.

A tolerance zone is a virtual region formed around the true feature [8]. Tolerances of orientation and location define zones within which all points of the toleranced feature have to be contained. Therefore, related geometrical tolerances also contain form deviation (Fig. 3.12). Related geometrical tolerances of axes or median faces, limit the form deviations of the axes or median faces, but not of the pertinent surfaces. The indication of form tolerances is not necessary when the related geometrical tolerance already limits the form deviations. Similarly, locational tolerances limit the location but also the orientation and the form of the toleranced feature. When applying a tolerance to the design, the tolerance symbol together with the tolerance frame is connected to the toleranced feature. The symbol indicates which type of tolerance it is. The information in the tolerance frame indicates the datum reference for the applied tolerance.

Understanding the causes and effects of dimensional and geometric variations is a major concern in the design and manufacture of mechanical products. In order to guarantee the interchangeability and quality of parts and assembly, and also to control the economical production processes, designers need to determine an

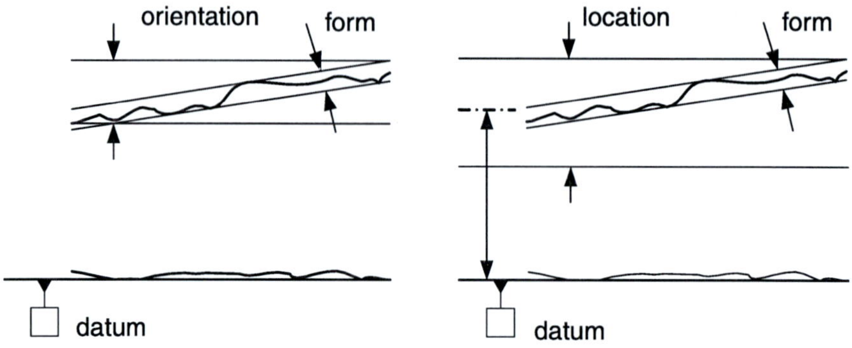

Fig. 3.12 Form deviation, orientational deviation, and locational deviation

acceptable range of dimensional and geometric variations of the shapes of the workpiece. The following GD&T issues are of essential concern to designers:

- Functionality and/or assemblability
- Tolerance Analysis: consequences of a proposed GD&T scheme
- Tolerance Allocation: determining how to distribute the allowable variation on the dimension of interest amongst all the independent contributors

Critical to these issues, is communication of the acceptable range that is comprehensible across the various divisions of the industry, such as design and manufacturing. Apart from unambiguous communication, automation is also critical for cost effectiveness and efficiency. All divisions in a manufacturing infrastructure that process tolerance specification information are shown in Fig. 3.13. The automation of these engineering processes with computer tools has been investigated for the past three decades. To accommodate these requirements, standards have evolved from the parametric tolerances to the modern geometric tolerance specification. Modern geometric tolerances provide the foundation for creating mathematical models that is essential for the development of computer tools.

The syntax of the current geometric tolerance specification is complex with fourteen basic symbols and eight modifiers. The meaning of the syntax—the semantics—has to be interpreted based on the feature under consideration, such as cylinder, sphere, cone, slot, tab, free-form surface, and their 2D counterparts (i.e., thin sheet metal parts). The tolerance zones have to be captured by computer tools for any processing. For example, in an automated inspection process, a point on an actual surface feature needs to be verified if it lies within the zone. The complexity of shapes, the numerous syntaxes and semantics based on the feature shapes, and the composition of tolerances of adjacent sides post difficulties in developing a universal method for capturing the tolerance zones.

The process of generating a tolerance zone for a part can be viewed as a morphological operation. However, individual features that constitute a part are specified with different tolerance values. Hence, the shape or the ratio of the dimensions of any feature with other features of a part is not necessarily preserved.

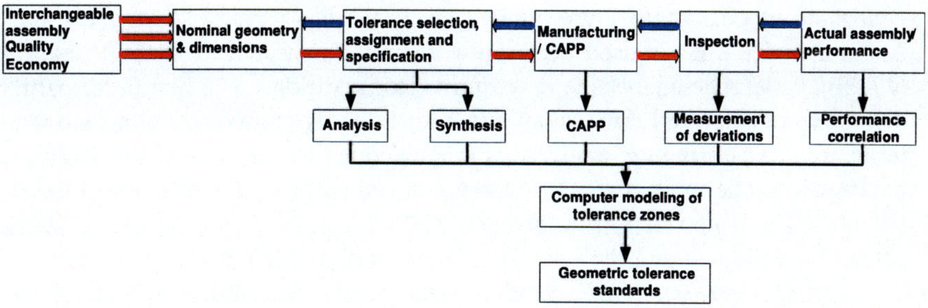

Fig. 3.13 Tolerance information utilization and transfer in a manufacturing enterprise

The composition of tolerance zones of adjacent features is left to the attention of the designer. Today's tools for assisting designers in allocating tolerances and identifying trade-offs during the design process can be grouped into two types:

- The manual procedure called Min/Max Tolerance Charting; it is consistent with ASME 14.5/ISO standards, but limited to 1D worst case analysis only.
- Commercial tolerance packages; these typically perform both worst case and statistical analysis, but based on point-to-point constraint solving. This makes them incompatible with the current tolerance standards that are based on tolerance zones, not point-to-point variations.

Comprehensive 3D analysis of tolerance stack-ups involving all types of dimensional and geometric variations is only possible if a mathematical model of such variations exists. But the current international standards are based on ad-hoc conventions collected from years of engineering practice, not on mathematical foundations. The attempt to "retrofit" an "official" math model to the tolerance standard is ongoing [24]. Shen et al. [16] classified GD&T models into six major categories listed below.

- *Attribute models* The basic characteristic of attribute models is that a tolerance is directly stored as an attribute of either geometric entities or metric relations in CAD systems [20, 23, 25–29]. The common deficiency of this approach is that they cannot do validation since GD&T semantics is not built into the model structure.
- *Offset models* In this approach, the maximal and minimal object volumes are obtained by offsetting the object by corresponding amounts on either side of the nominal boundary [25, 26]. Offset models can only represent a composite tolerance zone; they cannot distinguish between effects of different tolerance types, nor interrelations among tolerance specifications.
- *Parametric models* Tolerances are modeled as plus/minus variations of dimensional or shape parameters. Parameter values can be found by a set of simultaneous equations representing the constraints [28, 31, 32]. The parametric equations can be used for point-to-point tolerance analysis rather than zone based analysis. This is not consistent with GD&T standards.

- *Kinematic models* Entities are modeled in terms of "virtual" links and joints. A "kinematic link" is used between a tolerance zone and its datum features [27–30]. Tolerance analysis is based on vector additions. The first order partial derivative of analyzed dimension with respect to its component dimensions in terms of a transformation matrix was employed for tolerance analysis. Both the parametric model and kinematic model can represent all the tolerance classes, but not all the information involved in GD&T can be stored. Datum systems cannot be validated and the analysis is point based rather than zone based.
- *Degrees of Freedom (DoF) models* treat geometric entities (points, lines, planes) as if they were rigid bodies with DoFs [31–34]. Geometric relations (angular and linear) are treated as constraints on DoFs. ASME Y14.5 tolerance classes are characterized by how each DoF of each entity is controlled. Technologically and Topologically Related Surfaces (TTRS) models bear many similarities to DoF models [35, 36]. Later researchers have tried to express ASME Y14.5 tolerance classes in terms of TTRS but this is not fully achieved. Although mathematically elegant, TTRS models are indifferent to ASME Y14.5 Rule #1, floating zones, effects of bonus and shift, form tolerance, or datum precedence. DoF models facilitate the validation of DRF and tolerance types.
- *Hybrid models* Some hybrid models have been proposed to combine the good aspects of the different models. One of such hybrid models is the ASU GD&T Global Model [37], which is mainly a hybrid model of DoF model and Attribute model.

Commercial CAD software systems have certain tolerance analysis capabilities. The GD&T information is modeled in their proprietary data format. In order to realize interoperability for dimensional metrology systems, a neutral data format to represent GD&T information, best of which is the standard data model, is required. Up to now, the STEP standard is the only one that provides a semantic modeling of GD&T with its associated design geometry/features. In Sect. 3.3.4, the STEP GD&T data model is discussed in detail.

3.2.3 Measurement Features Information

As introduced in Sect. 3.2, feature is an item of application interest. The term "feature" in dimensional metrology can be defined as the individual measurable properties or surfaces of the workpiece being examined [38]. Identifying feasible measurement features from a design is one of the foremost tasks a dimensional metrology system needs to accomplish in order to generate a measurement plan. Dimensional measurement can occur at any stage of the life cycle of a product where checking for conformance with a design specification is required. At different product manufacturing stages, different measurement features may be identified.

The measurement feature identification process is closely related to the GD&T information and its associated design geometry and/or manufacturing feature at each particular stage when a measurement is needed on the workpiece. For post-process measurement, a workpiece has been manufactured before measurement operations are carried out. Therefore, measurement features can be identified directly from a CAD design by associating tolerances and the geometry elements controlled by these tolerances. Most CMM software systems are able to do such measurement feature recognition with certain human involvement. However for in-process measurement, some measurement operations are carried out during a manufacturing process. It is common that some measurement features locate on certain manufacturing feature surfaces. Therefore, more complicated analysis is needed by incorporating manufacturing feature, GD&T requirements, and measurement feature together. In today's industry, measurement feature identification for in-process measurement is mostly done through manual operation, in which an operator identifies the measurement features for each measurement operation.

In the above paragraph, two feature terminologies have been mentioned: manufacturing feature and measurement feature. The readers need to comprehend the differences between these two types of features. A manufacturing feature is a recognizable shape that has specific characteristics of part shape used in manufacturing. The purpose of manufacturing features is to facilitate the identification of manufacturing shapes that are human and computer interpretable. Whereas the measurement features are the shapes of a workpiece that have specific characteristics in measurement processes. Normally they are expressed in terms of the tolerances associated with the manufacturing features or primitive features. Primitive features are also called geometric features. They can be defined by the smallest recognizable canonical or primitive shape, which cannot be further decomposed otherwise, such as lines, points, planes, etc. Primitive features make up a large percentage of measurement features. In order to carry out a measurement operation, a datum system must be established first. A datum feature is a feature on a workpiece that is used to establish a datum.

A simple example is given in Fig. 3.14 to further explain the relation between manufacturing features, tolerances, measurement features, and datum. The workpiece shown in Fig. 3.14a has one machining feature—slot, to which a position tolerance and a parallelism tolerance are applied. The dimension and tolerance information is displayed in the design drawing shown in Fig. 3.14b. The parallelism tolerance is applied on one side of the slot surface, and position tolerance on the other. When a measurement operation is planned to check the parallelism tolerance, the planar surfaces, where the parallelism tolerance is applied onto, is the measurement feature shown Fig. 3.14c. Also, in order to measure any tolerance, a datum surface needs to be established first. The planar surface where Datum A is indicated is the datum feature.

A very significant characteristic of measurement features is that not all measurement features are directly connected to a tolerance. They are called the constructed measurement features, which are a constructed through the association of a tolerance and its controlled geometry. Figure 3.15 shows an example of

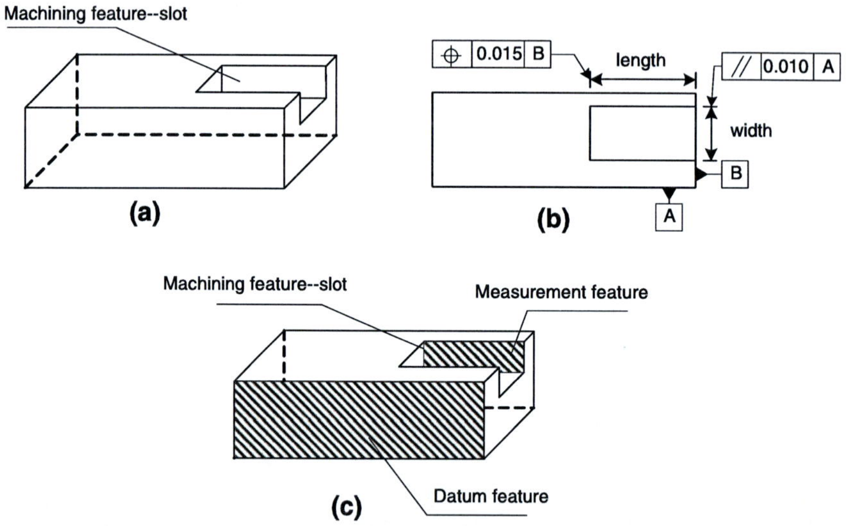

Fig. 3.14 Relationship between manufacturing feature, tolerances, and measurement features. **a** Workpiece with a manufacturing feature. **b** Top view of the workpiece. **c** A comparison of machining feature, measurement feature and datum feature

constructed measurement feature. A parallelism tolerance is applied to a cylinder to control the axis of the cylinder. However, it is impossible to measure the axis itself. The cylinder is, therefore, the measurement feature. After the cylinder is measured, the location of its axis is calculated and compared to the tolerance requirement.

Each tolerance has limited types of surface that it can control. Table 3.4 lists some possible combinations between tolerances, the surface they control, and the measurement feature.[2] The complete combinations are listed in Appendix A. Up to now DMIS and STEP AP 219 are the only standards that provide data models for measurement features and they are discussed in detail in Sect. 4.3.

3.3 Product Data Models and Standards

The most important set of information that is able to be generated from a product definition activity includes product geometry/topology information and GD&T requirements. These are the basic types of information for any type of product manufacturing. Computers and information technology were introduced into industry in an ad hoc manner to initially relieve particular bottlenecks in industrial processes. There was no need to think of the effect on the overall enterprise and the

---

[2] Courtesy of Mitutoyo America Corporation for providing this table. In the table S=Size, F= Form, O= Orientation.

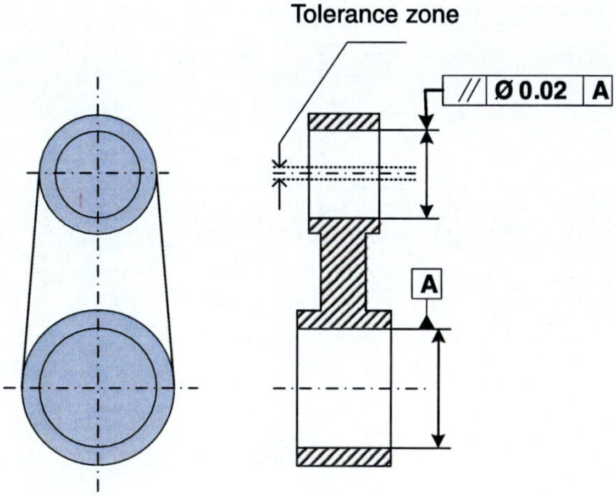

Fig. 3.15 An example of a constructed measurement feature

issue of systems integration. Any attempts to deal with data exchanges were also handled in an ad hoc manner. As computers are used more and more in all walks of an organization, in particular in the product development process, data exchange and sharing has now risen to the top of the agenda for many businesses.

As discussed in Sect. 2.3, a typical dimensional metrology system handles information from product definition, measurement process planning, measurement plan execution, and measurement result report and analysis. Each of these activities is supported by a number of commercial software systems, each of which imports/exports its own proprietary data format. To facilitate the data exchange, a neutral data format is crucial. This section introduces current product data models that are of direct relevance to dimensional metrology systems. The proprietary data models are discussed first followed by standardized neutral data models.

3.3.1 Types of Product Models

In order to support different scopes and methods, different aspects of a product must be represented in a product model. These aspects can be represented in different types of product models. Krause et al. [39] have identified an incomplete list of different types of product models:

- Structure-oriented product models,
- Geometry-oriented product models,
- Feature-oriented product models,
- Knowledge-based product models, and
- Integrated product models.

Structure-oriented product models represent some aspects of a product from a product breakdown perspective [39]. For instance, a product can be broken down

Table 3.4 Examples of tolerances and the surface they control

Type	Characteristic	Example	Feature(s)	Datum Type
S	Spherical Diameter S⌀	S⌀10±0.01	Sphere	N/A
S	Diameter ⌀	⌀10±0.01	Cylinder	N/A
S			Circle	N/A
S	Spherical Radius SR	SR10±0.01	Spherical Arc	N/A
S	Radius R	R10±0.01	Cylindrical Arc	N/A
S	Controlled Radius CR	CR10±0.01	Cylindrical Arc	N/A
S	Width	10±0.01	Parallel Plane	N/A
S			Parallel Points	N/A
F	Straightness ⏤	[⏤ ⌀0.1]	Cylinder	N/A
F			Derived Median Line	N/A
F		[⏤ ⌀0.1Ⓜ]	Cylinder	N/A
F			Derived Median Line	N/A
F		[⏤ ⌀0.1/10]	Cylinder	N/A
F			Derived Median Line	N/A
F		[⏤ 0.1]	Line	N/A
F			Parallel Planes	N/A
F			Derived Median Plane	N/A
F		[⏤ 0.1Ⓜ]	Parallel Planes	N/A
F			Derived Median Plane	N/A
F		[⏤ 0.1/10]	Line	N/A
F	Flatness ⏥	[⏥ 0.1]	Plane	N/A
F		[⏥ 0.1/10X10]	Plane	N/A
F	Circularity ○	[○ 0.1]	Sphere	N/A
F			Circle	N/A
F	Cylindricity ⌭	[⌭ 0.1]	Cylinder	N/A
O	Perpendicularity ⊥	[⊥ ⌀0.1 A]	Cylinder	Cylinder
O				Plane
O				Parallel Plane
O		[⊥ ⌀0.1Ⓜ A]	Cylinder	Cylinder or (Par)plane
O		[⊥ ⌀0.1Ⓛ A]	Cylinder	Cylinder or (Par)plane
O		[⊥ ⌀0.1Ⓟ 10 A]	Cylinder	Cylinder or (Par)plane
O		[⊥ ⌀0.1 AⓂ]	Cylinder	Cylinder
O				Parallel Plane
O		[⊥ ⌀0.1 AⓁ]	Cylinder	Cylinder
O				Parallel Plane

in functions and components in terms of, e.g. function trees, bill-of-materials, and assembly structures. Geometry-oriented product models represent the shape of a product. The shape of a product can be described in different ways depending on the purpose of the description including wire frame, surface, solid, and hybrid models. In addition, Feature-oriented product models represent a product in terms of features.

Knowledge-based product models formally represent accumulated knowledge about a product [39]. The accumulated knowledge can then be used to guide the design and constrain the design space. Consequently, a knowledge-based product model is used to guide and control a designer in her, or his, use of other types of product models, rather than representing some aspect of a product itself.

Integrated product models represent a product by combining different types of product models. An example of an integrated product model is ISO 10303-214 (STEP AP214), which combines structure, geometry, and feature oriented product models (ISO/TC184/SC4, 2001) [40].

3.3.2 Proprietary Data Models

There are a wide variety of CAD vendors to choose from. Each of these different types of CAD systems requires the operator to think differently about how he or she will use them and he or she must design their virtual components in a different manner for each. Table 3.5 represents a sample of commercial CAD companies and the products they offer.

There are many producers of the lower-end 2D systems, including a number of free and open source programs. These provide an approach to the drawing process without all the fuss over scale and placement on the drawing sheet that accompanied hand drafting, since these can be adjusted as required during the creation of the final draft. 3D wireframe is basically an extension of 2D drafting. Each line has to be manually inserted into the drawing. The final product has no mass properties associated with it and cannot have features directly added to it, such as holes. The operator approaches these in a similar fashion to the 2D systems, although many 3D systems allow using the wireframe model to make the final engineering drawing views.

3D "dumb" solids are created in a way analogous to manipulations of real world objects. Basic three-dimensional geometric forms (prisms, cylinders, spheres, and so on) have solid volumes added or subtracted from them, as if assembling or cutting real-world objects. 3D projected views can easily be generated from the models. Basic 3D solids do not usually include tools to easily allow motion of components, set limits to their motion, or identify interference between components.

3D parametric solid modeling requires that the operator use what is referred to as "design intent". The objects and features created are adjustable. Any future modifications will be simple, difficult, or nearly impossible, depending on how the original part was created. One must think of this as being a "perfect world" representation of the component. If a feature was intended to be located from the center of the part, the operator needs to locate it from the center of the model, not, perhaps, from a more convenient edge or an arbitrary point, as he could when using "dumb" solids. Parametric solids require the operator to consider the consequences of his actions carefully.

Table 3.5 Commercial CAD companies and their products

Company name	Product name	Product type
Autodesk	AutoCAD and Architectural Desktop	CAD
	Mechanical desktop	CAD
	Autodesk inventor	CAD
AutoDesSys	Form•Z (RenderZone Plus)	CAD
Bentley systems	MicroStation	CAD
CAMM	OmniCAD	CAD
Dassault systemes	CATIA	CAD/CAE/CAM
	SolidWorks	CAD
	ACIS 3D modeler	Kernel
DataSolid	CADdy++ mechanical design	CAD
	CADdy++ basic	CAD
	CADdy classic	CAD
	PowerSHAPE	CAD
DAKO	WorldCAT family	CAD
DpS CAD-center ApS	PCschematic ELautomation	CAE
Engineered software	PowerCADD	CAD
Fast AG	GraphiteOne	CAD
FastCAD	FastCAD	CAD
JVSG	IP video system design tool	CAD
Google	SketchUp	CAD
GrabCAD	GrabCAD	CAD
IMSI/Design	TurboCAD professional	CAD
	TurboCAD deluxe 2D/3D	CAD
	TurboCAD designer 2D	CAD
	DesignCAD 3D max	CAD
	DesignCAD express	CAD
Interfacial	Interfacial	CAD
Ironcad	IronCAD	CAD
Kubotek corporation	KeyCreator	CAD
	KeyMachinist	CAM
MacroVision Ltd.	Eagle eye	Kernel
Missler software	TopSolid design	CAD
Nemetschek	Allplan BIM	CAD
	Vectorworks	CAD
Open CASCADE SAS	Open CASCADE	Kernel
	SALOME	CAD/CAE
Parametric technology corporation	Pro/ENGINEER (Creo Elements/Pro)	CAD
	Pro/DESKTOP	CAD
	Granite	Kernel
	CoCreate OneSpace	CAD
PYTHA lab GmbH	PYTHA	CAD
RDCadd	CAD	CAD
Sescoi	WorkNC-CAD	CAD
Sigma design	Arris	CAD

(continued)

Table 3.5 (continued)

Company name	Product name	Product type
Siemens PLM solutions	NX	CAD/CAE/CAM
	Solid Edge	CAD
	Parasolid	Kernel
	I-DEAS	CAD
	NX I-deas	CAD
Trace software	Elecworks	CAD

Some software packages provide the ability to edit parametric and non-parametric geometry without the need to understand or undo the design intent history of the geometry by use of direct modeling functionality. This ability may also include the additional ability to infer the correct relationships between selected geometry (e.g., tangency, concentricity) which makes the editing process less time and labor intensive while still freeing the engineer from the burden of understanding the model's design intent history.

Most of these CAD systems have proprietary formats in which translation is required to move design models from one system to another. The largest problem that this lack of standardization presents to the industry is the potential for information loss or degradation during the translation process. For this issue a new market has emerged to support "before and after" translator model comparison and verification. In the following sections, some major standard data formats for the exchange of design data are discussed.

3.3.3 IGES

Design data translation based on IGES originated around the late 1970s. It is still one of the viable methods of transferring CAD data. The file format defined by this Specification treats the product definition as a file of entities. Each entity is represented in an application-independent format, to and from which the native representation of a specific CAD/CAM system can be mapped. The entity representations provided in this Specification include forms common to the CAD/CAM systems currently available and forms which support the system technologies currently emerging.

Entities are categorized as geometry and non-geometry. Geometry entities represent the definition of a physical shape. They include points, curves, surfaces, solids and relations. Relations are collections of similarly structured entities. Non-geometry entities typically serve to enrich the model by providing (a) a viewing perspective in which a planar drawing may be composed and (b) annotation and dimensioning appropriate to the drawing. Non-geometry entities further serve to provide specific attributes or characteristics for individual or groups of entities. The definitions of these groupings may reside in another file. Typical non-geometry entities for drawing definitions, annotations and dimensioning are the view, drawing, general note, witness line and leader. Typical non-geometry entities for attributes and groupings are the property and associated entities.

An IGES file consists of five sections: Start, Global, Directory entry, Parameter data, and Terminate. It may include any number of entities of any type as required to represent a product. Each entity occurrence consists of a directory entry and a parameter data entry. The directory entry provides an index and includes descriptive attributes about the data. The parameter data provides the specific entity definition. The directory data are organized in the fixed fields and are consistent across all entities to provide simple access to the frequently used descriptive data. The parameter data are entity-specific and are variable in length and format. The directory data and parameter data for all entities in a file are organized into separate sections, with pointers providing bi-directional links between the directory entry and parameter data for each entity.

IGES provides for groupings whose definitions will be found in a file other than the one in which they are used. Attributes for the geometric entity are defined in the directory segment; the corresponding data itself is defined in the parameter segment. The directory entry and the parameter portion contain all the information about the entity with linkages between the two segments. The connection between attributes and data segment is made with bi-directional pointers. Similar numerical identifiers are assigned for various finite element analysis entities and their post-processing entities.

IGES can also transfer both 2D and 3D finite elements for an FEA type of analysis. While IGES is a popular method of data transfer, it lacks a means of transferring solid objects. This leads to users spending more time to build the solid object on the receiving end. Although open-ended in terms of adding more entities, they are not standardized to be acceptable across all CAD systems. For more information, the readers are referred to the book by Bloor and Owen [41].

3.3.4 STEP

STEP is developed by the Sub-committee 4 (SC4) of ISO Technical Committee 184 (TC 184) Industrial automation systems and integration. STEP is intended to support data exchange, data sharing and data archiving. For data exchange, STEP defines the form of the product data that is to be transferred between a pair of applications. Each application holds its own copy of the product data in its own preferred form. The data conforming to STEP is transitory and defined only for the purpose of exchange. STEP supports data sharing by providing access to and operation on a single copy of the same product data by more than one application, potentially simultaneously. STEP is also suitable to support the interface to the archive. As in product data sharing, the architectural elements of STEP may be used to support the development of the archived product data itself. Archiving requires that the data conforming to STEP for exchange purposes is kept for use at some other time. This subsequent use may be through either product data exchange or product data sharing [42].

Another primary concept contributing to the STEP architecture is that the content of the standard is to be completely driven by industrial requirements. This, in combination with the concept that the re-use of data specifications is the basis for standards, led to developing two distinct types of data specifications.

- The first type—reusable, context-independent specifications. They are the building blocks of the standard.
- The second type—application-context-dependent specifications (application protocols).

This combination enables avoiding unnecessary duplication of data specifications between application protocols.

3.3.4.1 An Introduction of STEP

STEP consists of a large group of integrated resources, application protocols, and parts. A basic introduction of STEP standards is given before the detailed discussion of design data modeling in STEP application protocols.

1. *Components of STEP* The architectural components of STEP are reflected in the decomposition of the standard into several series of parts. Each part series contains one or more types of ISO 10303 parts. Figure 3.16 provides an overview of the structure of the STEP documentation.
2. *Description Methods* The first major architectural component is the description method series. Description methods are common mechanisms for specifying the data constructs of STEP. They include the formal data specification language developed for STEP, known as EXPRESS. EXPRESS is similar to programming languages such as PASCAL. Within a SCHEMA, various data types can be defined together with structural constraints and algorithmic rules. A main feature of EXPRESS is the possibility to formally validate a population of data types, i.e., to check for all the structural and algorithmic rules. Other description methods include a graphical form of EXPRESS known as EXPRESS-G, a form for instantiating EXPRESS models, and a mapping language for EXPRESS. EXPRESS-G, as a formal graphical notation for the display of data specifications defined in the EXPRESS language, supports only a subset of the EXPRESS language. EXPRESS-G is represented by graphic symbols forming a diagram. The detailed introduction of EXPRESS and EXPRESS-G data modeling language can be found in Sect. 2.2.3.
3. *Implementation Methods* The second major architectural component of STEP is the implementation method series. Implementation methods are standard implementation techniques for the information structures specified by the only

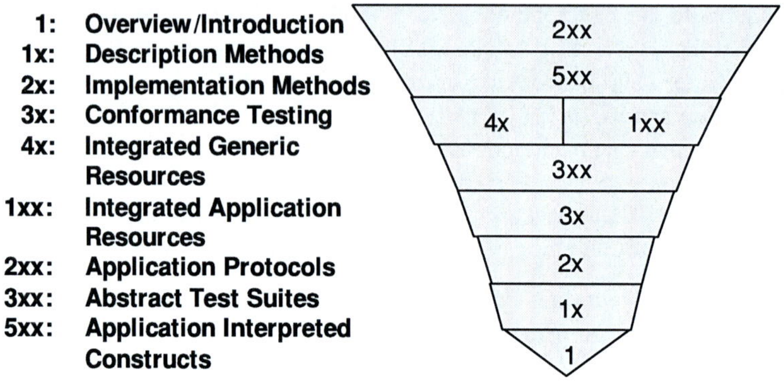

Fig. 3.16 STEP documents architecture

STEP data specifications intended for implementation, application protocols. Each STEP implementation method defines the way in which the data constructs specified using STEP description methods are mapped to that implementation method. This series includes the physical file exchange structure [43], the standard data access interface [44], and its language bindings [45–48].

4. *Conformance Testing* The third major architectural component of STEP is in support of conformance testing. Conformance testing is covered by two series of 10303 parts: conformance testing methodology and framework, and abstract test suites.

The conformance testing methodology and framework series provide an explicit framework for conformance and other types of testing as an integral part of the standard. This methodology describes how testing of implementations of various STEP parts are accomplished. The fact that the framework and methodology for conformance testing is standardized reflects the importance of testing and testability within STEP. Conformance testing methods are standardized in the ISO 10303-30 series of parts.

An abstract test suite contains the set of abstract test cases necessary for conformance testing of an implementation of a STEP application protocol. Each abstract test case specifies input data to be provided to the implementation under test, along with information on how to assess the capabilities of the implementation. Abstract test suites enable the development of good processors and encourage expectations of trouble-free exchange.

5. *Data Specifications* The final major component of the STEP architecture is the data specifications. There are four part series of data specifications in the STEP documentation structure, though conceptually there are three primary types of data specifications: integrated resources, application protocols, and application interpreted constructs. All of the data specifications are documented using the description methods.

Integrated application resources represent concepts related to a particular application context that supports common requirements of many other product data applications. Examples of application resource constructs include drawing sheet revision, drawing revision, and dimension callout. These constructs may be used by any application that includes drawings. Integrated application resources are standardized in the ISO 10303-100 series of parts.

Application protocols are the implementable data specifications of STEP. APs include an EXPRESS information model that satisfies the specific product data needs of a given application context. APs may be implemented using one or more of the implementation methods. They are the central component of the STEP architecture, and the STEP architecture is designed primarily to support and facilitate developing APs.

Many of the components of an application protocol are intended to document the application domain in application-specific terminology. This facilitates the review of the application protocol by domain experts. The Application Interpreted Model (AIM) is the component of the AP that is the normative, implementable information model in EXPRESS. Conformance classes are defined subsets of the AIM that may be used as a basis for conformance testing of implementations. Application protocols are standardized in the ISO 10303-200 series of parts.

Application Interpreted Constructs (AICs) are data specifications that satisfy a specific product data need that arises in more than one application context. An application interpreted construct specifies the data structures and semantics that are used to exchange product data common to two or more application protocols. Application protocols with similar information requirements are compared semantically to determine functional equivalence that, if present, leads to specifying that functional equivalence within a standardized AIC. This AIC would then be used by both application protocols and available for future APs to use as well. Application interpreted constructs are standardized in the ISO 10303-500 series of parts.

6. *A STEP File* In STEP instead of using numerals, text is used in identifying the entity. For example "Cartesian_point" is used as the identifier for points. These definitions are all given by the respective EXPRESS schema. The STEP file is generated conforming to the rules and format in the EXPRESS Schema. Unlike C or C++, EXPRESS is more like a formatted design language. Geometric objects are defined in terms of ENTITIES. An example of an EXPRESS file is listed below,

```
SCHEMA TEST_SCHEMA;

    ENTITY CARTESIAN_POINT;

          x_coordinate: REAL;
          y_coordinate: REAL;
          z_coordinate: REAL;

    END_ENTITY;

END_SCHEMA;
```

When the CAD model is compiled with an EXPRESS compiler and the data structure populated, a STEP file, whose format is defined in STEP Part 21, can be produced as shown below,

```
ISO-10303-21;
HEADER;
FILE_DESCRIPTION(('‘'), '1');
FILE_NAME('CARTESIAN-POINT',

        '2011-03-10T09:19:11-04:00',

            ('‘'),
            ('‘'),
            'STEPSTEP INTERFACE',
            'STEPSTEP DESIGN SYSTEM',
            '‘');

FILE_SCHEMA(('TEST_SCHEMA'));
ENDSEC;
DATA;
#1 = CARTESIAN_POINT(10.0,20.0,30.0);
#2 = CARTESIAN_POINT(5.0,10.0,15.0);
#3 = CARTESIAN_POINT(30.0,10.0,6.0);
ENDSEC;
END-ISO-10303-21;
```

3.3.4.2 AP 203 Edition 1 and 2

STEP Application Protocol (AP) 203 edition 1 [18] and edition 2 [3] (Configuration Controlled 3D Designs of Mechanical Parts and Assemblies) provides the data structures for the exchange of configuration-controlled 3D designs of mechanical parts and assemblies. AP 203 is but one part of the entire ISO 10303 product data standard. It was developed to represent one domain. AP 203 does not present itself as the data standard for configuration management of a product throughout its entire life cycle. The AP is centered on the design phase of mechanical parts. As STEP evolves, other APs (currently under development or proposed) will carry the data in AP 203 forward through the product life cycle. AP 203 edition 1 has fairly complete definitions of product design information. However, it does not provide semantic association between GD&T and design geometry. This has been improved in the second edition. The following information is within the scope of AP 203 edition 2, among which the bold and italic items represents the essential semantic association between GD&T and design geometry:

- products that are mechanical parts and assemblies;
- product definition data and configuration control data pertaining to the design phase of a product's development;
- representation of an instance of a part in an assembly through its usage in a sub-assembly;
- three dimensional shape representations of a part that includes:
  - geometrically and topologically bounded wireframe models;
  - geometrically bounded surface models;
  - topologically bounded solid models with facetted, elementary and advanced faces;
  - non surface bounded solid models including constructive solid geometry, curve swept and other swept solids, thickened face solid;
  - solids with construction history;
  - topologically bounded manifold surface and subsurface and non manifold surface models;
  - topologically bounded compound models;
- ***geometric validation properties to allow the translation of geometric shape representations (advanced boundary representation and faceted boundary representation solids) to be checked for quality;***
- ***geometric and dimensional tolerances applied to geometric shape representations;***
- materials and their composition of chemical substance;
- composite material structure and shape;
- catalogue data characterized by property value pairs;
- three dimensional presentation of product data:
- arranging geometric elements in layers and groups and assigning colors;
- presentation styles for points, curves, surfaces and sections, including hatching and tiling;
- saved views of particular camera positions and sections;
- textual annotation and notes applied to geometric elements;
- ***presentation of geometric and dimensional tolerances;***
- technical drawings as two dimensional presentation of product data.

In using AP 203, there are some constructs which have global applicability across all data in the exchange. These constructs relate to the file header for physical file exchanges, data definitions within the file related to the AP itself, and fundamental constructs which contain the information about people, organizations, dates, times, approvals, security classifications, and units of measure. Figure 3.17 illustrates the high-level entities in the AP 203 schema.

The application_context entity identifies the application which defined the data. The application attribute, based on its definition in ISO 10303-41 [49], should have the value "configuration controlled 3D designs of mechanical parts and assemblies" as this is the application domain AP 203 is meant to cover. The application_protocol_definition entity further identifies the AP. The application

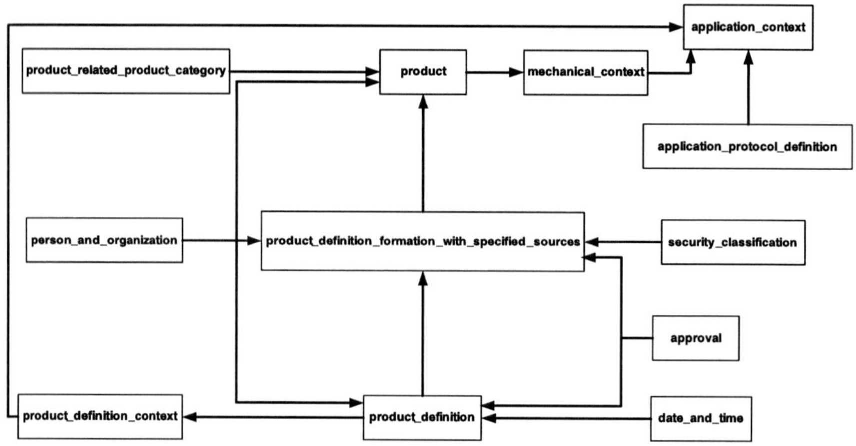

Fig. 3.17 High-level AP 203 entities

identified by the application_context entity is broken down into elements in STEP. In AP 203, these elements are contexts and the valid context entities are mechanical_context, product_definition_context, design_context, and product_concept_context. The mechanical_context entity is a subtype of the product_context entity which identifies from what engineering discipline's point of view the data is being presented. For AP 203 which uses the mechanical_context, the value for the discipline_type is restricted to be "mechanical". The mechanical_context entity will establish the viewing perspective and therefore the requirements source for product entities. It should be noted that this does not mean that AP 203 will only support purely mechanical parts. It actually means that any parts/products defined under AP 203 should have mechanical properties. It further means that these parts/products should be capable of being managed through the same configuration management processes that are used for mechanical parts.

The product_definition_context entity and its subtype design_context identify the life cycle stage or maturity of the data being presented. The product_definition_context entities will establish the viewing perspective and therefore the requirements source for product_definition entities. The product_concept_context entity also identifies what market segment or customers provided requirements for the data. This entity will establish the source of the requirements for product_concept.

AP 203 represents people and organizations as they perform functions related to other data and data relationships. A person in AP 203 must exist in the context of some organization. A person in an organization is then associated to the data or data relationship in some role indicating the function being performed. AP 203 represents dates and times to record when something occurred. In industry today, this is normally done with just a date. There are many constructs in AP 203 which require approvals. Approving in AP 203 is accomplished by establishing an approval entity and relating it to some construct through a cc_design_approval entity. AP 203 requires that certain constructs indicate their sensitivity to the owning organization. This is accomplished by establishing the security_classification entity and relating it to the construct via the cc_design_security_classification entity.

Among these high-level entities, product_definition is of particular importance to dimensional metrology because this is the entity that links to all shape and geometry definition entities.

AP 203 deals with all parts/workpieces as products. The part number for a part is stored in the id attribute. The nomenclature or name of the part is stored in the name attribute. If there is an expanded name or description of the part, this is stored in the description attribute. All STEP products must be founded in some product_context which identifies the engineering discipline from which the data is viewed. AP 203 uses two entities to form the link between the configuration management data for a part and the shape for a part. These two entities are product_definition_shape and shape_definition_representation (Fig. 3.18).

There must be only one product_definition_shape for each product_definition in an AP 203 exchange file. If there are multiple shape_definition_representation entities related to the product_definition_shape, these relationships describe alternate representations as shown in Fig. 3.18a. If the shape of the part is composed of shape constructs from multiple types of shape_representation to form the entire shape model, the main shape_representation shall be related to a shape_definition_representation which relates to the product_definition_shape as shown in Fig. 3.18b. The other shape_representations are related to the main shape_representation through a shape_representation_relationship. In some cases, the shape of a part is based on the shape of another part. This commonly occurs when the one part is the mirror image of the other. When this occurs, it is through a representation_relationship_with_transformation as shown in Fig. 3.18c.

Portions of a shape model can be designated as shape_aspects. This can be done just for internal model subdivisions or to attach specifications to portions of the shape. AP 203 edition 1 provides seven types of shape_representation which are grouped into five conformance classes. These conformance classes are [50]:

(1) geometrically bounded shape models which are represented by

- geometrically_bounded_wireframe_shape_representation, and
- geometrically_bounded_surface_shape_representation entities,

(2) wireframe with topology shape models which are represented by

- edge_based_wireframe_shape_representation, and
- shell_based_wireframe_shape_representation entities,

(3) manifold surface with topology shape models which are represented by

- manifold_surface_shape_representation entities,

(4) faceted boundary representation shape models which are represented by

- faceted_brep_shape_representation entities,

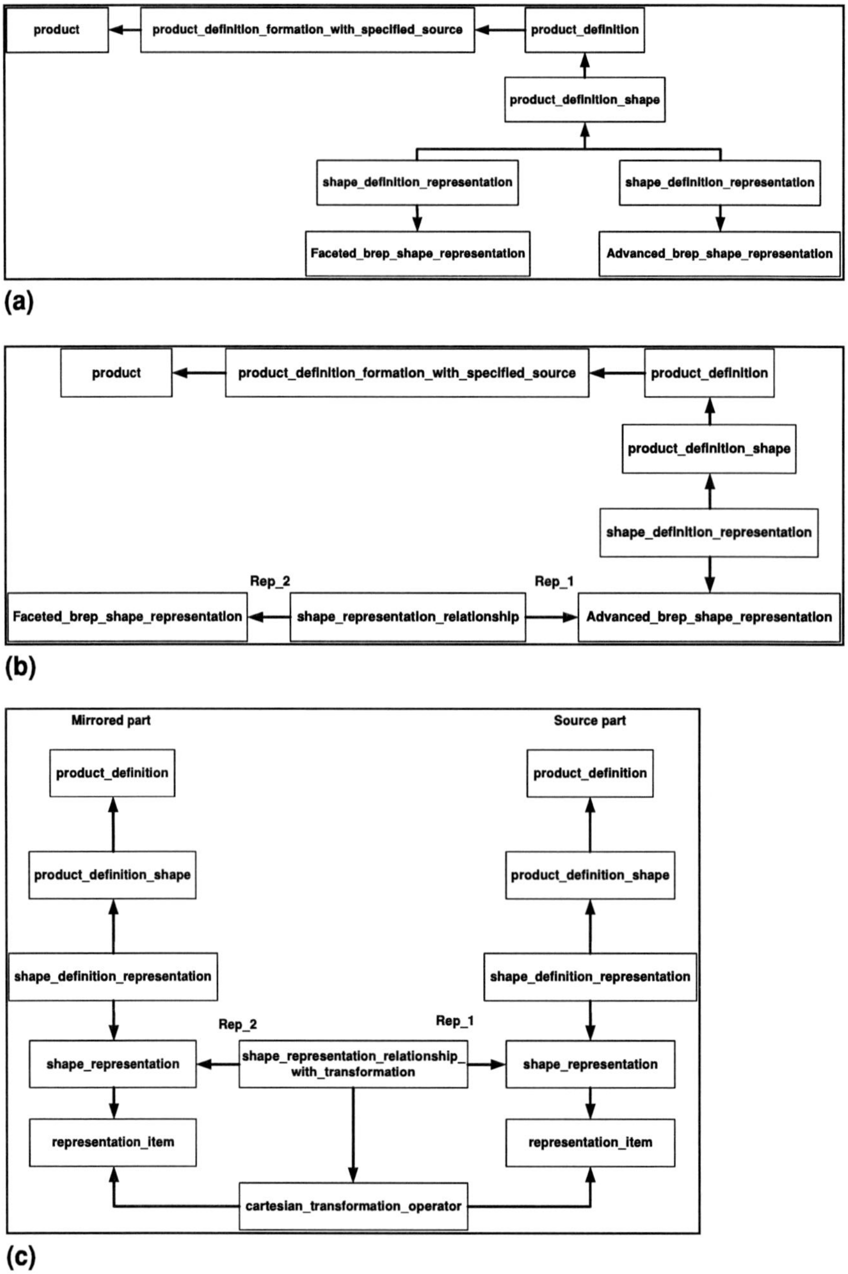

Fig. 3.18 Part and shape association in AP 203. **a** Alternative shape representation for one product. **b** Multiple shape representation for one product. **c** Shape that is mirrored from another shape on one product

(5) boundary representation models which are represented by

- advanced_brep_shape_representation entities.

A more sophisticated geometry model for the representation of the shape of a product has been developed in AP 203 edition 2, which can be categorized into six conformance options. They are:

(1) external model;
(2) wireframe models:

- geometrically_bounded_wireframe entities;
- edge_based_wireframe entities;
- shell_based_wireframe entities.

(3) geometrically bounded surface;
(4) topologically bounded solids:

- faceted_boundary_representation entities;
- elementary_boundary_representation entities;
- advanced_boundary_representation entities.

(5) solid geometry models:

- constructive_solid_geometry_3D entities;
- curve_swept_solid entities;
- swept_solid entities;
- thickened_face_solid entities;
- solid_with_local_modification entities;

(6) topological bounded models:

- manifold_surface entities;
- manifold_subsurface entities;
- non_manifold_surface entities;
- compound_shape_representation entities.

An AP 203 edition 2 Part 21 example file is listed in Appendix B.[3] In addition to these conformance options for geometric modeling, AP 203 edition 2 provides semantic association between GD&T requirements to the part shape/geometry representations. This association is illustrated in Fig. 3.19.

A geometric_tolerance is one of the following: angularity_tolerance, circular_runout_tolerance, circularity_tolerance, concentricity_tolerance, cylindricity_tolerance, flatness_tolerance, linear_profile_tolerance, parallelism_tolerance, perpendicularity_tolerance, position_tolerance, straightness_tolerance, surface_profile_tolerance, symmetry_tolerance, or a total_runout_tolerance. These geometric tolerance types are in accordance to those that are defined in the ASME Y 14.5 standard (refer to Sect. 3.2.2 for the GD&T information).

[3] Courtesy of STEP Tools Inc.

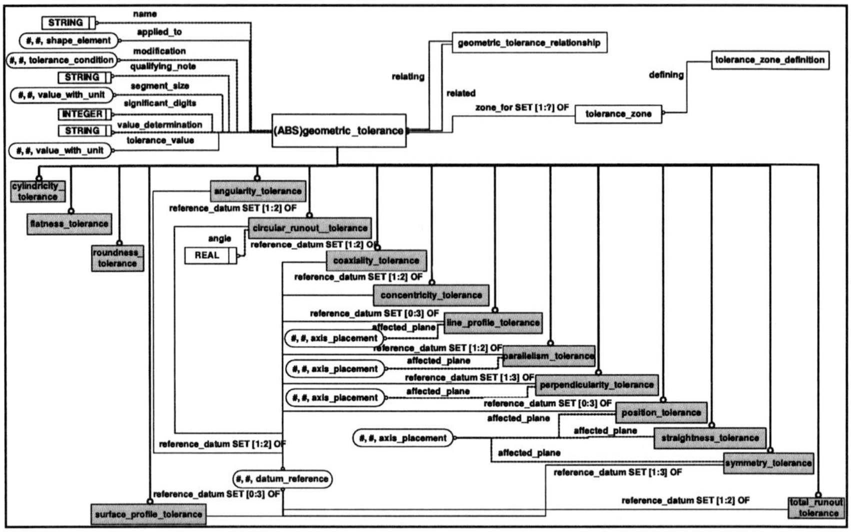

(a)

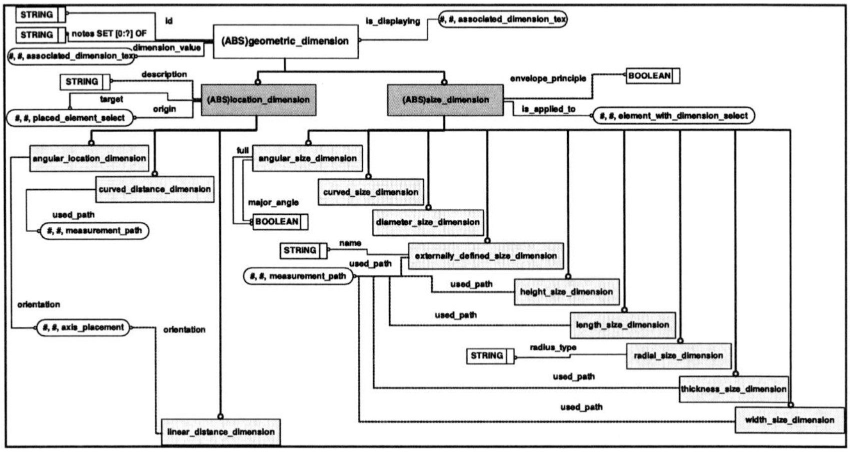

(b)

Fig. 3.19 GD&T definitions in AP 203 edition 2. **a** Geometric tolerance definitions in AP 203 edition 2. **b** Dimensional tolerance definitions in AP 203 edition 2

The data associated with a geometric_tolerance entity are:

- applied_shape—specifies the shape on a part that is being toleranced by a geometric_tolerance;
- geometric_tolerance_value—specifies the tolerance amount that a part is allowed to meet the required accuracy for proper fit;

- applied_to_multiple_datum_frames—specify the simultaneous or separate application of tolerances for different features tied to the same mobile datum reference frame;
- modifier_control—specifies the material condition which is applied to the shape being toleranced by the geometric_tolerance;
- significant_digits—specifies the number of decimal places indicating the accuracy of the tolerance;
- unit_of_measure—specifies the quantity of measure in which the value is given;
- zone_definition—specifies the tolerance zone that restricts the geometric_ tolerance.

```
ENTITY Geometric_tolerance
  ABSTRACT SUPERTYPE OF (ONEOF          (Angularity_tolerance,
                                         Circular_runout_tolerance,
                                         Coaxiality_tolerance,
                                         Concentricity_tolerance,
                                         Cylindricity_tolerance,
                                         Flatness_tolerance,
                                         Line_profile_tolerance,
                                         Parallelism_tolerance,
                                         Perpendicularity_tolerance,
                                         Position_tolerance,
                                         Roundness_tolerance,
                                         Straightness_tolerance,
                                         Surface_profile_tolerance,
                                         Symmetry_tolerance,
                                         Total_runout_tolerance));
   name :                                OPTIONAL STRING;
  applied_to :                           Shape_element;
  modification :                         OPTIONAL Tolerance_condition;
  qualifying_note :                      OPTIONAL STRING;
  segment_size :                         OPTIONAL Value_with_unit;
  significant_digits :                   OPTIONAL INTEGER;
  tolerance_value :                      Value_with_unit;
  value_determination :                  OPTIONAL STRING;
WHERE
    WR1: (NOT (EXISTS(segment_size))) OR
   ('AP203_CONFIGURATION_CONTROLLED_3D_DE
   SIGN_OF_MECHANICAL_PARTS_AND_ASSEMBLIES
   _ARM_LF.LENGTH_MEASURE' IN
   TYPEOF(segment_size.value_component));
    WR2: (NOT (EXISTS(tolerance_value))) OR
   ('AP203_CONFIGURATION_CONTROLLED_3D_DE
   SIGN_OF_MECHANICAL_PARTS_AND_ASSEMBLIES
   _ARM_LF.LENGTH_MEASURE' IN
   TYPEOF(tolerance_value.value_component));
    WR3: (NOT (EXISTS(significant_digits))) OR
   (significant_digits > 0);
   END_ENTITY;
```

Depending on the types of tolerance, datum information is required. The tolerance dependent datum information is defined with each type of geometry tolerance. For example angularity must have datum information. Datum information is linked to angularity_tolerance by reference_datum attribute. Therefore, it can be seen that the attribute applied_to associates a geometric tolerance to the shape/geometry that this tolerance is applied on.

```
ENTITY Angularity_tolerance
  SUBTYPE OF (Geometric_tolerance);
  reference_datum :                  SET[1:2] OF Datum_reference;
END_ENTITY;
```

Dimensional tolerances are also defined in AP 203 through entity geometric_dimension. A geometric_dimension is either a location_dimension or a size_dimension as shown in Fig. 3.19b. The geometric_dimension carries the following data:

- Id—specifies the name of this dimensional tolerance;
- dimension_value—specifies the value that has a dimension applied to it;
- note—specifies a qualifying note. There may be more than one note for a dimensional tolerance.

```
ENTITY Geometric_dimension
  ABSTRACT SUPERTYPE OF          (Location_dimension,
   (ONEOF                          Size_dimension));
  id : STRING;
  dimension_value :              dimension_value_select;
  notes :                        SET[0:?] OF STRING;
END_ENTITY;
```

The location_dimension and the size_dimension carry different data from each other. A location_dimension defines tolerances that are an allowable variation in location between an origin shape and a termination shape. The data associated with this entity are:

- Description—allows a string to be associated to this locational tolerance for description purposes;
- Directed—specifies a logical value designating the importance of direction for measuring a locational tolerance. If the value is TRUE, the locational tolerance is measured from point of origin to point of termination; if FALSE, an in tolerance result shall occur regardless of direction of measurement;
- Origin—specifies the shape on the part that defines the starting position for a locational tolerance;
- Target—specifies the shape on the part that defines the targeting position for a locational tolerance.

```
ENTITY Location_dimension
  ABSTRACT SUPERTYPE OF (ONEOF      (Angular_location_dimension,
                                     Curved_distance_dimension,
                                     Linear_distance_dimension))
  SUBTYPE OF (Geometric_dimension);
  description :                      OPTIONAL STRING;
  directed :                         OPTIONAL BOOLEAN;
  origin :                           placed_element_select;
  target :                           placed_element_select;
END_ENTITY;
```

A size_dimension defines the size dimension tolerance characteristic for a geometric element. The data associated with a size_dimension are:

- envelope_principle—specifies the envelope of the perfect shape corresponding to the maximum material that shall not be larger than the specified dimension tolerance.
- is_applied_to—specifies the physical shape of the part that is toleranced.

```
ENTITY Size_dimension
  ABSTRACT SUPERTYPE OF (ONEOF    (Angular_size_dimension,
                                   Curved_size_dimension,
                                   Diameter_size_dimension,
                                   Externally_defined_size_dimension,
                                   Height_size_dimension,
                                   Length_size_dimension,
                                   Radial_size_dimension,
                                   Thickness_size_dimension,
                                   Width_size_dimension))
  SUBTYPE OF
   (Geometric_dimension);
  envelope_principle :             OPTIONAL BOOLEAN;
  is_applied_to :                  element_with_dimension_select;
END_ENTITY;
```

To summarize, AP 203 edition 2 was upgraded with the GD&T model that is highly integrated with the existing model for geometry. Several iterations have already been made with significant feedback about functionality from the CAD vendors. The initial GD&T models were developed for several different AP's with different scopes so each was slightly different. A new harmonized model for all the known GD&T requirements was produced in September 2004 and was incorporated into AP 203 edition 2. The necessary data sharing has been achieved by developing a highly intricate data model with many intertwined data definitions. This makes implementing the GD&T model almost as challenging as implementing the original geometry models that were very object oriented with many inheritance relationships. Consequently, there are questions as to whether or not the new model can be implemented. Hence, early implementation projects are being formed to show that such implementation is feasible and valuable. Major CAD vendors (i.e., CATIA, NX, Pro/Engineer) have participated in the validation of AP 203 edition 2, and significant progress has been made through these validation tests. As of 2010, this edition of AP 203 has been circulated in the ISO, and it has been approved for registration as a final draft international standard waiting to be voted for formal publication.

3.3.4.3 AP 214 and AP 224

STEP AP 214 [51] was developed for the exchange of information between the applications that support the development process of the mechanical aspects of automotive vehicles. This application protocol was developed for automotive manufacturers and their suppliers. The products supported by this AP include parts, assemblies of parts, tools, assemblies of tools, and raw materials. Eight types of representation of the shape of a part or a tool were defined in this AP. They are:

(1) 2D–wireframe representation;
(2) 3D–wireframe representation;
(3) geometrically bounded surface representation;
(4) topologically bounded surface representation;
(5) faceted–boundary representation;
(6) boundary representation;
(7) compound shape representation;
(8) constructive solid geometry representation.

However, compared to AP 203, AP 214 did not receive wide acceptance by CAD vendors. Most CAD vendors are able to export design files in AP 203 edition 1 but not AP 214. This book aims to discuss interoperability for information exchange in dimensional metrology systems. Therefore, only a very brief discussion of AP 214 is given.

As introduced in Sect. 3.2, feature-based design is one of the three main design approaches in today's industry. While AP 203 and AP 214 provide design geometry and topology information of a product, AP 224 was developed for the definition of product data for mechanical product definition for process planning using machining features. Defined at the top level of this application protocol is the manufacturing_feature which contains the information necessary to identify shapes that represent volumes of material to be removed from a part by machining. A Manufacturing_feature is defined as a Machining_feature, a Replicate_feature, or a Transition_feature. Figure 3.20 shows a simplified EXPRESS-G diagram of the AP 224 manufacturing feature taxonomy.

A Replicate_feature is defined by a basis shape and the arrangement of identical copies of that base shape. Each base shape is a Machining_feature oriented to the first defined position of a pattern. The patterns describe how to replicate that feature for different placements on the part. A Replicate_feature can be a Circular_pattern, a General_pattern or a Rectangular_pattern. A Transition_feature defines a transition area between two surfaces. It can be a Chamfer, an Edge_round or a Fillet.

A Machining_feature identifies a volume of material that is to be removed to obtain the final part geometry from the initial stock. A Machining_feature may be one of the following,

- Multi_axis_feature
- Revolved_feature
- Outer_round
- Spherical_cap
- Thread
- Knurl
- Marking
- Compound_feature

The remainder of this section briefly discusses the above eight machining features. Multi_axis_feature is a type of milling feature; it may not be turned on a lathe. There are eleven types of Multi_axis_features,

- Boss (Circular_boss, General_boss, Rectangular_boss)
- General_removal_volume
- Hole (Counterbore_hole, Countersunk_hole, Round_ hole)
- Rounded_end
- Planar_face
- Pocket (Cutout, General_pocket, Recess, Rectangular_closed_pocket, Rectangular_open_pocket)
- Profile_feature (General_outside_profile, Shape_profile)
- Protrusion
- Rib_top
- Slot
- Step

Revolved_feature is the result of sweeping a planar shape by one complete revolution about an axis. The planar shape needs to be finite in length, coplanar with the axis of revolution, and should not intersect the axis of revolution. The axis of revolution shall be the same as the Z-axis of the feature. The Revolved_feature may be either an outer shape of a part or a volume removal, depending on the material direction. A Revolved_feature can be a General_revolution, Groove, Revolved_flat, or Revolved_round.

In order to generate an AP 224 file, a feature recognition process is needed. The feature recognition process examines the topology and geometry of a part (i.e., AP 203 or AP 214) and matches them with the appropriate definition of predefined and domain-specific features (i.e., AP 224). This way, a model of lower-level entities is converted into a model of higher-level entities. Many research works have been published in the field of feature recognition and various approaches have been adopted. The main advantage of these approaches is that they do not impose any constraints on designers and the method may be independent of CAD systems and CAD data formats.

Section 3.2.3 gave a discussion of measurement features and the identification of measurement features. Such information is closely related to measurement process planning especially high-level planning, the detailed introduction of measurement feature data models is, therefore, presented in Chap. 4.

3.4 Product Lifecycle Management Information

The above sections discussed the most important set of information that is able to be generated from a product definition activity such as product geometry/topology information, GD&T requirements, etc. These are the basic kinds of information for any type of product manufacturing. However, dimensional metrology systems require more than just the basic product design information. Product lifecycle related information represents another set of information that is crucial to generate an efficient dimensional measurement plan for product and/or enterprise quality control. This section provides a detailed discussion of this information.

In the 19th century, American inventor Eli Whitney championed the beginning of the Industrial Revolution with a focus on "interchangeable parts". It was during this time that the groundwork in the American system of manufacturing was created.

The concept of "interchangeable parts" was considered that a complex product could be assembled from independently manufactured parts according to an accurately documented process. The use of parts that were manufactured independently and integrated in final assembly according to precise specifications contrasted greatly with the previous practice of artisan industry, where one highly skilled individual would personally control the entire process of small-scale design and production.

This paradigm shift in manufacturing strategy introduced significant gains in efficiency for the newborn industrial world. The utilization of interchangeable parts prescribed the use of precise technical documentation that was to be created and shared by multiple parties in the design and manufacturing process [52].

From here the beginning of configuration management through engineering control measures planted the seed for what would become Product Lifecycle Management (PLM). Configuration management is essentially the process for controlling documentation release and proper versioning. PLM can be described as basic configuration management with the addition of a higher-level informational organization of a manufactured part's progress through the design, manufacturing, measurement, analysis and aftermarket lifecycle. PLM systems manage the release status for perhaps tens of thousands of individual parts and their associated information all managed at the enterprise level for concurrent engineering purposes.

System control is established through two basic implementation guidelines:

(1) A centralized persistent storage repository exists with appropriate security user access control.
(2) A well-defined process exists that ensures correct handling of the data and information contained within, including standardized design, quality control, manufacturing and metrology engineering disciplines

In a recent research study of supply chain technology the Gartner Group AMR Research organization defined five core components of PLM [49]:

(1) *Product Data Management (PDM)*—A generic term for the archival system that manages revision and configuration of specifications to provide a single version of product design.
(2) *Collaborative Product Design (CPD)*—Online conferencing and design or visualization applications that support distributed development teams.
(3) *Direct Material Sourcing (DMS)*—Automates request for quote and supports part or supplier re-use to reduce downstream design complexity in the supply chain.
(4) *Customer Needs Management (CNM)*—Captures and manages customer requirements through development to ensure the voice of the customer is heard.
(5) *Product Portfolio Management (PPM)*—Provides visibility to the NPI pipeline status and supports the business decisions to prioritize product funding.

The primary strategic value of PLM is to give manufacturers a competitive edge through faster time to market, and often results in a company's ability to command premium pricing and gain increased market share. While strategic benefits should always be emphasized in any business case, they must always be complemented by more tangible operational benefits in order to provide value. For example, in an AMR research, General Motors has attributed more than $1 billion in savings to PLM, which allowed the company to simplify its IT infrastructure. GE Aircraft credits a 33% improvement in engineering cycle time to using digital design methodologies [49].

The economic validity is now well proven as early adopters realize the cost savings of centralized well controlled enterprise engineering software systems. However, the value of these systems is not universally implemented, for instance, Burkett and Smith state that "PLM adoption differs across industries and that discrete manufacturers—including aerospace, automotive, and high-tech—are the more mature users" [53]. This leaves a large part of the industrial market open to new initiatives and potential for interoperability.

3.4.1 Product Data Management

In the area of dimensional metrology, it is the PDM information that is of most interest. It is within this pillar of the PLM system that product nominal and tolerance information is kept. PDM systems focus on change management.

Targeted largely at manufacturing users, change management provides tools for requesting changes after designs are released to production. Intended to break down some of the barriers between engineering and production, change management plays an important part in serial product development environments where many design changes occur after final design release. Focused change management implementations rank among the most successful uses of PDM today [54].

PDM systems also provide solid document management for manufacturing teams. These applications store documents under version control, and provide access rights for design teams. Created to help make sense of concurrent CAD assembly modeling, these tools also extend to basic product structure modeling and may record CAD bills of materials. Databases that are compliant to the STEP standards are gaining popularity in this group of applications, because they must provide a single source for CAD data.

Although document management is a core function of product data management systems, the value of a collaborative enterprise environment cannot be overstated. Not only do these systems support a check-in/check-out baseline for version control and history tracking of toleranced design models, manufacturing plans and work instructions, they also provide the foundation in which teams can operate efficiently. PLM systems typically support full workflows that include subscription to events such as email alerts for document changes. They are also well designed with object oriented hierarchies to provide for structures such as master modeling through part families and parent–child references that promote the reuse of standard component artifacts that can establish an enterprise product digital inventory.

Rules can be established that support standardization of product build up. Process efficiencies are in turn gained through establishing the enterprise digital domain. For example, best practices can be facilitated through various system enforced policies such as completeness and semantic validation.

In decentralized operations in which geographically dispersed engineering teams must work together concurrent engineering plays a vital role. Product Lifecycle Management allows these teams to reduce workflow bottlenecks by allowing people to work on elements of the design, manufacture and metrology steps when the upstream step is complete. For example, when a given design model is ready the machining instructions can be generated by the process engineers. Likewise, the quality control plan can be created in order to develop the measuring plans and the subsequent low level inspection programs that will ultimately be used to inspect parts when the product is physically produced. All of this is done well ahead of any actual physical operation.

Despite the potential for maximizing efficiency in the product lifecycle, it has been identified that the multitude of terms and abbreviations along with the mix-up between the discipline itself and the supporting information systems has led to some confusion regarding the substance of PLM. It is this confusion that has subsequently impeded the standardization of the field, which may have otherwise seen widespread implementation in the workplace [52].

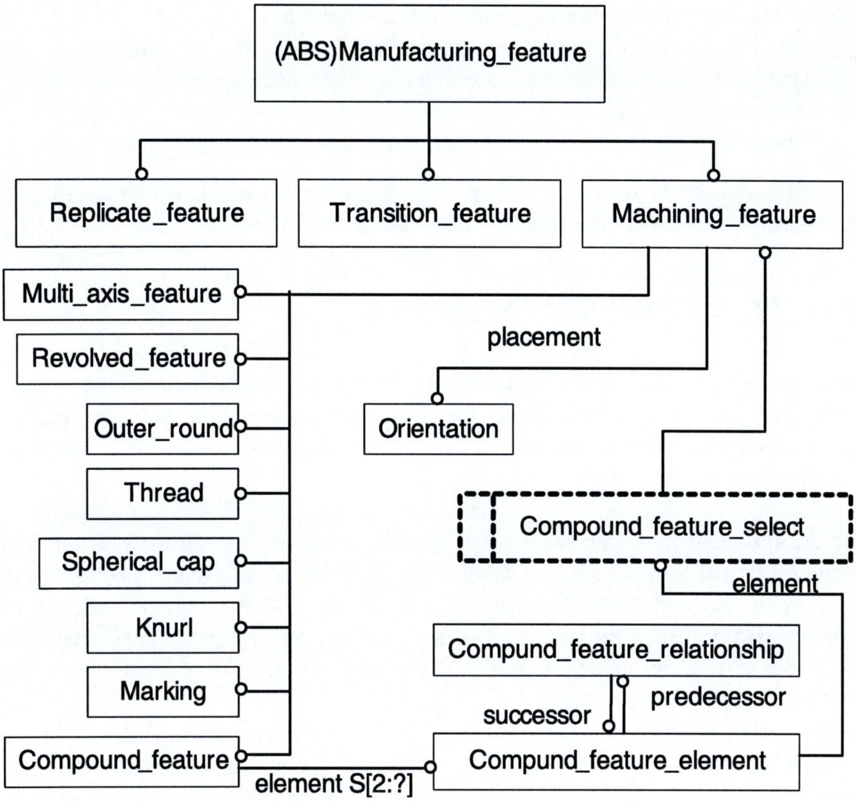

Fig. 3.20 Simplified manufacturing feature taxonomy in AP 224

Commercial PLM software systems tend to have two orientations:

(1) *CAD* Systems that were originally intended to manage the engineering work-in-process design files and approval processes, as are typically found in organizations with large mechanical engineering departments, expand into the cross-functional domain by including engineering change order process capabilities and interfaces with enterprise resource planning.
(2) *Bill of Materials* (*BOM*) Systems that focus on product-level configuration management and cross-functional change control processes that begin to offer improved connectivity to the CAD world.

In summary, PLM systems are used as one of the three cornerstones of enterprise systems in manufacturing organizations, along with Enterprise Resource Planning (ERP) and Customer Relationship Management (CRM). Regardless of the roots of a particular system, most PLM systems include the following basic functionality [52]:

(1) Secure vault for all product-related documentation
(2) Item and document classification, including part number generation
(3) Approved manufacturers and manufacturer item management
(4) BOM management
(5) Change control processes—engineering change requests and orders

Additional areas of functionality include compliance management, project management, costing and enterprise resource planning [54].

3.4.2 Key Characteristic Management

Over the last decade Key Characteristic (KC) methodologies and tools have been studied and practiced in several domains of the product lifecycle, and many world-class companies have introduced KCs considerations into their product development practices [55].

PLM is the most important systematic strategy and enabling technique to realize this emerging paradigm shift in the manufacturing industry. PLM is a strategic business approach that applies a consistent set of business solutions in support of the collaborative creation, management, dissemination, and use of product definition information across an extended enterprise. The approach is applied from concept to the end of life and it integrates people, processes, business systems, and information [56].

In particular, for complex products such as aircraft and automobiles, it is not economically or logistically feasible to control thousands of parameters (e.g. material properties, dimensions, and tolerances) and processes. Instead, companies must devote most of their attention to critical product characteristics and focus their efforts in collaborative and global product development by communicating, sharing and coordinating the identification and control of KCs. An increasing number of commercial firms are making KCs a non-negotiable technical requirement in their product development activities [57].

Currently there is no unique definition for a KC. Some typical definitions are given as follows. The Boeing advanced quality system standard D1-9000 defines a KC as a feature whose variation has the greatest impact on the fit, performance, or service life of the finished product from the perspective of the customer [57]. D1-9000 has been adapted as Aerospace Standard AS9103. The AS9103 defines a KC as a feature of a material, process, or part/assembly whose variation has a significant influence on product fit, performance, service life, or manufacturability [58]. As a result of their common background, the two definitions are almost the same. KCs are usually identified and marked on drawings or in specifications. A unique identifying number or label should be assigned to each KC so that related data can be tracked and mapped to the production processes that create the KCs.

Different companies or industries adopt different symbols. In the automotive industry, the inverted delta (∇) is used to designate safety and government regulated KCs, while a diamond (◇) is usually used to mark-up performance, fit, or appearance KCs. The designation symbol ‘|KEY>’ is adopted for mark-up of a KC in the aerospace and defense industries [57]. Although the KC terminology, definitions, and implementation schemes may vary between corporations, the organization-specific methods have common goals, i.e., to identify a small set of critical features for an organization to focus on during design and manufacturing [59].

During production and testing phases, the most important task is to identify and isolate the root causes of faults caused by process variations for a production or manufacturing system/line [60, 61]. The use of KCs is a powerful tool to help identify and reduce sources of variability. Reduction of variability can eventually result in greater product performance, fewer defects, and lower manufacturing cost. There are two major sources of variability in technical processes [62].

(1) The inherent variability of manufacturing processes. Every factor in a manufacturing process possesses inherent variability.
(2) The inherent variability of measurement systems. Several large manufacturing firms believe that variability in their measurement system initially contributed 20–25% to the problems and defects found in their shop floor.

Most research and applications of KCs are focused on variation reduction (VR) and variation risk management (VRM) during production and measurement and testing [61]. The major case studies come from aircraft and automotive industries, for which the final product quality is assured and improved. As described in Sect. 4.3, to date the most important application of KCs has been focusing on VR and VRM during the production phase. Such VR efforts fall into four areas [59].

(1) Data collection through measurement or testing during production operations to monitor process performance and initiate preventive actions.
(2) The implementation of process improvements during manufacturing activities.
(3) Assessment of feedback received from users and support personnel, and product reliability data.
(4) Implementation of design enhancements to improve quality, performance, manufacturability, and affordability.

Closing the gap between the product definition and the actual manufacturing production activities within the enterprise is one of the key priorities in digital manufacturing [63]. As a result, key characteristic and related variation information flow must propagate from design to production and ideally be implemented using closed-loop and bidirectional relationships rather than by current open-loop and unidirectional associations. Historically, key characteristic production data have not typically been collected and fed back to up-stream phases. Measurement and metrology information and knowledge (e.g. dimension and error data, process capability data, process FMEA knowledge) needs to be integrated with product and process design, particularly in assembly design [64].

3.4.3 Product Lifecycle Management Data Models

Although the information introduced above is very important for efficient dimensional metrology systems, the research effort of developing a comprehensive data model to represent product lifecycle management information is still fairly new. One of the reasons is the complexity of such data model. PLM integrates people, data, process, knowledge, and business systems together. A proper PLM data model needs to incorporate information from different sectors of a corporation. However, in the past ten years more and more research efforts have been put in this area. The following sections provide an overview of current data models of product lifecycle management.

3.4.3.1 NIST Core Product Model

The National Institute of Standards and Technology (NIST) has put forth an effort to lay the foundation of a product information modeling framework for lifecycle management [65]. It is recognized that PLM systems form the highest level of the corporate software hierarchy and depend on subsidiary systems for detailed information capture and dissemination. PLM systems tend to rely on PDM systems for managing the information describing the product itself. For many manufacturers, only the geometric description of products generated by CAD systems is managed directly. These companies would in turn rely on PDM subsystem CAD referencing for managing product descriptions.

CAD representations tend to arise only at later stages of design, after a form has been assigned to the product concept; therefore, PLM systems tied only to CAD representations of products cannot be used before the form is assigned. In order to realize PLM's potentials, PLM systems need to interact with product information used in the early stages of conception and ideation, where designers and planners deal with the function and performance of products, and not yet with their form.

The Product Engineering Program at NIST has as its goal to "establish a semantically-based, validated product representation scheme as a standard that supports seamless interoperability among current and next generation CAD and between CAD systems and other systems that generate and use product data. Specifically, the primary needs for the next generation of CAD/CAM/CAE software systems are interoperability among software tools, collaboration among distributed designers and design teams, integration of data and knowledge across the product development cycle (from design to analysis to manufacturing and beyond), as well as knowledge capture, exchange and reuse" [65].

The conceptual information architecture under development at NIST has the following key attributes:

(1) It is based on formal semantics, and will eventually be supported by an appropriate ontology to permit automated reasoning;
(2) It is generic: it deals with conceptual entities such as artifacts and features, and not specific artifacts such as motors, pumps or gears;
(3) It is to serve as a repository of a rich variety of information about products, including aspects of product description that are not currently incorporated;
(4) It is intended to foster the development of novel applications and processes that were not feasible in less information-rich environments;
(5) It incorporates the explicit representation of design rationale, considered to be as important as that of the product description itself; and
(6) There are provisions for converting and/or interfacing the generic representation schemes into a production-level interoperability framework.

An interoperability framework resulting from the application of the conceptual information architecture will:

(1) Provide a generic depository of all product information at all stages of the design process;
(2) Serve all product description information to the PLM system and its subsidiary systems using a single, uniform information exchange protocol; and
(3) Support direct interoperability among CAD, CAE, CAM and other interrelated systems where high bandwidth, seamless information interchange is needed.

The primary objective of the Core Product Model (CPM) is to provide a base-level product model that is open, non-proprietary, generic, extensible, independent of any one product development process and capable of capturing the full engineering context commonly shared in product development. The CPM model consists of two sets of classes, called object and relationship, equivalent to the UML class and association class, respectively [66, 67].

Figure 3.21 illustrates the entities comprising the CPM. All entities are specializations of the abstract class CommonCoreObject. The CoreEntity class is intended to abstract Artifact and Feature information and the CoreProperty class is intended to abstract Function, Form, Geometry, and Material information.

3.4.3.2 Open Assembly Model

The aim of the Open Assembly Model (OAM) is to provide a standard representation and exchange protocol for assembly and system-level tolerance information. The main schema structure of OAM is shown in Fig. 3.22. OAM is extensible; it currently provides for tolerance representation and propagation, representation of kinematics, and engineering analysis at the system level [66].

The assembly information model emphasizes the nature and information requirements for part features and assembly relationships. The model includes both assembly as a concept and assembly as a data structure. For the latter it uses the model data structures of ISO 10303—the STEP standards.

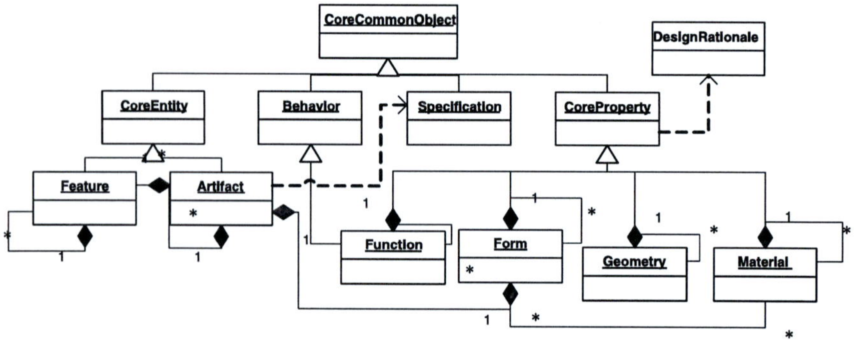

Fig. 3.21 Entities in the core product model

3.4.3.3 Design-Analysis Integration Model

Computer-Aided Design of a product's geometry and Computer-Aided Engineering for the analysis of its behavior are in common use today. However, the integration of the efforts of the professionals in the two disciplines is not as complete as it should be, resulting in the limited interoperability of the two sets of tools. Typically, a product's behavior needs to be analyzed in several functional domains (e. g., structural, thermal, kinematics, and economics) and the results of the analyses may suggest design changes for improving or optimizing the behavior [66].

The Design-Analysis Integration Model (DAIM) is a conceptual data architecture that provides the technical basis for tighter design-analysis integration than is possible with today's tools and information models. It is also intended to make analysis-driven design (often referred to as form-to-function reasoning) more practical. Figure 3.23 shows the main schema of the DAIM model.

3.4.3.4 Product Family Evolution Model

Many manufacturing concerns develop product families so as to offer a variety of products with reduced development costs [68]. The Product Family Evolution Model (PFEM) represents the evolution of product families and of the rationale of the changes involved [69]. The model consists of three sub-models: family, evolution, and evolution rationale shown in Fig. 3.24.

Family Evolution consists of two aspects: Family Derivation and Design Evolution. Family Derivation refers to the set of precedence relationships between derivative series and versions in the evolution of the product line. Design Evolution contains the design information that changed between particular series or versions and their predecessor(s). The design evolution driving factors are the justifications of the changes in the design and are reflected in the Rationale model.

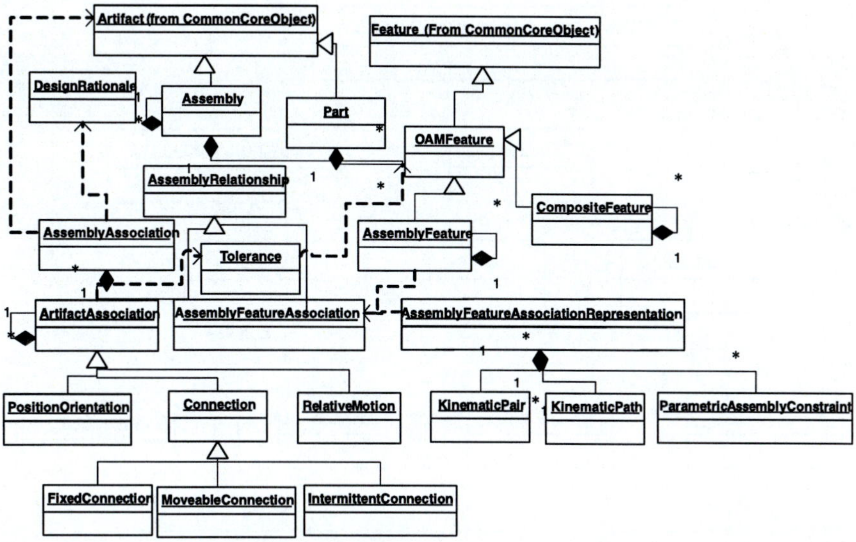

Fig. 3.22 Main schema for open assembly model

3.4.4 PDM and ERP

Enterprise Resource Planning (ERP) software evolved from earlier Material Requirements Planning (MRP) systems for inventory control and later Manufacturing Resource Planning (MRP II) technology for shop-floor scheduling and co-ordination. ERP controls and manages the entire manufacturing facility in areas including not only production but also purchasing, finance, and engineering. By coordinating the manufacturing operation for increased efficiency, ERP has become a principal tool for manufacturers to reduce manufacturing time and cost as well as facilitate teamwork and collaboration.

PDM systems manage product-related information throughout the enterprise including design geometry, engineering drawings, project plans, part files, assembly diagrams, and product specifications. However, PDM is moving beyond the product design department to support enterprise-wide business processes and the management of all product-related information and documents, including those on the shop floor and in manufacturing engineering departments.

At the same time, ERP has begun to support portions of engineering. ERP vendors continue to release a range of capabilities required for PDM. These include features such as component classification, configuration management, extended part information, document archiving, process workflow, and program management [70]. With the huge overlap in data and functionality in the two systems, many companies are starting to realize linking the two has the potential to provide an extremely powerful information tool. All PDM vendors without

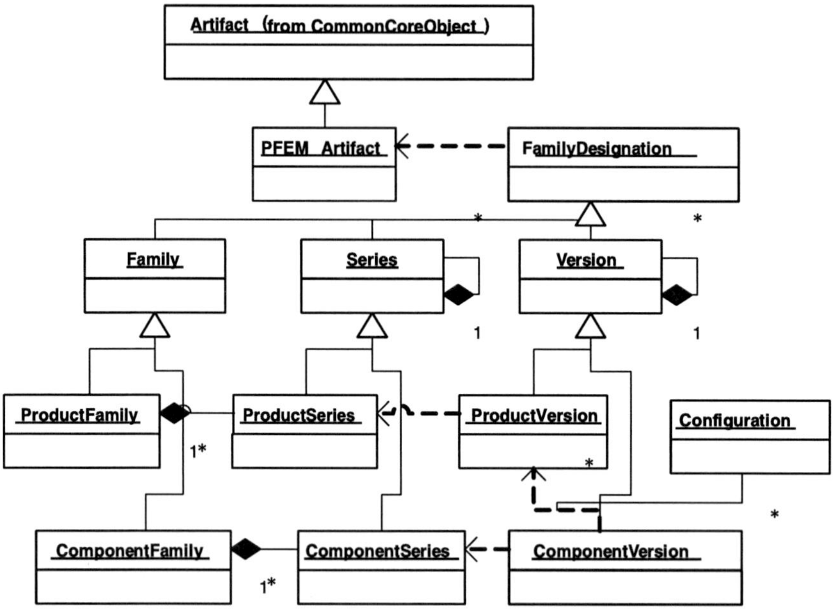

Fig. 3.23 Design-analysis integration model

exception are now actively pursuing links to ERP. Also, a growing number of ERP suppliers are engaged in providing some level of PDM functionality.

Most experts and vendors agree the technology of linking PDM to ERP has the potential to be a straightforward process. Perhaps the greatest difficulty is getting people to agree on how the two systems should be linked and what department should control what information. Many of the challenges arise from the fact that PDM and ERP have different origins and functional design. Historically, PDM systems have been championed by and controlled by engineering departments while ERP systems have generally been considered a manufacturing or even an overall business operations responsibility. As both types of systems have proliferated and increased their scope, problems have appeared, particularly in areas where they overlap.

Many experts agree there is no single correct answer to the question of how the systems should be integrated. However, one possible scenario is that ERP is a slave to PDM; PDM tells ERP what to do. In this scenario, ERP is viewed as a static system which receives downloads of released information. Bidirectional transfer of information can be achieved, for example, when PDM released information goes to ERP and feedback from ERP, like change requests, may be sent to PDM. PDM is the design group's "play area" and the ERP system really owns the configuration from that point on. In this situation, the engineering bill of material and production schedule is created and managed from the ERP point of view. Specific points of integration will need to be further identified as these two software disciplines converge. The majority of PDM/ERP integrations start through

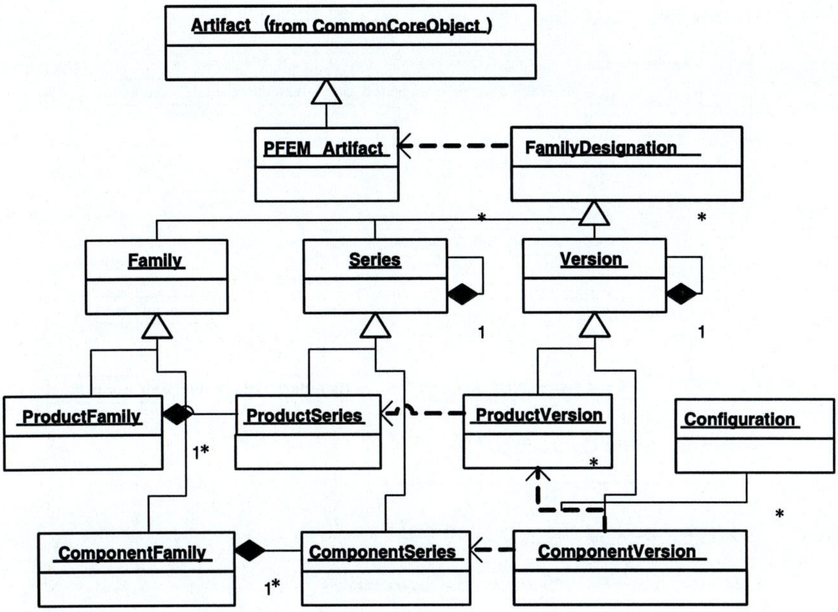

Fig. 3.24 Product and component families

the product structure or BOM. Product structures are at the heart of ERP systems, defining parts and how they are put together on the shop floor. Likewise, product structures are central to PDM but are more functionally oriented toward product capabilities and how products are defined [70].

Since product structures are common and key to both ERP and PDM, companies often use product structures as the main link between the two systems. This saves time and duplication of work largely due to the fact that data already entered in one system will not need to be re-created in another.

3.4.5 Traceability Information

Traceability can be defined as the set of practices that can be adopted by any production sector to make available all essential information about a product [71], or as defined in ISO 9000/2000 quality procedures: "as the ability to trace the history, application or location of an entity by means of recorded identifications" [72]. Tracking and tracing represents the historic information on properties of objects in the object systems [73]. The objective of "manufacturing traceability" is to provide the information about the manufacturing process needed to be able to react against defects or wrong behaviors in final products that originated during manufacturing. With the review of the

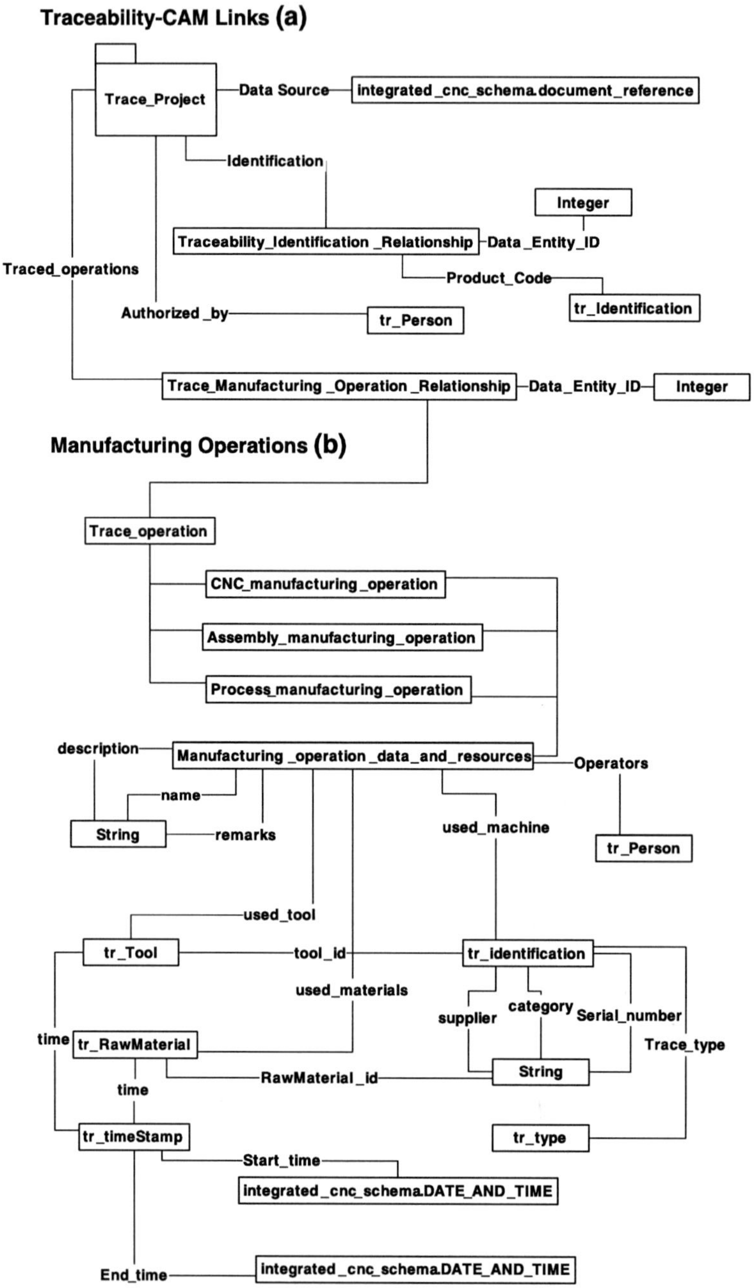

Fig. 3.25 A STEP compliant traceability model

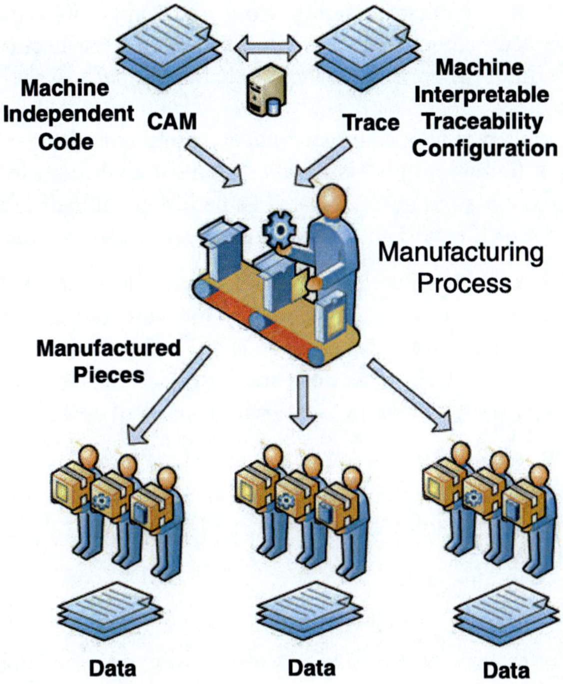

Fig. 3.26 Shop floor traceability

traceability data it may be possible to find, for instance, the lot identification of all components used to make a specific piece (Trace Back or Tracing). Then if a component lot is identified as suspected, the manufacturer can identify the final products made with components from that lot [74].

Traceability is the ability to retain the identity of the product and its origin, and usage nowadays concerns both government and business. It is the answer to the needs of several arenas of activities like Quality Management, Information Management, Risk Management, Logistical Flow, Commercial advantage, Evaluation of Management Demand [75].

The usage of traceability is not just restricted to applications such as recall, proof-of-quality, proof-of originality, or etc. In fact, in finding the roots of problems, remedy activities, and continuous improvement, the traceability system is a matter of essential assistance [76].

Three main problems are identified:

(1) The "interoperability" problem: or the ability of plant-level production applications and business systems to share information, exchange services with each other based on standards, and to cooperate using the information and services.

(2) The availability problem, coming from the temporary character of the relations between enterprises. Traceability data may be necessary after long periods of time, when some supply chain company may be no longer reachable.
(3) The integration problem: common model points out the need for common traceability information models for the extended enterprise, to overcome the use of company custom models. Data can be lost or difficult to merge because companies use different data formats and organize information differently.

Traceability information requirements and specifications about how the traceability data should be registered may vary a lot from one product to another. However, some general traceability data should always be present to allow answering essential traceability questions also known as the 6W-question approach. It should allow answering, among other, basic traceability questions [77, 78]:

- What and with what has been done?
- How it has been done?
- When it has been done?
- Who has done it?
- Where it has been done?
- Why it has been done?

It is important to consider the linkage between product traceability and product structure (CAD-CAM data). The STEP standard [79] defines products as hierarchical aggregations of assemblies and parts with features or attributes. These entities are uniquely identified and can be referenced across STEP APs and files. The use of a STEP compliant data model for traceability, besides guaranteeing interoperability, provides the implicit mechanism to link traceability data and product structure data. The proposed STEP compliant traceability model incorporates these links as references contained in the "Trace Manufacturing Operation Relationship" entity from Fig. 3.25 part A.

This entity can reference parts or assemblies (Assembly Manufacturing Operations entity), features (CNC Manufacturing Operations entity) or any kind of part attributes (Process Manufacturing Operations entity). Figure 3.25 part B shows part of the STEP compliant data model which handles this information. This has been modeled for a sample CAD-CAM STEP-NC manufacturing environment, and includes data like: raw materials, parts, subassemblies and operations to transform them into final products, data about used tools, operators, time, etc. [80].

An efficient traceability implementation has to provide exactly the right information at exactly the right time, and this means the value of information is dependent on its relevance (fit for the purpose), its currency (timeliness), its accuracy, its availability and its accessibility (ease of use).

Traceability information is most often collected for post measurement data analysis as shown in Fig. 3.26. It is quite common in Statistical Process Control (SPC) to use traceability information to perform comparative studies between process parameters such as Shift, Operator, Material, etc.

In addition to process parameters, other types of traceability data may be captured for different purposes. For example, parts may be serialized and the serial numbers are captured alongside measurement data for safe keeping in the event that product failures in the field may identify a lot or batch of product that must be recalled. In all cases, the definition and capture of manufacturing process and product traceability is an absolute requirement in most quality control programs, especially as it relates to dimensional metrology.

3.5 Summary

Product definition is the process in which a part is designed using CAD design software based on customer requirements. One of the key activities in any product design process is to develop a geometric model of the product from the conceptual ideas, which can then be augmented with further engineering information pertaining to the application area. In today's product design industry, there are three popular design approaches. They are feature-based design, parametric and variational design.

In the product definition process, three categories of information are very important to dimensional metrology. They are design geometry and feature representation information, GD&T information, and measurement feature information. Features can be thought of as 'engineering primitives' suited for some engineering tasks. They originate in the reasoning processes used in various design, analysis and manufacturing activities, and are therefore often strongly associated with particular application domains. Apart from features, another type of essential information generated through the product definition process is tolerances. Tolerances treat the uncertainty with which the realized shape or measurements of a real manufactured object compare to their design ideals.

The method of classifying features for the product design process is largely dependent on the shape representation schemes and the application domains of the feature data. There are six shape representation schemes for solid modeling in today's industry. They are CSG, sweep representation, B-rep, and combinations of these three.

The information required for GD&T and a symbology to communicate it on a part drawing have been standardized by the ISO committee as a set of standards. The geometric tolerances can be categorized into form, orientation, location, and run-out tolerances. The size variations of a product are controlled by the dimensional tolerances. The standardized GD&T symbology communication provides a means for specifying the shape requirements of, and the interrelationships between, part features. Major 3D GD&T models can be categorized into six types. They are attribute model, offset model, parametric model, kinematic model, degree of freedom model, and hybrid model.

Measurement features are the individual measurable properties or surfaces of the workpiece being examined. Identifying feasible measurement features from a design is one of the foremost tasks a dimensional metrology system needs to accomplish in order to generate a measurement plan. The measurement feature identification process is closely related to the GD&T information and its associated design geometry and/or manufacturing feature at each particular stage when a measurement is needed on the workpiece.

There are a wide variety of CAD vendors to choose from. Each of these different types of CAD systems requires the operator to think differently about how he or she will use them and he or she must design their virtual components in a different manner for each. Each of the commercial CAD systems has its own proprietary data format making the data exchange between different CAD software systems extremely difficult. This imposes one of the key interoperability issues among computer-integrated manufacturing systems. There are a few standardized data models for the exchange of design data. IGES is one of the earliest efforts in standardizing product design data exchange between CAD systems. An IGES file consists of five sections, Start, Global, Directory Entry, Parameter Data, and Terminate. IGES provides for groupings whose definitions will be found in a file other than the one in which they are used. However, it does not include proper GD&T definitions and product management information.

STEP standards have been developed to support data exchange, data sharing and data archiving. For data exchange, STEP defines the form of the product data that is to be transferred between a pair of applications. Each application holds its own copy of the product data in its own preferred form. The data conforming to STEP is transitory and defined only for the purpose of exchange. AP 203 editions 1 and 2 provide the data structures for the exchange of configuration-controlled 3D designs of mechanical parts and assemblies. AP 203 edition 1 has fairly complete product design geometry/topology information. AP 203 edition 2 was upgraded with more sophisticated geometry representation definitions. More importantly, AP 203 edition 2 provides semantic GD&T definitions with their associated geometry.

STEP AP 214 was developed for the exchange of information between the applications that support the development process of the mechanical aspects of automotive vehicles. This application protocol was developed for automotive manufacturers and their suppliers. It provides geometry based design information, however, it did not receive main acceptance among CAD vendors. AP 224 was developed for the definition of product data for mechanical product definition for process planning using machining features.

Product Lifecycle Management presents another set of information useful in dimensional metrology systems. Most of PLM related information is closely related to product definition activity and it gives manufacturers a competitive edge through faster time to market. Within PLM information, the product data management and key characteristic management are most closely related to efficient dimensional metrology systems. A number of recent PLM data models are discussed such as NIST CPM model, OAM model, DAIM model, etc. Traceability is an important element for data management in manufacturing; many industries require traceability capture for recall purposes as well as for data analysis.

Dimensional metrology systems depend heavily on traceability information to track and trace product and process errors. STEP standards have provided a standardized way of representing traceability information for parts and assemblies.

References

1. IMTI (2006) A roadmap for metrology interoperability. Integrated Manufacturing Technology Initiative (IMTI, Inc.)
2. Xu X (2009) Integrating advanced computer-aided design, manufacturing, and numerical control : principles and implementations. Information Science Reference, Hershey
3. ISO (2009) ISO 10303-203:2009: industrial automation systems and integration—product data representation and exchange—Part 203: Application protocol: configuration controlled 3D design of mechanical parts and assemblies
4. Shah JJ, Mäntylä M (1995) Parametric and feature-based CAD/CAM: concepts, techniques, and applications. Wiley-Interscience, Hoboken
5. van 't Erve AH, Kals HJJ (1986) XPLANE, a generative computer aided process planning system for part manufacturing. CIRP Ann—Manuf Technol 35(1):325–329
6. Wingård L (1991) Introducing form features in product models: a step towards CAD/CAM with engineering terminology. Department of Manufacturing Systems, Royal Institute of Technology
7. Shen Y, Shah JJ (1994) Feature recognition by volume decomposition using half-space partitioning. In: 20th Design Automation Conference, vol 69(1). American Society of Mechanical Engineers, Design Engineering Division (Publication) DE, pp 575–583
8. ANSI (1994) ASME Y14.5M-1994: Dimensioning and tolerancing
9. ISO (2004) ISO 1101:2004: Geometrical product specifications (GPS)—geometrical tolerancing—tolerances of form, orientation, location and run-out
10. ISO (2010) ISO 14405-1: Geometrical product specifications (GPS)—dimensional tolerancing—Part 1: Linear sizes
11. Dixon JR, Cunningham JJ, Simmons MK (1987) Research in designing with features. Workshop on intelligent CAD, pp 137–148
12. Pratt MJ, Wilson PR (1985) Requirements for support of form features in a solid modeling system-final report. CAM-I Report R-85-ASPP-01
13. Wozny MJ, McLaughlin HW (1986) A taxonomy of form features for solid modeling. In: Wozny MJ (ed) Geometric modeling for cad applications: selected and expanded papers from the IFIP WG 5.2 working conference, North-Holland
14. Giacometti F, Chang TC (1990) Object-oriented design for modeling parts, assemblies and tolerances. In: Proceedings technology of object oriented languages and systems (TOOLS), pp 243–255
15. Brändli N, Mittelstaedt M (1989) Exchange of solid models: current state and future trends. Comput Aided Des 21(2):87–96
16. Shen Z, Shah J, Davidson J (2008) Analysis neutral data structure for GD&T. J Intell Manuf 19(4):455–472
17. Kim J et al (2008) Standardized data exchange of CAD models with design intent. CAD Comput Aided Des 40(7):760–777
18. ISO (2007) ISO 10303-203: Industrial automation systems and integration—product data representation and exchange—Part 203: Application protocols: configuration controlled 3D design
19. ISO (2006) ISO 10303-224: Industrial automation systems and integration—product data representation and exchange—Part 224: Application protocol: mechanical product definition for process planning using machining features

20. Pratt MJ (2008) Introduction to ISO 10303-the STEP standard for product data exchange. J Comput Inf Sci Eng 1(1):102–103
21. Henzold G (1995) Handbook of geometrical tolerancing: design, manufacturing, and inspection. Wiley, West Sussex
22. Shah JJ et al (2007) Navigating the tolerance analysis maze. Comput-Aided Des Appl 4(1–6):705–718
23. Zhao YF (2009) An integrated process planning system for machining and inspection. Department of Mechanical Engineering. Ph.D. thesis, University of Auckland
24. Kramer TR et al (2001) A feature-based inspection and machining system. CAD Comput Aided Des 33(9):653–669
25. Johnson RH (1985) Dimensioning and tolerancing-final report. R84-GM-02-2, CAM-I
26. Ranyak PS, Fridshal R (1988) Features for tolerancing a solid model. ASME Comput Eng Conf 1:262–274
27. Maeda T, Tokuoka N (1995) Toleranced feature modeling by constraint of degree of freedom for assignment of tolerance. In: Proceedings of 4th CIRP Design Seminar, pp 89–103
28. Tsai JC, Cutkosky MR (1997) Representation and reasoning of geometric tolerances in design. Artif Intell Eng Des Anal Manuf: AIEDAM 11(4):325–341
29. Requicha AAG (1983) Toward a theory of geometric tolerancing. Int J Robotics Res 2(4):45–60
30. Hoffmann P (1982) Analysis of tolerances and process inaccuracies in discrete part manufacturing. Comput Aided Des 14(2):83–88
31. Krishnan KK, Eyada OK, Ong JB (1997) Modeling of manufacturing processes characteristics for automated tolerance analysis. Int J Ind Eng: Theory Appl Pract 4(3):187–196
32. Turner JU (1993) Feasibility space approach for automated tolerancing. J Eng Ind 115(3):341–346
33. Gao J, Chase KW, Magleby SP (1995) Comparison of assembly tolerance analysis by Direct Linearization and modified Monte Carlo simulation methods
34. Chase KW, Gao J, Magleby SP (1997) Tolerance analysis of 2-D and 3-D mechanical assemblies with small kinematic adjustments. In: 21st annual Design Automation Conference, vol. 82(1). American Society of Mechanical Engineers, Design Engineering Division (Publication) DE, pp 353–360
35. Zhang BC (1992) Geometric modeling of dimensioning and tolerancing. Department of Mechanical and Aerospace Engineering, Arizona State University
36. Wu YY (2002) Development of mathematical tools for modeling geometric dimensioning and tolerancing. Department of Mechanical and Aerospace Engineering, Arizona State University
37. Kandikjan T, Shah JJ, Davidson JK (2001) A mechanism for validating dimensioning and tolerancing schemes in CAD systems. CAD Comput Aided Des 33(10):721–737
38. Clement A, Rivière A, Serre P (1997) A declarative information model for functional requirement. In: Proceeding of the 5th CIRP seminar on computer aided tolerancing, Toronto, pp 3–16
39. Krause FL et al (1993) Product modelling. CIRP Ann—Manuf Technol 42(2):695–706
40. Nielsen J (2003) Information modeling of manufacturing processes: information requirements for process planning in a concurrent engineering. Department of Production Engineering, Royal Institute of Technology
41. Ungerer M, Buchanan K (2002) Usage guide for the STEP PDM schema V1.2. PROSTEP AG and ADL/PDES,Inc
42. Kemmerer SJ (1999) STEP: the grand experience. National Institute of Standards and Technology
43. ISO (2002) ISO 10303-21: Industrial automation systems and integration—product data representation and exchange—Part 21: Implementation methods: clear text encoding of the exchange structure

44. ISO (1998) ISO 10303-22: Industrial automation systems and integration—product data representation and exchange—Part 22: Implementation methods: standard data access interface
45. ISO (1998) ISO 10303-23: Industrial automation systems and integration—product data representation and exchange—Part 23: Implementation methods: C++ language binding of the standard data access interface
46. ISO (1998) ISO 10303-24: Industrial automation systems and integration—product data representation and exchange—Part 24: Implementation methods: C language binding of the standard data access interface
47. ISO (1998) ISO 10303-27: Industrial automation systems and integration—product data representation and exchange—Part 27: Implementation methods: Java TM programming language binding to the standard data access interface with Internet/Intranet extensions
48. ISO (2002) ISO 10303-28: Industrial automation systems and integration—product data representation and exchange—Part 28: Implementation methods: XML representations of EXPRESS schema and data
49. AMR, AMR Research Inc (2010) http://www.gartner.com/technology/supply-chain/amr-research.jsp Accessed 7 Dec 2010
50. PDES (1998) Recommended practices for AP 203
51. ISO (2001) ISO 10303-214: Industrial automation systems and integration—product data representation and exchange—Part 214: Application protocol: core data for automotive mechanical design processes
52. Rosen J (2010) Product lifecycle management and you. Ind Eng 42(1):44–49
53. Burkett M, Smith A (2008) Is PLM right for your business? Industry week
54. Collier W (1996) Managing the product lifecycle: the changing role of enterprise PDM. Comput Graph World 19(9):112–116
55. Zheng LY et al (2008) Key characteristics management in product lifecycle management: a survey of methodologies and practices. Proc Inst Mech Eng Part B J Eng Manuf 222(8):989–1008
56. Li WD, Qiu ZM (2006) State-of-the-art technologies and methodologies for collaborative product development systems. Int J Prod Res 44(13):2525–2559
57. AQSD1-9000 (1998) AQS D1-9000: advanced quality system tools. The Boeing Company
58. AS9103, Variation Management of Key Characteristics (2001) Society of automotive engineers. Pennsylvania
59. Kiener G (2008) Manufacturing developing guide. Wright-Patterson Air Force Base
60. Thornton AC (2004) Variation risk management: focusing quality improvements in product development and production. Wiley, Hoboken
61. Ceglarek D, Shi J (1995) Dimensional variation reduction for automotive body assembly. Manuf Rev 8(2):139–154
62. Motley B (2005) Introduction to variability and variation reduction. Defense AT&L, pp 53–55
63. Chryssolouris G et al (2009) Digital manufacturing: history, perspectives, and outlook. Proc Inst Mech Eng Part B J Eng Manuf 223(5):451–462
64. Maropoulos PG et al (2007) Key digital enterprise technology methods for large volume metrology and assembly integration. Int J Prod Res 45(7):1539–1559
65. Sudarsan R et al (2005) A product information modeling framework for product lifecycle management. CAD Comput Aided Des 37(13):1399–1411
66. Fenves Steven J (2004) A core product model for representing design information. NIST Internal Report, 6736
67. Booch G (2005) The unified modeling language user guide. Addison-Wesley, Boston
68. Ho T-H, Tang CS (1998) Product variety management: research advances. International series in operations research & management science. Springer
69. Wang F et al. (2003) Towards modeling the evolution of product families
70. Kempfer L (1998) Linking PDM to ERP. Comput-Aided Eng 17(10):58–64

71. Cheng MJ, Simmons JEL (1994) Traceability in manufacturing systems. Int J Oper Prod Manag 14(10):4–16
72. Wilkinson G, Dale BG (2002) An examination of the ISO 9001:2000 standard and its influence on the integration of management systems. Prod Plan Control 13(3):284–297
73. Töyrylä I (1999) Realising the potential of traceability—a case study research on usage and impacts of product traceability. Department of Industrial Engineering and Management, Helsinki University of Technology
74. ECR (2004) Using Traceability in the supply chain to meet consumer safety expectations. Efficient Consumer Response Europe
75. van Dorp CA (2002) Extending ERP with recipe and material traceability. Eight Americas Conference on Information Systems
76. Sohal AS (1997) Computerised parts traceability: an implementation case study. Technovation 17(10):583–591
77. Jansen-Vullers MH, Van Dorp CA, Beulens AJM (2003) Managing traceability information in manufacture. Int J Inf Manag 23(5):395–413
78. Chiu M-L, Lan J-H (2005) Information and IN-formation: information mining for supporting collaborative design. Autom Constr 14(2):197–205
79. Peng TK, Trappey AJC (1998) A step toward STEP-compatible engineering data management: the data models of product structure and engineering changes. Robotics Comput-Integr Manuf 14(2):89–109
80. Campos JG et al. (2006) e-Traceability: traceability for collaborative spread CAD-CAM-CNC manufacturing chains. In: Proceedings of the 5th WSEAS International Conference on E-ACTIVITIES, Venice, Italy

Chapter 4
High-Level Dimensional Metrology Process Planning

Dimensional metrology systems consist of a number of software and hardware systems. Section 2.3 introduced the idea that a typical dimensional metrology system can be divided into four major elements: product definition, measurement process planning, measurement process execution, and analysis and reporting of quality data. Chapter 3 has discussed the product definition activity that generates product design and lifecycle management information for the downstream manufacturing and measurement activities. In order to effectively measure products and have efficient quality control of a manufacturing process, a measurement process planning activity is always needed. The measurement process planning, also commonly called inspection process planning in academic research, produces process plans to measure products so that the functionality of the product is ensured during or after the manufacturing process.

This chapter first discusses what a typical measurement process planning activity is and the functionalities it should accomplish. This is followed by an overview of Computer-aided Inspection Process Planning (CAIPP) research in the past three decades. The review separates the CAIPP research efforts into early research, which is prior to mid-1990s, and recent research (mid-1990s up to date). These research efforts had different focuses and contributed to the development of different sections of CAIPP systems. The remainder of the chapter presents the current data models for CAIPP systems including a brief overview of proprietary data models and some emerging new standard data models to enable the exchange of high-level measurement plans.

4.1 Computer-Aided Inspection Process Planning Activities

Measurement process planning is an integral part of the design and manufacturing activities. In industry and academic research, measurement process planning is often referred as computer-aided inspection process planning. In this book,

Y. Zhao et al., *Information Modeling for Interoperable Dimensional Metrology*,
DOI: 10.1007/978-1-4471-2167-1_4, © Springer-Verlag London Limited 2011

measurement process planning is the same as CAIPP. A measurement process planning system determines what characteristics of a product are to be inspected, where and when. Modern manufacturing is increasingly characterized by low volume, high variety production, tight tolerance, and high quality and more complex products. Part and product inspection is evolving to be an important module of integrated manufacturing. This requires fast yet accurate inspection as well as effective integration with the product model and relevant database. The need for more automated measurement process planning and better decision support tools increases as the complexity and variety of products increase and the product development cycle decreases.

Decisions made in the course of process planning (both manufacturing and measurement process planning) have a significant effect on the resulting product quality, in addition to the production time and cost. Some manufacturing methods and sequences selected during process planning may be more prone to errors and inconsistencies due to a large number of setups or improper choice of datums and references. Coupling manufacturing process planning with measurement process planning leads to the closure of the desired quality assurance loop and, when taken in the wider context of concurrent engineering, will ensure that quality is "designed-in" from the start, and reduce costly rejected and/or reworked parts [1].

The most important functions of CAIPP activities are:

- to extract or accept as input (from the product definition activity, dimensional measurement equipment specifications, etc.) all the information necessary to generate a complete measurement process plan (called the high-level process plan)
- to generate a device-independent process plan containing the necessary information to execute the part measurement process.

To generate the measurement process plan, information such as part material, machine accuracy, and measuring constraints that needs to be considered to support the following decision making:

- what measurements to make and in what order
- which features need to be measured
- what are the measurands (quantities being determined by measurement)
- what are the measurement purposes
- what are the measurement methods
- how to handle outliers and filter measurement results
- among available measurement resources, which (measurement device, sensor) pairs, if any, will successfully accomplish the measurement

As measurement is an integral part of manufacturing systems, it is important to view the CAIPP activity from a broader point of view—the process planning activity for both machining and measurement. Generally speaking, the measurement process serves the following three functions for the entire manufacturing process:

- to inspect a part for the purpose of determining whether the part is within the required tolerances. This is usually carried out after the part has been manufactured. The purpose of this is usually to determine if the part is usable, but the result could also be used to change the product design or manufacturing process in the future.
- to inspect a batch of parts to determine whether the batch meets quality requirements according to statistical criteria. A sampling plan is needed for this case.
- to inspect a part during manufacturing for the purpose of process control. This is done during manufacturing. Operations carried out after inspection would vary depending on the result of the inspection.

In order to plan a measurement process, additional tolerances and manufacturing information need to be defined for different manufacturing environments. Based on the manufacturing information (such as manufacturing features, manufacturing operations, etc.), tolerance requirements, and measurement purposes, measuring features, measurement datums, and measurands can be determined. For example, a pocket is a machining feature but it is not a measuring feature. Depending on the tolerance applied on the pocket, measuring features are different. Following the high-level (also known as the macro) process planning activity, low-level (also known as the micro) process planning is carried out to generate a detailed measurement program, including precise measurement paths and measurement points, along with instructions for recording and reporting. In general, there are two types of correlations between machining and measurement process planning: machine-then-inspect and inspect-while-machining. The three types of measurement of dimensional metrology introduced in Sect. 1.2 can be categorized into these two types of process planning systems. Remote measurement (also known as post-process measurement) belongs to the former type; while in-situ measurement and in-process measurement belong to the latter one. Figures 4.1 and 4.2 depict the activities within these two types of process planning systems.

It can be seen in both figures that machining and measurement process planning are divided into high-level and low-level planning activities. In Fig. 4.1, the high-level machining process planning activity (A21) generates high-level machining process information such as machining feature and operation sequence, machine tool selection, etc. The machining process information generated from this activity is used by the high-level measurement process planning activity (A23) for traceability purposes, which can be later on used for statistical machining process analysis and control. Apart from this connection, the machining process planning activities (A21 and A22) and the measurement planning activities (A23 and A24) are not associated with each other. In industry, most post-process measurements are planned separated from the machining process planning system.

In Fig. 4.2, high-level machining and measurement process planning activities (A21' and A22') produce the same type of information as those high-level process planning activities in Fig. 4.1. The difference here for inspect-while-machining processes is the integrated machining and measurement process

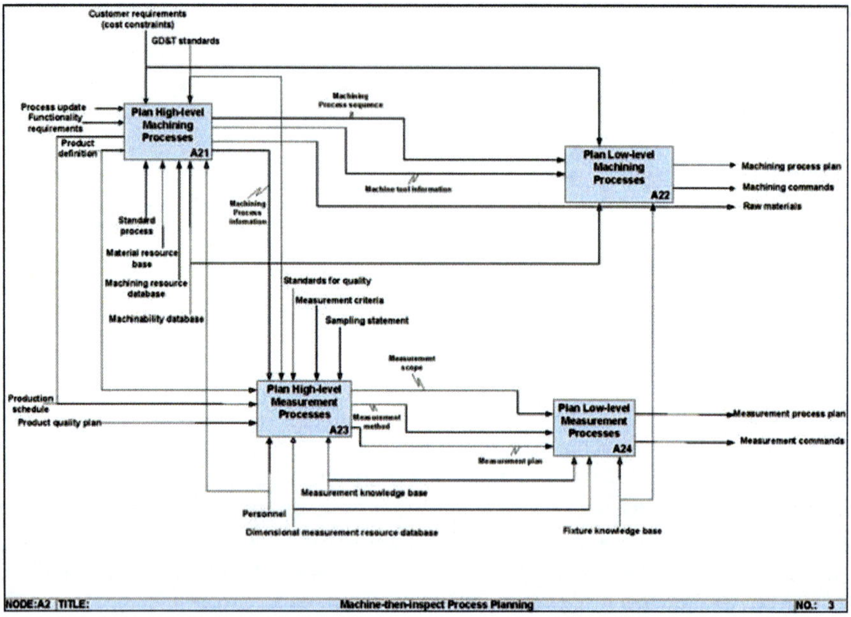

Fig. 4.1 IDEF0 activity model of machine-then-inspect process planning

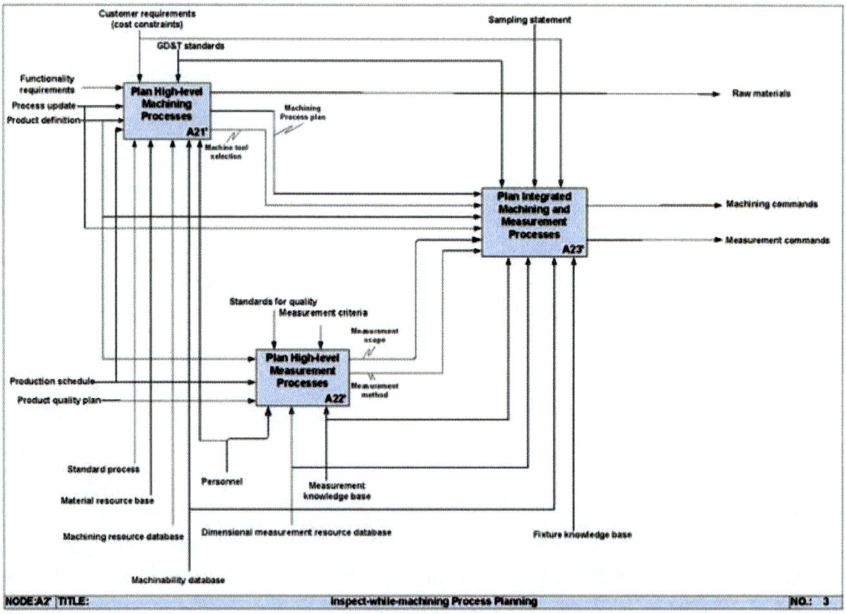

Fig. 4.2 IDEF0 activity model of inspect-while-machining process planning

planning activity (A23'). The purpose of this activity is to appropriately select critical GD&T characteristics that need to be controlled throughout the machining process and assign limited yet effective measurement operations to measure these characteristics. This activity is not the simple sum of low-level machining and measurement process planning. It decides what machining features are controlled by these GD&T characteristics, when these machining features are made, and what surfaces to measure. Then, the low-level machining and measurement process planning activities are carried out.

To summarize, the overall CAIPP activity is to develop a measurement process plan. This activity is normally divided into high-level and low-level process planning. The high-level process planning defines the measurement scope, a dimensional measurement equipment list, a sequence of high-level measurement operations, etc. The low-level process planning activity decides the number of measurement points, allocation of these points, measurement path, etc. For the measurement process that is carried out throughout the machining process, certain machining process information needs to be considered before generating the low-level measurement process plan. Most of academic research in CAIPP systems does not separate these activities as they are here. It is also hard to draw a strict line between high-level and low-level measurement process planning activities. However, the fundamental tasks each CAIPP system should achieve remain the same. In the next section, CAIPP research works in the past three decades are reviewed and these tasks are discussed.

4.2 Computer-Aided Inspection Process Planning Research

A CAIPP system may include automated or semi-automated modules capable of identifying and recognizing the dimensional inspection features along with the associated inspection constraints. It should be able to recommend an inspection method for each dimensional inspection feature. The resulting inspection operation also needs to be integrated into an overall inspection plan [2].

Automatic inspection planning for dimensional and geometric inspections can be at a high level or a low level. The high-level planning is concerned with producing a collection of setups. Each setup is related to accessibility of the features to be inspected, the probes to inspect each type of feature and the relative orientation of the part. Attempts are made to group the features, the types of tolerances and the type and size of probes to be used. The low-level planning primarily addresses the issue of point selection, path generation, and generation of executable code. Although much of the inspection carried out in industry continues to be conducted using conventional metrological equipment, most previous work on CAIP systems has been directed towards inspection operations performed on CMMs.

Research on CAIPP systems started from the early 1980s. Before the mid-1990s, most of the research works remained on conceptual-level CAIPP systems. These systems can be categorized into two groups:

1. tolerance-driven inspection process planning systems, and
2. geometry-based inspection process planning systems.

The research in the first category focused on planning inspections for those features that have specific tolerance requirements. The research in the second category focused on planning the inspection process to obtain a complete geometric description of a manufactured workpiece using the inspection data. Thus, comparison can be made with the design model for a complete geometry inspection.

From the middle of the 1990s, research on CAIPP systems started to shift to one or some of the modules that a typical CAIPP system has, such as sampling strategies and probing path planning strategies. At the same time, non-CMM measurement devices, such as 3D optical scanners, have attracted more and more attention. Therefore, CAIPP system research for non-CMM measurement methods has become another major characteristic of the research trend during this period. The following sections provide a detailed review of the research prior to 1995 in the two aforementioned categories respectively, followed by reviews on recent CAIPP research according to the modules that each research category focused on.

4.2.1 Early Research (Prior to 1995) on CAIPP

Early research (prior to 1995) on CAIPP systems is reviewed briefly. The focus is on CAIPP systems for 2½D features. Free-form surface inspection is quite a different research area. Interested readers are referred to the review article by Li and Gu [3].

4.2.1.1 Tolerance-Driven CAIPP Systems

One of the earliest CAIPP systems was developed by EIMaraghy and Gu [4]. It used a knowledge-based approach to generate inspection tasks. The system was developed in PROLOG and used a feature-oriented modeling approach. It took into account the characteristics of the CMMs, the function and geometry of the inspected part as well as the geometric and dimensioning standards and theories. It was the first system to group inspection features according to their datum, assign inspection priority based on the nature and magnitude of the assigned tolerance and check feature accessibility in a given part orientation. Figure 4.3 shows the planning logic which resulted in a recommended inspection feature sequence, probe selection and part orientation sequence. The system has a modular structure and features serve a key role.

Helmy [5] developed a feature recognition module that extracts the data of a component from its B-Rep geometric model, and then uses the data to generate a DMIS [6] inspection program. An attributed Adjacency Graph (AAG) was used to

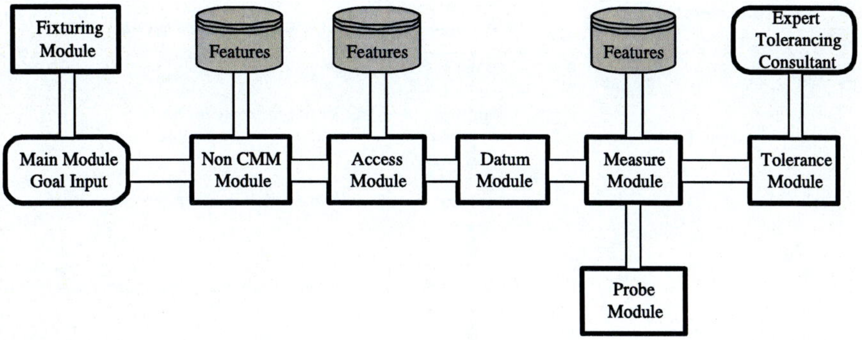

Fig. 4.3 An expert inspection planning system model

group the inspection features. AAGs were introduced by Joshi and Chang [7] to enable machined feature recognition for machining process planning. The recognition approach includes procedures for each different manufacturing feature such as steps, slots and cylindrical holes. Using these recognition procedures, together with the AAG representation and a wireframe visualization interface, the features of a component to be inspected are selected interactively. The implementation of the system requires the user to enter the machine coordinate system, the number of measurement points required, and the tolerances to be measured.

Hopp and Hocken [8] presented an approach using an inspection control hierarchy to generate control codes for CMMs (Fig. 4.4). After the user selects the required tolerance from a CAD database, the scope of the inspection is determined and the characteristics of the tolerance are identified. The surfaces involved in the characteristics are then selected for inspection and probing. Next, probing points, path planning, machine motion, and servo commands are carried out sequentially. A CMM inspection program is then generated. Some commercial systems such as Valisys [9] and Audimess [10] use a similar approach.

Medland and Mullineux [11–13] tried to integrate a CMM with a manufacturing system. The inspection plan is created automatically from a feature-based model, which contains information about the features, their significance (i.e., importance of their dimensional accuracy for the acceptance of the part), and the requirements on different probe types and orientations to reach the feature. The developed system is modular and based on a manufacturing network where communication is achieved through files exchanged within an integrated manufacturing environment. The measuring activities are controlled by a combination of dedicated programs and a constraint modeling system.

The system developed by Merat et al. [14] was part of a large effort to develop a Rapid Design System (RDS). The objective is to reduce the time from design to manufacture and inspection. In this system, tolerances are represented as features. An overall inspection plan consists of fragments, each of which relates to how toleranced geometry of a given feature is to be inspected. These Inspection Plan Fragments (IPFs) are generated based upon rules and methods used in industrial

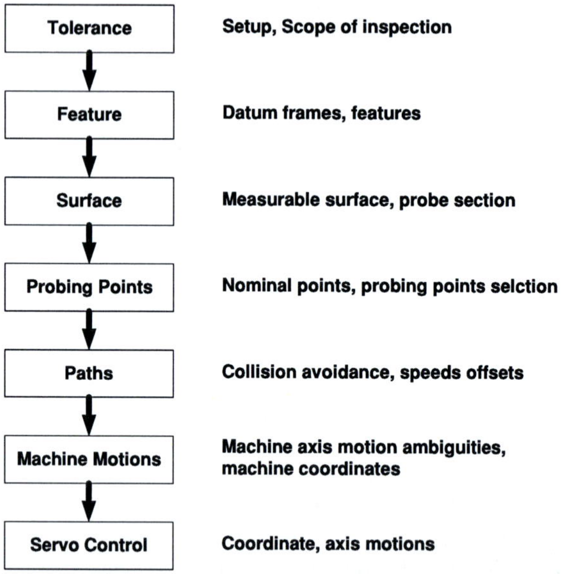

Fig. 4.4 An inspection control hierarchy

practices. Inspection planning is the selection of appropriate IPFs which result in an overall time efficient plan. IPF is generated by a macro called the IPF Generator. For each tolerance it generates a corresponding IPF with a suitable CMM probe, probing orientations and any required inspection tools other than CMMs such as depth micrometers. Feature accessibility analysis is not included and the inspection steps for various features are not prioritized or clustered to generate an optimal sequence.

The CAIPP system developed by Yau and Menq [15–17] consists of five modules: (1) inspection specification, (2) automatic inspection planning, (3) CMM verification, (4) CMM execution, (5) comparative analysis. The core of the system is a knowledge-based inspection planner that monitors process flow and assists decision-making. The main function of the inspection specification module is to translate functional requirements, tolerances, manufacturing parameters and CMM constraints into inspection specifications. The results of the specification module are used by the planning module to generate the probe path. The manufacturing accuracy and tolerance specification are taken into consideration. The generated path is then verified to ensure a collision-free path. The execution module carries out the inspection and generates the data. The measurement data together with the design model and inspection attributes are processed by the comparative analysis module to generate an inspection report.

Tannock et al. [18] developed a measurement planning system. They classified their measurement workpiece via a feature-based approach, and established measurement planning data through inquiries.

Brown and Gyorog [19] discussed a prototype system named IPPEX (Inspection Process Planning EXpert system) for the development of a generative process

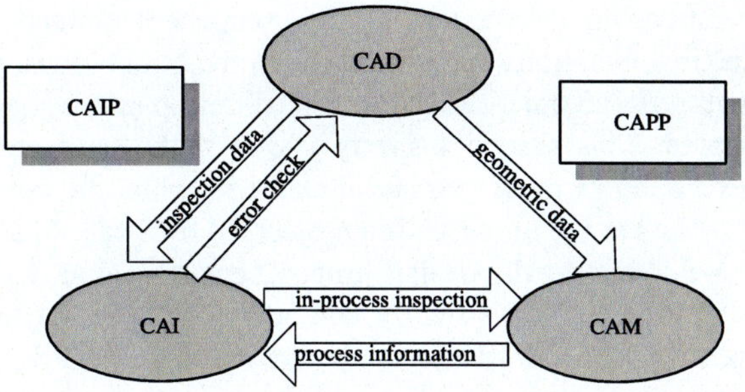

Fig. 4.5 The role of CAIP and the inter-relations between CAD/CAI/CAM

planning expert system for dimensional inspections. IPPEX uses a product geometric modeler coupled with a dimensional and tolerance modeller to generate inspection instructions in the form of an operation plan and as a part program in compliance with the DMIS standard.

All the above reviewed research works focused on developing conceptual level CAIPP systems for CMMs. Most of these CAIPP systems need an inspection operator's input for either selecting inspection features, or tolerances that need to be inspected. This type of research has laid a good foundation for the later-stage CAIPP research.

4.2.1.2 Geometry-Based CAIPP Systems

Unlike a tolerance-driven CAIPP system, geometry-based CAIPP systems largely ignore tolerance information, but focus on geometry-matching between a machined part and its designed shape. Duffie et al. [20] developed a technique to obtain a measured database for a machined part and then compared it with a CAD database. Inspection features were defined by operators. The inspection of part surfaces is carried out automatically using a tactile sensor. This inspection process results in the collection of a database of measured coordinates on the part surface. This measured database is compared with a CAD database defining the desired part geometry, and then results in a determination of the error between the actual measured part and the desired part geometry at each measured point.

Menq et al. [21] developed an optimal match scheme that aligns measurement data with design data during the CAD-directed dimensional inspection. Cho and Kim [22] developed a flexible 3D inspection system for sculptured surfaces by employing CMM, CAD database and vision system technology. The proposed system (shown in Fig. 4.5) performed optimum inspection planning, recognition of the workpiece, and compensation for alignment errors. The recognition/localization database was generated from the CAD database based on a new concept called Z-layer. Then, a 3D shape of the object on the table of the CMM was constructed by using a vision guided CMM.

Corrigall and Bell [23, 24] at Loughborough University of Technology, UK developed a system for code generation for CMMs using geometric data and relationship information of the component defined in a product model. Datum setting operations, measuring and probe orientations, probing points and safe rapid paths are automatically determined, and part programs for a CMM are also generated. This system inspects 100% of the geometry of a component with the exception of those geometric elements which lie beyond the capacity of CMM.

The Design to Inspection project led by Sira [25] aimed to develop methods that would support the design process, ensuring that designs could be manufactured and inspected consistently and sufficiently. Prototype software, known as Computer Aided Validation Expert System (CAVES), was developed to validate designs. The project identified the limitation of current geometric modelers and concluded that a powerful product modeling system is required if product validation is to be achieved in an automated fashion.

Geometry-based CAIPP systems have not received as much attention as tolerance-driven CAIPP systems. In comparison with a tolerance-driven CAIP system, a geometry-based system tries to measure the entire part; a process that is time consuming.

4.2.2 Recent CAIPP Research for On-Machine Measurement and CMM

From the CAIPP research reviewed above, it is found that a complete CAIPP system must have modules for the following tasks:

1. inspection feature selection and sequencing;
2. measurement/sampling points selection and optimization;
3. collision-free probing path planning and generation (including probe accessibility and orientation);
4. inspection execution commands generation.

From the mid-1990s, CAIPP research shifted its focus to one or some of the above modules. Also, as new measuring devices and measuring technologies, such as non-contact measuring devices, became mature and available for CAIPP systems, inspection process planning for using different types of measuring devices became another research trend. Some of these new measuring devices include laser scanning devices, optical measuring sensors, pneumatic measuring devices, etc. Compared with traditional gages and touch probes, non-contact probes are able to provide large amounts of data in a relatively short time. Therefore, the inspection process planning strategies for these new measuring devices are different from those for traditional touch tactile sensors.

Bogue [26] discussed the limitations of contact-probe-based CMMs scanning and described a new, laser-based 3D geometrical scanning system developed jointly by Metris and Volvo for assembling purposes. Vezzetti [27] presented a selective sampling acquisition approach for boundary definition in reverse

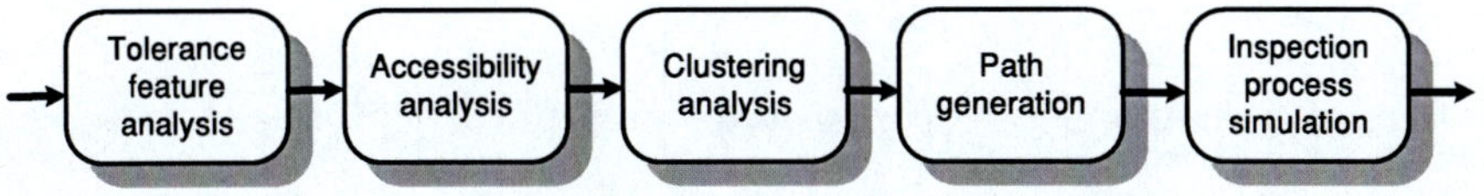

Fig. 4.6 Flowchart of the feature-based inspection process planning system

engineering. The proposed approach is developed for optical scanning devices. Minoni and Cavalli [28] proposed an optical measuring probing system, which can be used to perform on-line measurement. However, this optical equipment also has stringent requirements with respect to the measuring environment. For example, mist, unclean workpiece surfaces, reflective surfaces, and temperature lead to measurement errors. Aguilar et al. [29] analyzed the accuracy and error mechanisms of laser scanning probes using simulations and experiments. Several tests have been carried out with a laser scanning probe mounted on a CMM to determine the main error sources. The research on CAIPP for optical measurement devices is still limited.

In the following section, the relevant CAIPP research works for CMMs are reviewed in the order of the above modules.

4.2.2.1 Inspection Feature Selection and Sequencing

Inspection features are rooted in dimensions and tolerances that have a significant influence upon the functionality of the component. Determination of these inspection features used to rely upon the skill and experience of inspection engineers. Most of the research works reviewed in Sect. 4.2.1 (the early research) either required the user to specify each and every face to be probed for inspection, or automatically selected machining features that were previously recorded and controlled for inspection. Therefore, the degree of automation was severely limited. Recently, the research focus has been to develop CAIPP systems that can recognize/extract inspection features directly from a CAD model and sequence them automatically. When a workpiece is measured on a CMM, most of the machining operations have finished already before measurement. The functionality of inspection is merely an acceptance check. The inspection features selection and sequencing for this type of inspection process are more related to the probe accessibility and probe orientation. Therefore, inspection feature grouping and clustering is the main focus of most related CAIPP research.

Zhang et al. [30] proposed a feature-based inspection process planning system for CMMs. The proposed system is a prototype designed to produce an inspection process plan directly from a CAD model. The inspection process planning prototype system includes five functional modules (Fig. 4.6): tolerance feature analysis, accessibility analysis, clustering analysis, path generation and inspection process simulation. The tolerance feature analysis module is used to parse tolerance information and establish relationships between tolerance information and

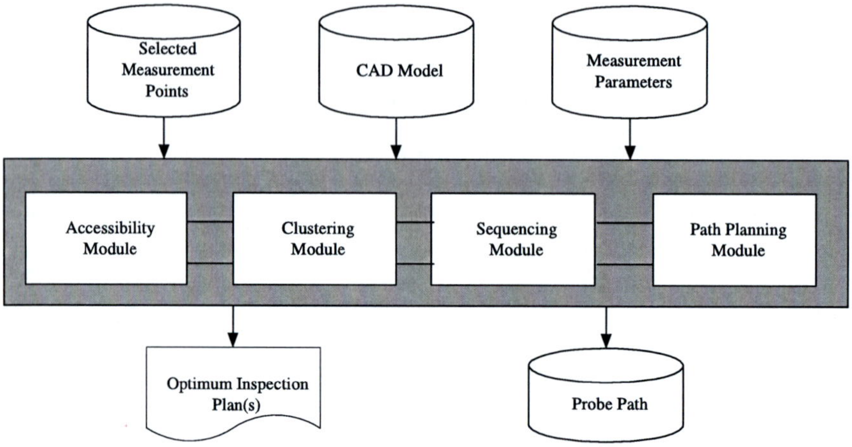

Fig. 4.7 CATIP system structure

surface features. The accessibility analysis module evaluates all accessible probe relationships between tolerance information and surface features. The clustering analysis module groups inspection probe and surface features into inspection groups so that time for inspection probe exchange and calibration can be reduced to a minimum. The path generation module determines the number of measurement points, their distribution and their inspection sequences. The inspection process simulation module provides an animated display of the inspection probe path and checks whether a collision occurs between the part and the inspection probe.

Vafaeesefat and ElMaraghy [31] proposed a methodology to automatically define the accessibility domain of measurement features and group them into a set of clusters. The methodology uses the CAD model of the workpiece and tolerance information as input to an algorithm for defining feature accessibility. The CAD model is first converted to the STereo Lithography (STL) or Virtual Reality Model Language (VRML) format for the Probe Orientation Module (POM). The user chooses probes and measurement features, defines coordinate systems, specifies tolerances and datum for measurement points. Then, probe paths are generated automatically.

Limaiem and ElMaraghy [32] proposed a Computer-Aided Tactile Inspection Planning (CATIP) system. Inspection features are selected based on a CAD model and the tolerance requirements. These inspection features are the input of the system, which contains four modules (Fig. 4.7). Inspection features are sequenced based on their accessibility and minimization of probe orientation.

Hwang et al. [33] proposed a CMM inspection planning system for the purpose of minimizing the number of part setups and probe orientations. Inspection features are selected based on the tolerance specifications by the users. After receiving the inspection feature information, the proposed system firstly analyzes the accessibility of each feature. Then, the feature accessibility information is used to derive the required part setup and probe orientation. Based on a proposed decision rule that aims to minimize the number of changes of part setup and probe orientation, the sequence of inspection features was decided.

From the above review, it can be concluded that probe accessibility and probing orientations are the major considerations for CMM-based inspection feature grouping and sequencing. The CAIPP systems for CMMs, apart from the different focuses of each research system, mostly analyze the accessibility of each inspection feature and the necessary probe orientation changes in order to decide the sequence of inspections. The effort is to minimize the change of probe orientations which contribute to the bulk of CMM inspection time.

4.2.2.2 Measurement/Sampling Points Selection and Optimization

The inspection processes carried out on CMMs often use touch-type probes to perform point-to-point motions when recording 3D coordinates of a workpiece. The more measurement points (or sampling points) that are chosen, the more reliable are the results. However, since an increase of the number of measurement points usually leads to an increase in measuring time, the appropriate number of measurement points has to be determined for each feature and tolerance to be measured. This section reviews related research on touch-type probes. Since scanning probes collect measurement points by dragging along the measurement surface, a large amount of data can be collected in a relatively short time. The measurement points allocation and probing path planning for scanning probes is distinctively different from that required for touch-type probes. Limited research was carried out in this area.

Elkott et al. [34] reviewed research works on sampling strategies for CMM inspection. Based on this review, the authors summarized the literature review of sampling for inspection planning (Table 4.1). The brief review of the research works in the table is represented in the following paragraphs. Some useful methods have been proposed to decide proper measurement points for each inspection feature by considering tolerance levels, geometric characteristics, and desired confidence levels.

Menq et al. [35] developed a method based on a given design tolerance and machining accuracy to determine the optimum number of measurement points. Dowling et al. [36] discussed the statistical issues that arise when CMMs are used. They carried out research and simulation on commonly used methods for estimating a feature's deviation range—the orthogonal least-squares and minimum-zone methods. Hwang et al. [37] proposed a knowledge-based inspection planning system for CMMs. This system integrates part geometry information, tolerance information and heuristic knowledge of experienced inspection planners to determine the number and position of measurement points. Based on their previous research, Lee et al. [38] and Cho et al. [39] proposed a similar fuzzy system for determining the optimum number of measurement points for their proposed OMM system. The surface area of the target surface, the grade of design tolerance and the volumetric error of the machine tool used to produce the workpiece are used as

Table 4.1 Summary of research on measurement points sampling

	Prismatic and conical surfaces	Free form surfaces
Sampling optimization	Woo and Liang [40]	Menq et al. [17, 35]
	Zhang et al. [41]	Jiang and Chiu [43]
	Cho, Lee et al. [39, 42]	Elkott et al. [34]
	Jiang and Chiu [43]	Cho, Lee et al. [39, 42]
Sample size	Woo and Liang [40]	Menq et al. [17, 35]
Alternate sampling plans	Hocken et al. [44]	Menq et al. [17, 35]
	Fan and Leu [45]	Pahk et al. [48]
	Lee and Mou [46, 47]	Elkott et al. [34]
Sample location	Woo and Liang [40]	Menq et al. [17, 35]
	Hocken et al. [44]	Pahk et al. [48]
	Fan and Leu [45]	Kim and Ozsoy [52]
	Zhang et al. [41]	Edgeworth and Wilhelm [53]
	Lee and Mou [46, 47]	Elkott et al. [34]
	Oray et al. [49]	
	Kim and Raman [50]	Cho, Lee et al. [39, 42]
	Fang et.al. [51]	
	Cho, Lee et al. [39, 42]	

input parameters. The Hammersley's algorithm is used to locate the measurement points on the target surfaces. At the same time, the non-contact measurement point problem is handled to relocate the measurement points. Since the decomposed primitives may contain holes, slots and/or pockets where some measurement points may lie, these measurement points should be relocated.

The algorithm developed by Hwang et al. [37] was applied to relocate these non-contact measurement points. The effect of selecting a particular measurement sampling strategy has been recognized as a major component of measurement uncertainty. This effect is due to the systematic and pseudo-random errors contained in the measurement system.

Elkott et al. [34] stated that the previous research emphasized the sampling of primitive shapes, i.e., conical shapes, spheres, cylinders and planar surfaces. Researchers who worked on the sampling of free-form surfaces often adopted a uniform sampling pattern. Others who applied surface features-based methodologies developed algorithms that require large sample sizes to inspect free-form surface features. Moreover, while a few developed methodologies attempt to optimize sample size, they do not seek the optimal locations of the sample points. Most methods depend to a great extent on the skills of the users of those systems. To overcome these shortcomings, the authors proposed a sampling system that combines several sampling solutions. The system is able to automatically select a sampling algorithm that best suits the surface being inspected.

Jiang and Chiu [43] developed a statistical method for the determination of the number of measurement points for 2D rotational part features. The authors proposed a feature-based technique to determine a sufficient number of measurement points for CMMs. To use a feature-based approach in determining the number of measurement points, an acceptable error amount must be provided as the decision criterion. However, the errors caused by the measurement and the part dimension deviation from the norm are normally not separated. For form features, it is logical to use form tolerances as the acceptable error amount since it best represents the limit of the sum of all possible error sources. Regression and least-square methods were used for checking if the number of selected measurement points satisfies the requirement.

4.2.2.3 Probing Path Planning and Generation

After measurement points are generated, the main task for a probing path planning and generation module is to evaluate measurement points' accessibility, avoid collision, and optimize probing paths. Most of the research for CMM probing path generation focused on generating collision-free probing paths. It is assumed that the inspection features have already been sequenced previously for these research works.

Albuquerque et al. [54] used an iterative method of point placement and collision avoidance for multiple, interacting features to automatically generate probing paths (Fig. 4.8). A list of surfaces to be measured is obtained from the overall inspection planner. For each of these surfaces an initial set of points is generated, constrained only by the desired minimum configuration and number of inspection points on each surface. The system then addresses the mapping and subdivision techniques for each set of point placements. Each set of measurements is checked for measurability after transforming the inspection point coordinates into the CMM workspace coordinates. This process is followed by iterative replacement of points in accessible regions. After a sufficient number of measurable points have been placed during the iteration process, a collision-free path is generated. This research considered many requirements such as flexible and accessible point placement, feature intersecting, and probing path optimization.

Ainsworth et al. [55] developed a probe path generation system that utilizes interactions between CAD systems and users. The system has three stages, path generation, modification, and verification. The order in which measurement points are negotiated must be adapted to the geometry in question. With each inspection feature being essentially sampled over a grid of points, the measurement may be performed in unidirectional or bi-directional scans. The former is generally better suited to closed and/or highly folded surfaces, and the latter is more suited to relatively flat, open surfaces. By using a CAD model and the generated sampling points as input, the implemented path planning software initially generates a measurement path for each selected entity, based on the default parameters set by the user. The path is displayed as a set of line segments, together with the 3D model of the part. Following this, the system allows the user to modify interactively any of the path parameters. Finally, the defined measurement path is post-processed into machine executable programming code.

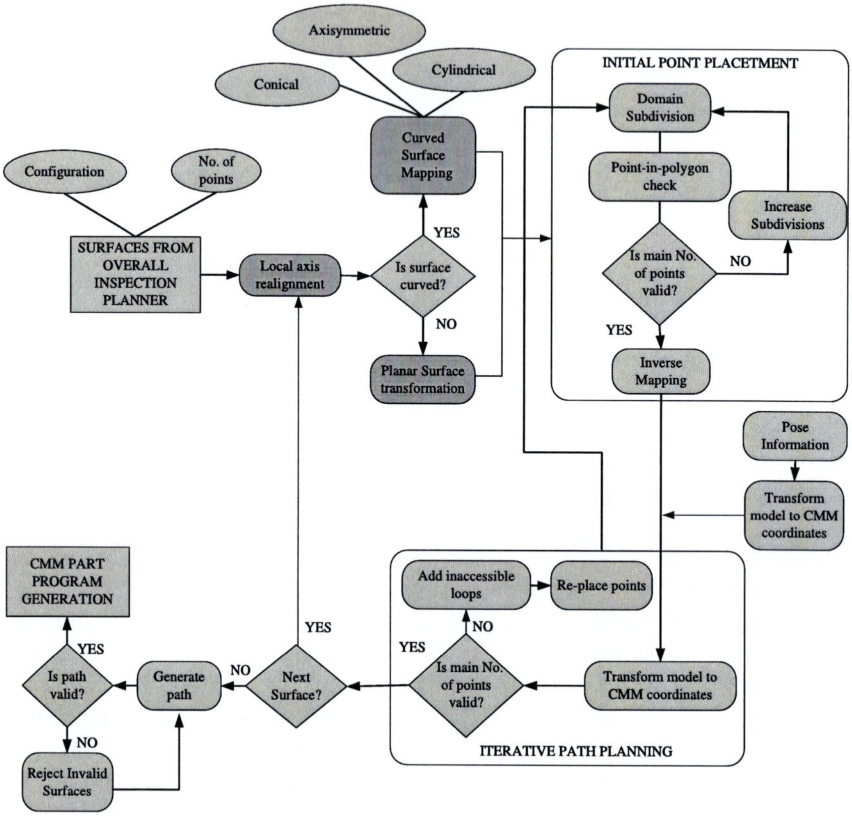

Fig. 4.8 Flowchart of the collision free probing path generation

Lin and Murugappan [56] proposed a framework for automatic CMM inspection probing planning. A three-phase approach was taken:

1. developing a general algorithm for path generation;
2. selection of a CAD system with an API (application programming interface);
3. implementation of the algorithm.

The main objective of this work is to develop a general algorithm for CMM inspection path generation, which can be implemented with any CAD system API. The algorithm assumed that the CMM probe is a point object. This helps convert collision detection of the moving probe with the part, into the simpler detection of collision of a single point with the part. Fixtures are not considered in this research.

4.2.3 Review of CAIPP Systems for OMM

All the reviewed CAIPP research in the above sections is for CMMs only. CAIPP systems for OMM operations received very little attention before the mid-1990s.

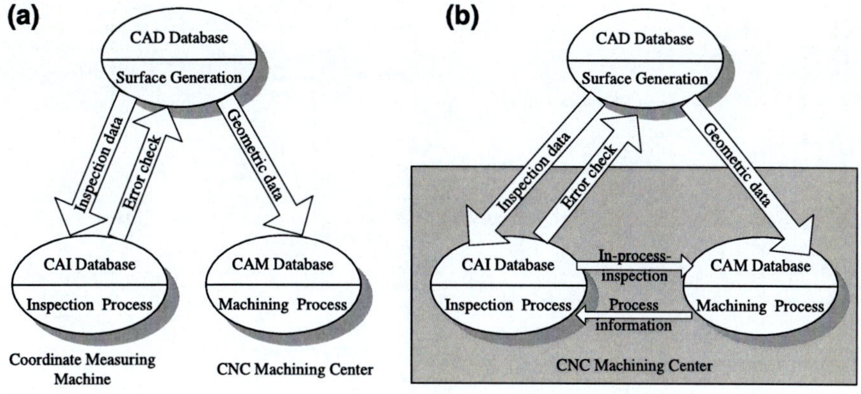

Fig. 4.9 CAIPP system using **a** CMM and **b** OMM

The main reason for this is that CNC machines were not able to provide high enough accuracy to carry out acceptable OMM operations. OMM was treated as a delay of production resources. However, this situation started to change when new generations of high accuracy and high performance CNC machines became available. Industry and researchers realized that certain OMM operations can be carried out in the machine center to provide real-time, in-process measurement. With proper process planning, this type of in-process measurement can largely reduce scraps. Measurement data can also be used for evaluating machine performance and providing statistical data for quality and machine maintenance control.

Successful implementation of OMM in machining centers, however, requires robust and reliable hardware, software, and reliable data. A multi-tool capacity machine tool is often a must. An open architecture controller is also essential for inclusion of any additional probing software that may be needed. The measuring system, which may be comprised of different probes, sensors and electronic elements, is needed for implementing the OMM process on the machine tool. The feedback mechanism needs to be in place and in real-time.

CAIPP systems for OMM and CMMs are different. In the research carried out by Cho and Seo [57], the differences of inspection planning strategy for the OMM and CMM in CAD/CAM/CAI environment were analyzed. Figure 4.9 shows the inspection process planning comparison between OMM and CMMs.

The inspection feature selecting/sequencing module has the most differences. For those systems that use CMMs, the inspection feature selecting/sequencing module focuses on accessibility and collision detection, probing approach direction, etc. For OMM operations, whereby a part is machined and inspected on the same machining center, machining feature sequence is the main consideration for inspection feature sequencing. The inspection feature selecting/sequencing module in CAIPP systems for OMM tends to focus on grouping inspection features according to the machining feature sequence. Probing accessibility and probing

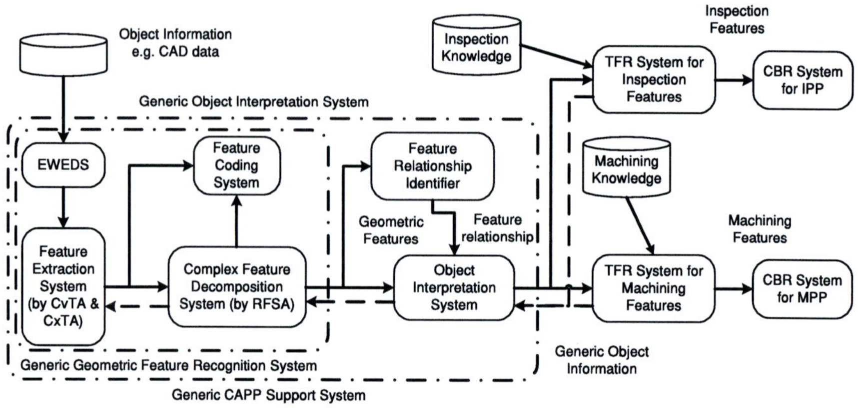

Fig. 4.10 Framework of GCAPPSS

orientation are of minor significance for these systems. The related research is reviewed in the following paragraphs.

Wong et al. [2, 58] proposed a feature recognition approach for non-CMM inspections, based on the environment of the Generic Computer-Aided Process Planning Support System (GCAPPSS) proposed by Yuen et al. [59]. Figure 4.10 shows the GCAPPSS system. A key feature of GCAPSS is the Generic Object Information System (GOIS), which consists of a generic geometric feature recognition system, a feature relation identifier, and an object interpretation system. The GOIS accepts the object information from a CAD model data in the Extended Winged Edge Data Structure (EWEDS), and processes them through a feature extraction system to identify simple and complex features. The output of the generic geometric feature recognition system contains features that are different from the machining features. The feature relationship identifier receives this geometric feature information and establishes feature relationship information for the object interpretation system. This, in turn, provides machining and inspection process planning systems generic object feature information respectively.

This research classifies the most frequently occurring dimensional inspection measurands into the following seven cases:

1. The distance between two parallel faces which can be length, width, gap, slot, fin, height, protrusion, depth, recess or thickness. The actual measurement process depends on the shape, size and orientation of the pair of faces of interest.
2. The diameter of a complete cylinder/hole.
3. The diameter or radius of a partial cylinder/hole or a cylindrical face.
4. The distance between a cylinder/hole and a parallel face.
5. The distance between a pair of cylinders/holes.
6. Coordinate measurement (or profile measurement) of a curved surface (free-form or otherwise) with respect to a bounded reference plane.

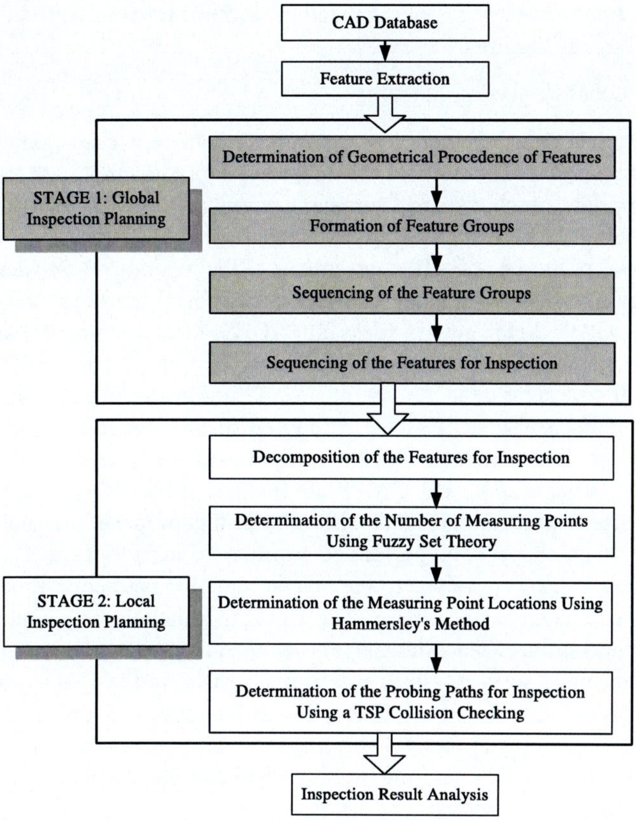

Fig. 4.11 Overall schematic structure of OMM process planning

7. A combination of the above. A wide range of measuring equipment and length standards may be used during this stage.

One problem of the proposed algorithm is that it often generates enormous numbers of different inspection options. The authors have proposed a knowledge-based technique—by using a series of "filters" to subject each inspection process.

Lee et al. at Inha University [38], Korea proposed an optimal inspection planning strategy (Fig. 4.11) for workpieces comprising many primitive form features. This is a two-stage process:

- Stage I: Global inspection planning

  At this stage, an optimum inspection sequence is determined. First, the geometrical precedence of the features is determined by analyzing their nested relations, and then the features are grouped according to the extracted characteristics. Next, the inspection sequence of the feature groups is determined, and then the sequence of the features in each group is determined to generate a global inspection plan. The planning procedure is represented as a series of the heuristic rules developed. The application of the rules results in an inspection sequence of the features.

- Stage II: Local inspection planning

  Each feature is then decomposed into its constituent geometric elements such as planes, circles, etc. The tasks of this local inspection planning are to determine the appropriate number of measurement points, their locations, and the optimum probing paths to minimize measuring errors and times.

Chung [60] proposed a CAIPP system for OMM operations on free-form surfaces. An Initial Graphics Exchange Specification (IGES) translator was developed to translate CAD/CAM output files into IGES files. Trimmed Non-uniform Rational B-spline (NURBS) surfaces are extracted through the IGES translator. Measurement codes are generated by means of coordinate transformation and the uniform sampling software (which is proposed in this research) linked with the IGES translator. The same techniques were used in the research carried out by Cho and Seo [57], where CAM and CAIPP are integrated by taking into account the geometric information of machined surfaces. For this purpose, the analysis of the machined surface shapes was performed in order to carry out the CAIPP effectively. This analysis corresponds to the machining error prediction process, which predicts the machined surface shape. The key is to simulate the geometrical form of the machined surface. Machining errors can then be predicted by comparing this simulated machined surface with the designed surface in the CAD system.

For the rest of the CAIPP modules, the techniques used for CMMs process planning can be employed for OMM operations. Techniques for selecting proper measurement points and generating probing paths for OMM operations are mostly "borrowed" from the research for CMMs.

4.2.4 Review of STEP Enabled CAIPP Systems

STEP has, by far, the most comprehensive data model for product related information. However, among STEP-enabled research efforts, CAIPP systems received the least attention. Very limited research has been carried out on developing STEP compliant CAIPP systems, most of which started after the mid-1990s. Among the reviewed research, CMMs were mostly chosen to carry out measurement operations. Multiple STEP APs have definitions related to measurement process and operations. Nevertheless, these standards overlap with each other and are still incomplete in supporting CAIPP. The international metrology communities and STEP standards committees are aware of these problems and are addressing some emerging issues through a series of meetings.

The NIST in the U.S. [61, 62] proposed a Feature-Based Inspection and Control System (FBICS) for machining and inspecting mechanical piece parts. FBICS controls a machining center or a CMM for inspection. It uses a feature-based description of the shape of the object to be made or measured as the principal input

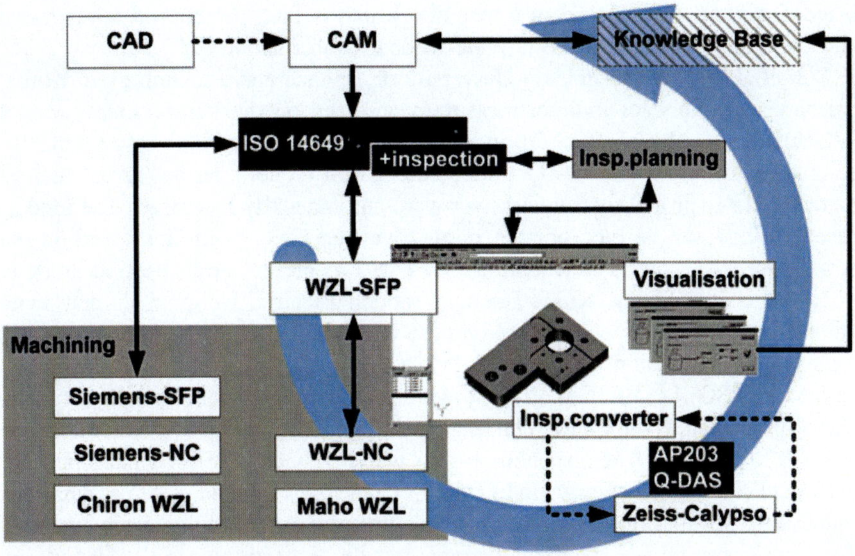

Fig. 4.12 Prototype implementation of the STEP enabled CLM

for machining and/or inspection. ISO 10303 AP 224 predefined machining features are used for defining an inspection node in the process plan. DMIS is used to control the CMM. For each AP 224 feature type, corresponding DMIS features are defined to represent the AP 224 feature for inspection operation; and then, measurement commands are generated in DMIS. The characteristics of FBICS are a tightly integrated open architecture, hierarchical tasks and control, standard data representation with clearly defined modules, command interface and data interfaces.

Brecher et al. [63] in the Laboratory for Machine Tools and Production Engineering (WZL) at Aachen University in Germany, developed a system for a closed-loop process chain which integrated inspections into the STEP-NC machining information flow. The research presents a system that supports milling and CMM-employed inspection operations for feature-based, closed-loop machining. A prototype implementation (Fig. 4.12) was carried out for the proposed system. Two STEP-NC-based controllers were used for the implementation: (1) Sinumerik 840D+STEP-NC enabled ShopMill controlling of a Chiron machine tool, and (2) a WZL-NC controlled Maho 600E machine tool. The milling operations are defined in an ISO 14649 data file including inspection workingsteps. The data file is processed by the WZL-SFP program, which provided graphics of the workpiece and manual tolerance and inspection workingstep definition input. Hence, machining commands for the aforementioned machine tools are generated for machining and inspection commands are generated for CMM (Zeiss CMM with Calypso software in this implementation) to carry out inspection operations.

Inspection results are stored in a text file, which is then parsed and reintegrated into the STEP-NC data file to provide feedback information.

Suh et al. [64] at the Pohang University of Science and Technology in Korea, presented a method of indirect measurement based on the Virtual Gears Model (VGM), obtained by NURBS fitting of the surface points measured by a CMM. By comparing the VGM with CAD model (soft-master model), various errors such as tooth profile error and tooth trace error were automatically measured. The model-based method can be incorporated in an advanced CNC controller based on the new CAM–CNC interface scheme of STEP-NC as an on-line inspection module.

In the United States, NIST, Boeing, General Electric, Unigraphics and some other industry partners collaborated on a STEP-enabled CLM scenario including probing activities using ISO 10303 AP 238. Two of the three probing operations defined in ISO 14649 Part 10 [65] were tested in this demonstration (workpiece_probing and workpiece_complete_probing). The workpiece was measured on a CMM machine. The inspection results were fed back to the original input (AP 238) data file for necessary modification. Then, the modified AP 238 data was tested. Due to possible CNC machine axis misalignment, offsets were coupled. This demonstration was the first attempt to test the inspection operation definitions in the existing ISO 14649 standards.

Ali et al. [66] developed an inspection framework for closing the inspection loop through integration of information across the CAx process chain. The proposed STEP-compliant inspection framework works with an inspection workplan, workingstep, and a mechanism to feed back inspection results across the total CAx process chain. STEP-NC (ISO 14649-16), DMIS and ISO 10303 AP 219 (application protocol for dimensional inspection information exchange for CMMs) [67] were used as the basis for representing the product and manufacturing models. This research mainly focused on utilising a CMM.

ISO committees and major inspection related research groups have been holding joint meetings in the effort to further develop and harmonize existing inspection standards such as DMIS, AP 219, and ISO 14649 Part 16 [68] (data for touch probing based inspection for CMMs). In 2006, the Automotive Industry Action Group (AIAG)'s MEtrology Project Team (MEPT) started to explore STEP-NC enabled solutions in conjunction with the work on Dimensional Markup Language (DML) and the new Quality Measurement Data (QMD) standard at the ISO TC184/SC4 WG3-T24 STEP-Manufacturing Hershey meeting, US. At this meeting, Airbus presented its requirements for tolerances in next generation CNC machining. Boeing presented a test result of using AP 203 edition 2 for a CLM machining process. During the meeting, it became clear that the several standards/specifications under the oversight of the MEPT, namely, I++ DME, DMIS, DML, QMD, and Scan Data, should generally fit well within the context of the appropriate STEP APs.

In 2007, a STEP-enabled on-machine inspection demonstration was carried out at a STEP meeting in Ibusuki, Japan. At attendance were NIST, STEP Tools Inc., Boeing, Airbus, and other major industry companies. An on-machine inspection operation was carried out on a fish-head shaped workpiece (provided by Airbus).

Although inspection path planning and measurement points optimization were not considered in this demonstration, it was the first physical demonstration of STEP-enabled on-machine inspection. At the same meeting, a High-level Inspection Planning (HIPP) system was proposed for conducting STEP-enabled inspection tasks. A new edition of AP 238 was also proposed with new inspection feature definitions and changes to some related existing entities.

To summarize, the above sections first reviewed CAIPP research in the past twenty years. The research on CAIPP started from mid 1980s. Before the mid-1990s, CAIPP research is more at a conceptual level that is to investigate what modules should be included in a CAIPP system. Four modules have been identified through these research efforts. They are the inspection feature selecting/sequencing module, the measurement/sampling point selection and optimization module, the probing path generation module, and the inspection execution module. These four modules can be further divided into smaller modules. Recent CAIPP research (after the mid-1990s) focused on one or some of these modules and their sub-modules. In the meantime, non-CMM inspection operations started to attract attention from industry and researchers. CAIPP systems for non-conventional measuring devices and operations became another research trend. As a consequence of all these research efforts on CAIPP systems, greater automation has been achieved in today's inspection process planning.

All the reviewed research focused on developing and testing CAIPP systems for single-device measurement systems. This is a common practice in academic research, which is also known as "reductionism" or the practice of decomposing a problem into smaller, more manageable pieces. This kind of research usually has some implicit assumption that the collection of solutions to these smaller problems would somehow combine to yield a solution to larger problem. In industrial product development, however, developing new technology is only part of the solution. For commercial CAIPP software systems, on one hand they need to receive design information from commercial CAD software, on the other hand, they also need to generate measurement information for device specific operations. Industry is faced with numerous proprietary inspection languages and interfaces. This causes overhead problems associated with maintaining multiple systems or locks users to one vendor. In particular, there are no adequate standard systems for linking coordinate measurement machines and other types of automated inspection systems with systems that analyze and track dimensional inspection results.

STEP standards have been developed to facilitate seamless information exchange in manufacturing systems including dimensional metrology systems. There is very limited research on STEP compliant CAIPP systems, most of which are very preliminary research in validating existing inspection related data models in STEP standards. Apart from STEP, there are some other standards efforts in modeling information for high-level measurement plan exchange. In the next section, these standard data models as well as proprietary data models in commercial systems are discussed.

4.3 Information Modeling of High-Level Dimensional Metrology Process Plans

It has been introduced that a CAIPP system can generally be divided into high-level and low-level process planning systems. The low-level process plan activity is closely associated with the chosen measurement devices; therefore, there is a large overlapping between low-level measurement process planning and measurement plan execution activities. Often the measurement device manufacturers provide very limited low-level measurement process plan capabilities. For example with a given measurement feature, the measurement device control system is able to generate the location of measurement points and probing paths. It is the exchange of high-level measurement plan information that is presenting an interoperability barrier in manufacturing today. The dimensional metrology industry needs a standard data model for the exchange of high-level measurement plans between different CAIPP systems.

This data model should consist of the information that bridges design intent to quality conformance testing and measurement plan execution. It needs to contain sufficient information to create a detailed inspection routine for any given manufactured part that could be inspected by any dimensional measurement source or other type of probe or sensor. The high-level measurement process planning system should also interface with enterprise production planning systems such as PLM and ERP (refer to Sect. 3.4.4). Therefore, the high-level measurement plan data model not only needs to define measurement plan information (i.e., measurement features, key characteristics, measurement operations, etc.), but also should define high-level production plan information for traceability purposes. This information is summarized in the following.

- Workpiece information—the information specific to the manufactured part to be measured. This information also contains data pertaining to workpiece ID, name and location of master CAD model, and setup construction (including part coordinate system and data reference frames).
- CAD design information—the CAD data model that refers to the label assigned to the specific CAD file used for generating the measurement plan (either explicitly or implicitly).
- Quality information—the information specific to any quality plans associated with the manufactured part. This includes both sampling plan and corrective action plan information.
- Measurement cycle time—the maximum time per unit allowed to measure a product. This information is generated based on ERP systems.
- Sample size—the number of pieces to be measured from a larger population within a manufacturing production run. This information is for batch production quality control and commonly generated based on production requirement information.

- Traceability information—the information needed during inspection of the manufactured part. This information consists of items within lists that can be associated with the measured part for downstream analysis.
- Standard reference information—the information that describes a "Master Artifact" that can be used as a reference for calibrating measurement equipment once a low level execution program is generated from a measurement plan.
- Material information—the information about the specific material of which the part to be measured is made. Examples include; Plastic, Aluminum, Titanium, etc.
- Manufacturing operation information—the information that specifies the manufacturing operations in a multi-stage process in which the part is to be measured; for example rough grinding, polishing, etc.
- Operator information—the information about the specific person that measured the part.
- Assignable Cause—the special causes of variations that are separate and distinct from common variation in the manufacturing process such as broken tool, dull tool, operator error, etc.
- Corrective action—the action that can be performed if an assignable cause of variation is found in the manufacturing process. There can be multiple corrective actions associated with any specific assignable cause.
- Time and date information—the information that refers to the time and date the measurement plan is generated.
- Measurement information—the information about the characteristics that need to be measured. This constitutes the core section of the high-level measurement plan file and includes items such as measurement feature, GD&T, etc.
- Characteristic—also known as measurand or dimension. The characteristic defines what is to be measured. Measurement plans can contain both variable characteristics (diameter, circularity, flatness, distance, location, etc.) and attribute characteristics (burrs, scratches, dents, discoloration, etc.).
- Nominal values—also known as the target value. This is the design specification for a perfectly manufactured part.
- Upper tolerance limit—also known as the upper specification limit. This is the design specification for the upper threshold of dimensional conformance.
- Lower tolerance limit—also known as the lower specification limit. This is the design specification for the lower threshold of dimensional conformance.
- Measurement feature—the specific geometric entity that contains the characteristic(s) to be measured such as a hole, a planar face, etc.
- Criticality information—the information that specifies the level of importance of a non-conformance to the specific characteristic to be measured.
- Event notification—the type of communication or alarm that is to occur in the event of a non-conforming characteristic measurement.
- Measurement operation information—the information to be used in the measurement process such as: staging the part, zeroing the gage, changing part orientation, etc.

The information summarized here is in a general form. Each item can be further divided into more detailed information entities. It is upon the data model designer to associate these entities in a hierarchical and semantic way.

4.3.1 Information Flow in Commercial CAIPP Systems

For the last several decades the primary inspection planning technique used by quality engineers on complex prismatic manufactured parts was based on the artful interpretation of 2D drawing key characteristic and blueprint tolerances through joystick programming based on heuristic approaches. Under this method, the CMM probe and stylus moved with human guidance over a sequence of approach and sensor contacts to inspect the entire workpiece. The method is very labor intensive, requires special skill and experience. The task typically results in long programming cycles for complex part inspection.

Today, there is a growing number of commercial inspection planning solutions on the market that can generate a part program in minutes instead of hours or days. All CMM manufacturers offer standard learn and repeat inspection software and to varying degrees have developed expert automatic programming solutions. In addition, there are several independent software vendors that have developed inspection planning software that can run on most manufacturers' CMMs via DMIS part program creation.

Although these commercial systems have many differences due to development strategy or target market focus they all have basic fundamental similarities (shown in Fig. 4.13).

- Import CAD model

  In order to automatically generate a part program, the inspector must have a design model to begin. In most cases a solid model with semantically validated data reference frames and geometric dimensions and tolerances is preferred. Some systems use a direct CAD interface that reads and acts upon the model with the native model interfaces. In other instances the CAD model may be brought into the inspection planning environment through the use of interoperability translators. This may be done either because the native programming interfaces are not made available by the CAD vendor or the cost prohibitive nature of direct interfaces since licensing costs for translator or CAD kernel add to the cost of the system.

- Augment design model

  The design model will already contain feature nominal information, but in many cases may not contain the tolerance information required to judge the part to design conformance during the measurement process. The model must be augmented with the appropriate information in order to validate conformance to specification. Most systems allow the user to embed inspection data directly into the CAD model. The part coordinate system must be set, the features must be identified for inspection,

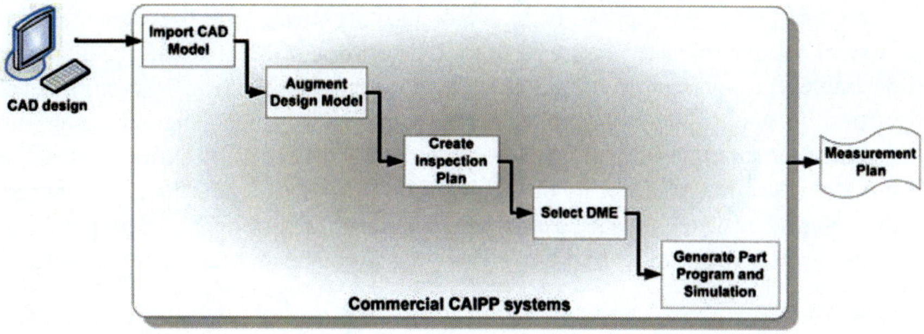

Fig. 4.13 Information flow in commercial CAIPP systems

and the tolerances must be established for quality control purposes. Depending on the intelligence of the system the user may have access to point and click geometry selection, including datum construction and validation for lineage based GD&T buildup. Some expert systems even provide intelligent feature extraction from the design model nominal geometry and generate rule based GD&T including feature control frame creation for display purposes.

- Create inspection plan

Once the CAD model is brought into Inspection Planning software and the coordinate systems, datum reference frames, features and characteristic tolerances are established, the basic information is available for offline inspection planning. The advantages to this approach are many, including the ability to perform concurrent engineering. The fully augmented design model may be available months before the first part is manufactured which allows the quality engineering department to create the inspection plan before the first part comes off the production line. It also provides the manufacturer the ability to avoid non value added "online" programming time in which the CMM is not available for quality inspection.

An inspection plan is essentially the sequence of features and characteristics to be measured. However, it is important that the model have semantically correct information identified in the design model in order to ensure a meaningful quality measurement plan. At this stage the user may be able to modify the inspection plan to remove certain unimportant features or change the sequence based on holistic knowledge such as part orientation or fixture restrictions that may inhibit sensor probing in some places of the workpiece. Once the high level inspection plan is created the measurement plan may be stored within the model as an attribute segment or associated with model, as a proprietary formatted XML file, for example.

- Select dimensional metrology equipment (DME)

The next step is to prepare for the CNC CMM part program creation. The part program is the set of instructions sent to the CMM controller that physically executes the probe head movements in order to acquire coordinate information used to calculate actual characteristic observations.

At this point it is necessary to select the target machine for measurement. Several equipment criteria may be used to determine if the machine is suitable or desirable for generation of the low level part program from the high level inspection plan. Machine volume, machine availability, machine accuracy, and machine sensor types, including probe articulation and scanning ability must all be used to determine the viability of the inspection plan to that machine. In addition, this same information must be used when generating the low level part program due to the special commands used when changing probe posture or scanning paths.

- Generate part program and simulation

Once the target DME is qualified and selected the part program can be generated. First, the part coordinate system must be aligned to machine coordinates before the probe path can be determined. Most systems provide proprietary methods for probe path optimization, including automation of clearance moves and determination of collision safety distance.

Kinetic simulation and program verification with collision detection and avoidance algorithms are extremely important features of any measurement program generating software. Probe systems can be very expensive (tens of thousands of dollars) and the parts they inspect can be even more so. Any collision can result in the loss of substantial sums of money as well as significant downtime for quality measurements.

Although the benefits of commercial expert automated inspection planning systems are vast, the challenges are also many. For example, there is no standard for product manufacturing information representation that supports passing the design model from vendor to vendor. This inhibits the portability of techniques from one CAD system to another. This also prevents portability from one planning system to another. As these commercial systems mature it should be expected to see more development in the use of templates for speed and repeatability to be used as corporate standards or industry best practices as well as rules based feature selection for consistency.

Another challenge is the limited availability of machine libraries for full kinematic representations of dimensional metrology equipment and the integration of master model technology and PLM techniques to notify and provide ability to modify inspection plans and regenerate inspection programs. In any case, in today's market vendors claim to be able to save 50–90% part programming time with intelligent offline CAD w/GD&T automation which is a huge advancement from traditional inspection planning strategies.

4.3.2 Dimensional Inspection Information Exchange Data Model (STEP AP 219)

ISO 10303 AP 219 specifies an application protocol for the exchange of information resulting from the dimensional inspection of solid parts, which includes administering, planning, and executing dimensional inspection as well as analyzing and archiving the results. The focus of this AP is the analysis and reporting activity for dimensional inspection. The measurement process itself is not within the scope of this AP. The primary benefit will be a link between dimensional inspection programs, provided by ISO 22093 (DMIS 4.0), Web-based analysis and reporting practices, and standard information models for manufacturing provided for example, by ISO 10303 AP 224 and ISO 10303 AP 238. The information provided by DMIS will be mappable into entities of this part of ISO 10303 and transportable into other ISO 10303 based implementations [67]. In addition this part of ISO 10303 captures the digital representation of GD&T requirements in standards ISO 1101, and ISO 5459 developed by ISO TC 213 on Geometrical Product Specifications and Verification.

AP 219 consists of the following Units of Functionality (UoF). Each of the UoFs has a list of entities. The high-level relationship between these UoFs is illustrated in Fig. 4.14. In the figure, each box represents a UoF and the entities this UoF contains.

- Administrative data

  Administrative data contains the information used in the management of product data, such as dates, organizations, etc. The administrative information is associated to the dimensional measurement program that is to be executed.

- Dimensional measurement analysis

  Dimensional measurement analysis defines a collection of possible analysis choices for calculating tolerance parameters and feature parameters from measured data. The information defined in this UoF is associated to dimensional measurement parameter entities.

- Dimensional measurement documentation

  This UoF provides the ability to specify documents that are directly related to the definition of a product. These documents may be specific to an operation on the part being manufactured or to a property of the part at a particular stage in the manufacturing process. The information defined in this UoF is associated with shape representation, functional limitations, manufacturing features, and part property.

- Dimensional measurement execution

  This UoF contains dimensional measurement execution program information and the initial data points gathered from a dimensional measurement operation on a workpiece. This UoF is associated with dimensional measurement parameters, functional limitations, part properties, and execution programs. The dimensional measurement parameters stores the measured data collected from the executed program. The GD&T requirements measured through this execution are traceable through functional limitations.

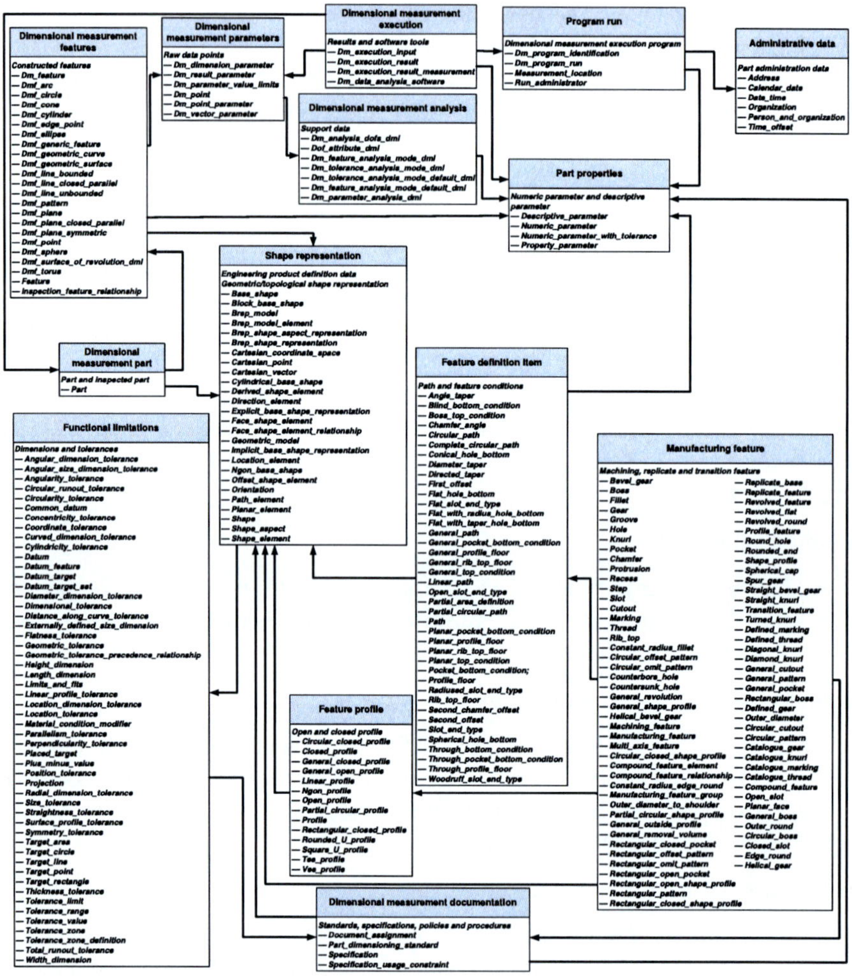

Fig. 4.14 High-level data model defined in STEP AP 219

- Dimensional measurement feature

  The dimensional measurement feature UoF consists of measurement feature information. It associates with dimensional measurement part UoF (for workpiece related information), shape representation UoF (for geometric/topology information), dimensional measurement parameters UoF (to connect the measured results to the measurement feature), and part property UoF. Dimensional measurement feature definitions are considered to be one of the most important UoFs in AP 219. Figure 4.15 shows the EXPRESS-G diagram of entities of the dimensional measurement feature UoF. The correlation between measurement features and manufacturing features is established through entity inspection_feature_relationship. A dimensional measurement feature (dm_feature) can be a pattern feature (dmf_pattern), a surface revolution feature (dmf_surface_of_revolution_dml), or one of the 18 basic geometry shapes (i.e., dmf_point, dmf_arc, dmf_line_unbounded, etc.).

- Dimensional measurement part

  The dimensional measurement part UoF contains the information necessary to identify the workpiece that is to be measured. This UoF also provides the property information that is associated with the workpiece. It is associated with the shape representation UoF so as to link the workpiece with its geometric shape information.

- Dimensional measurement parameters

  This UoF contains the information to record measurement results. The measurement results can be raw points, vector information, or analyzed results such as a value with tolerance.

- Feature definition item

  The feature definition item UoF contains the information necessary to create a machining feature. It also identifies the relationship between machining features and shape aspects, which is the geometry/topology information.

- Feature profile

  The feature profile UoF contains the information to identify profile features. A profile feature is created by sweeping its 2D projection along a path.

- Manufacturing feature

  The manufacturing feature UoF contains the information to identify manufacturing features. This UoF as well as the feature profile and the feature definition item UoFs are referenced from AP 224, AP 238 and other integrated resources of STEP standards.

- Functional limitations

  The functional limitations UoF contains the information to identify the important characteristics that need to be measured. These characteristics include dimensional and geometric tolerances. This UoF also provides datum information and the association between tolerances and the shapes controlled by tolerances. This UoF references other STEP parts such as Part 519 and AP 224.

- Part properties

  This UoF contains the description of parameters associated with the workpiece that is to be inspected. These parameters can be either numerical parameters or descriptive parameters.

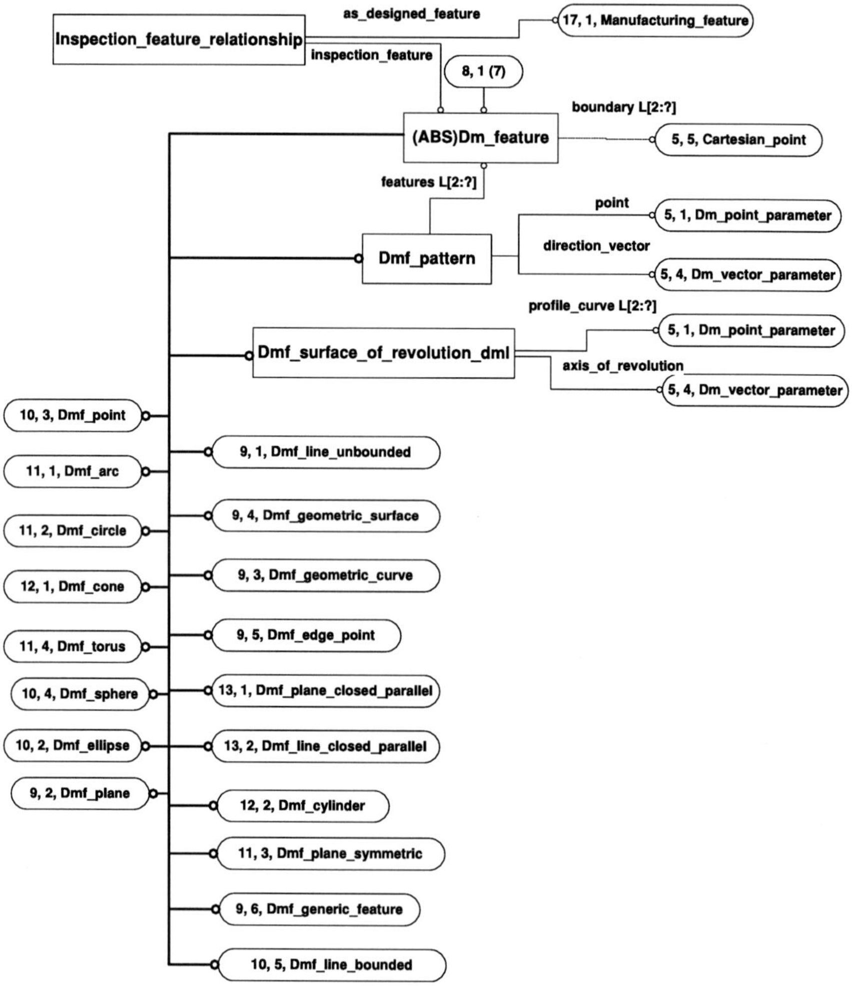

Fig. 4.15 Dimensional measurement features defined in AP 219

- Program run

  The program run UoF specifies the program used by the measurement system to direct measurement activities and gather measurement data.

- Shape representation

  The shape representation UoF contains the information used to define shapes such as manufacturing features, measurement features, etc. The shapes defined in this UoF are the primary geometry and shape elements. Integrated resources such as Parts 42 and 514 were referenced by this UoF.

The design purpose of AP 219 overlaps the purpose of other standards such as DML and DMIS. DML was developed to capture dimensional measurement result data in XML format. It defines limited types of measurement features and it was developed for the use of CMMs only. The DMIS standard also has the capability of storing dimensional measurement results. Compared with DML and DMIS, AP 219 does not have obvious advantages in storing measurement data due to its complex data model. Therefore, it was not well received by the industry. However, AP 219 is the first and only standard effort trying to provide semantic associations between tolerances, measurement features, dimensional measurement results and analysis. As part of the ISO 10303 standards, this application protocol also connects the measurement process with manufacturing features, which is able to further connect to manufacturing process information from other APs. Nevertheless, with the obvious intent AP 219 is inadequate in providing complete definitions of dimensional measurement features, dimensional measurement results collections and analysis methods. There are many entities in AP 219 that were left empty for further development. Without industry's interest and involvement, it is hard to see the future development of AP 219.

4.3.3 High-Level Inspection Process Planning Data Model (STEP AP 238)

In STEP standards, dimensional measurement information is defined not only in AP 219 but also in AP 238 [69]. AP 238 is also known as STEP-NC. It is the application of STEP methods to NC machines. Its title is "STEP Data Model for Computerized Numerical Controllers". It represents NC programming data. STEP-NC has been and continues to be a global effort with the goal of providing a data model for a new breed of intelligent CNC controllers. Within ISO, two different subcommittees (SC1 and SC4) of TC 184 have been contributing to the development of this standard. SC1 focuses on the control of machines and the standard developed by SC1 is ISO 14649, while SC4 focuses on industrial data and its developed standard is AP 238. Since numerical control programs for machining are data for the control of industrial machines, there is a natural overlap between SC1 and SC4.

The ISO 14649 set of standards, which are subtitled "Data model for computerized numerical controllers," were developed by SC1. The models are written in EXPRESS and are ARM type models, in that they use domain terminology to describe machining. ISO 14649 has the following Parts that became international standards in 2004.

- ISO 14649-1: Overview and fundamental principles [70];
- ISO 14649-10: General process data [65];
- ISO 14649-11: Process data for milling [71];
- ISO 14649-12: Process data for turning [72];
- ISO 14649-111: Tools for milling [73];
- ISO 14649-121: Tools for turning [74];

These Parts are arranged hierarchically, in that Part 11 uses Parts 10 and 111, while Part 12 uses Parts 10 and 121. Part 10 provides a set of basic capabilities for process planning for machined parts. Parts 11 and 12 specialize these capabilities for milling and turning, respectively. ISO 14649 was adopted by the SC4 committee as the ARM for AP 238. Both ISO 10303 AP 238 and ISO 14649 are commonly referred to as "STEP-NC". Unlike ISO 14649, which is divided into separate Parts as described above, AP 238 incorporates the equivalent of all the Parts of ISO 14649 (except Part 1) with a few modifications in a single model. The model is then mapped to the STEP integrated resources to obtain an implementation model—the AIM model. Although ISO 14649 uses the EXPRESS language as the data modeling language, the full inheritance model of EXPRESS was not employed by the SC1 committee in developing the ISO 14649 data model [75]. The data modeling rules in the EXPRESS language were only lightly used in ISO 14649. The integration between ISO 14649 and STEP integrated resources was not planned by SC1. However, the SC4 team continued all the research and demonstrations in the integrated way by using the AIM model. The SC4 integrated resources are normalized to make them easily extendible. If specific weaknesses can be identified then they should be extended for manufacturing. However, the editors of the STEP APs such as AP 224, AP 219 and AP 240 in addition to AP 238 have not yet identified any weaknesses. The difference between the ISO 14649 and ISO 10303 AP 238 data models is illustrated most clearly by the link between features, geometry and tolerances. In the AP 238 data model, the tolerance data is defined by the Geometric and Dimensional Tolerancing (GD&T) model developed for AP 203 edition 2, AP 214 and AP 224. This allows an application program to traverse the data from a feature, to the faces in that feature, to the design tolerances that apply to those faces, to the datum that define the tolerances, to the plane that defines each datum, to another feature that when machined defines that datum plane and so on.

Dimensional measurement related information is defined in Part 16 of ISO 14649. However, Part 16 is in a very primitive stage with very limited GD&T and inspection operation definitions. The metrology standard development groups including ISO, the AIAG group, and the MEPT team have joined to address the gaps and overlaps between different inspection data models through a harmonized STEP inspection data model. The incompleteness of the inspection-based STEP data model has been realized and the need to harmonize STEP/STEP-NC with some widely used and emerging interface specifications such as DMIS and I++ DME has been recognized.

Both ISO 14649 Part 16 and ISO 10303 AP 219 are incomplete and undergoing changes. ISO 14649 Part 16 does not have the definition of inspection features and geometric tolerances, whereas ISO 10303 AP 219 does not specify the inspection operations and strategies for corresponding inspection features. Both of these standards support only inspection operations carried out on CMMs. Hence, these standards lack definitions of measurement machine functions and technologies, metrology device information, and measurement strategy information. The latest joint meeting across these committees was the 53rd ISO TC184/SC4 meeting in

Dallas, USA, 2007. NIST proposed a newly developed AP 238 ARM model for the high-level inspection process planning system at the Ibusuki meeting. The ARM model, named as the High-level Inspection Process Planning (HIPP) data model, combined the information requirement models for machining defined by ISO 14649 Parts 10–12, 111, and 121. The ARM model was also augmented with product data management information and information necessary to harmonize the inspection feature descriptions with the STEP manufacturing application protocols and link to the aforementioned data. The data model is still under development. The objectives of HIPP data model include:

- to provide a standard means of transmitting high-level metrology objectives from one party to another (e.g. from an automobile manufacturer to a supplier). A HIPP file should say what needs to be inspected and what to report without specifying how the inspection should be carried out. Call a plan of this sort an "objective HIPP"
- to provide a standard means of embodying a detailed high-level metrology process plan that (1) can be translated into a machine program in a language such as DMIS or (2) can be executed directly by a smart machine controller. Call a plan of this sort an "executable HIPP". The executable HIPP approach is conceptually the same as the approach to machining already taken in AP 238.
- to make the executable model fit with machining models so that it is feasible to write process plans that include both machining and dimensional measurement on the same machine, using the same dimensional measurement entities in plans of this type as in plans for machines that do only dimensional measurement. Several aspects of the model implement the third objective.

A very important contribution this data model brought to the standardization of dimensional metrology information exchange is a harmonized dimensional measurement feature definitions from major dimensional metrology data models—DMIS 5.1, AP 219, and FB Meas® [76]. This EXPRESS data model for the dimensional measurement features was prepared by modeling the taxonomy and inserting attributes either from AP 219 EXPRESS model or from DMIS 5.1 model. The modeling takes advantage of the placement of a feature; although every type of feature may be placed anywhere in a part coordinate system, fewer parameters are required in the model than in some other models of the same features. The dimensional measurement features (named as dm_feature in the HIPP model) were categorized into two groups: dm_simple_feature and dm_composite_feature (shown in Fig. 4.16). Each of these has multiple subtypes identifying most of the common features in inspection process planning, such as 3D measurement features—tuboid, cone, cylinder, and composite measurement features—pattern features and compound features. Some of the dimensional measurement feature definitions in EXPRESS are shown as follows.

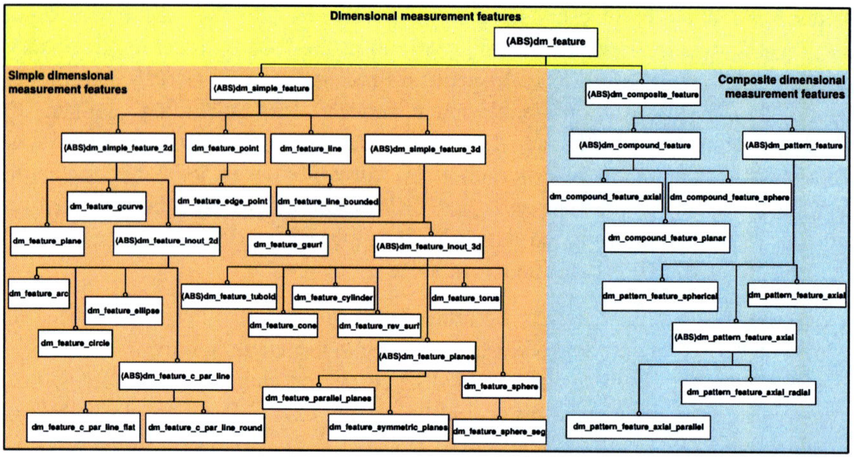

Fig. 4.16 Dimensional measurement feature classifications in HIPP

```
ENTITY dm_feature
  ABSTRACT SUPERTYPE OF (ONEOF (dm_composite_feature,
  dm_simple_feature));
  its_id                                  : identifier;
  its_workpiece                           : workpiece;
  its_related_machining_operations        : OPTIONAL SET [0:?] OF
                                            machining_operation;
  its_operations                          : OPTIONAL SET [0:?] OF
                                            dm_operation;
  feature_placement                       : axis2_placement_3d;
END_ENTITY;

ENTITY dm_compound_feature
  ABSTRACT SUPERTYPE OF (ONEOF (dm_composite_feature,
                               dm_pattern_feature))
  SUBTYPE OF (dm_feature);
END_ENTITY;

ENTITY dm_simple_feature
  ABSTRACT SUPERTYPE OF (ONEOF           (dm_feature_point,
                                          dm_feature_line,
                                          dm_simple_feature_2d,
                                          dm_simple_feature_3d))
  SUBTYPE OF (dm_feature);
  explicit_representation                : OPTIONAL SET [1:?] OF
                                           explicit_item;
END_ENTITY;
```

```
ENTITY dm_simple_feature_3d
  ABSTRACT SUPERTYPE OF (ONEOF (dm_feature_inout_3d,
                                dm_feature_gsurf))
  SUBTYPE OF (dm_simple_feature);
END_ENTITY;

ENTITY dm_feature_inout_3d
  ABSTRACT SUPERTYPE OF (ONEOF          (dm_feature_cone,
                                         dm_feature_cylinder,
                                         dm_feature_planes,
                                         dm_feature_rev_surf,
                                         dm_feature_sphere,
                                         dm_feature_torus,
                                         dm_feature_tuboid))
 SUBTYPE OF (dm_simple_feature_3d);
 inout                                  : inner_outer;
END_ENTITY;

ENTITY dm_feature_cone
  SUPERTYPE OF (dm_feature_cone_seg)
  SUBTYPE OF (dm_feature_inout_3d);
  whole_angle                           : plane_angle_measure;
END_ENTITY;
```

The dimensional measurement features are currently half associative in the HIPP data model. That is, there is a pointer (the explicit_representation attribute) from a dimensional measurement feature to the same portion of the shape of the part represented by geometry in the CAD model of the part, but there is no pointer from a dimensional measurement feature to the same portion of the shape of the part represented by a manufacturing feature. A means for providing associativity to manufacturing features should be added. This is particularly critical for in-process measurement done on a machine that can both cut metal and measure. This disadvantage has been improved by Zhao [77, 78] in her proposed integrated EXPRESS data model. In this data model a mechanism was established to link any of the following elements: machining operation, workingstep, machining features and tolerances (shown in Fig. 4.17) by a pointer from machining_feature (the its_applied_tolerances attribute) to tolerances. For example, a machining workingstep has its corresponding machining operation and machining feature. A machining feature has one or more machining operations. The tolerance applied on this machining feature is connected through the optional attribute indicated by dotted line in Fig. 4.17.

In the HIPP data model, the placement of simple (i.e., not compound) dimensional measurement features is with respect to the nominal part coordinate system. The explicit representation of a simple dimensional measurement feature is also given in the nominal part coordinate system. This model does not yet include

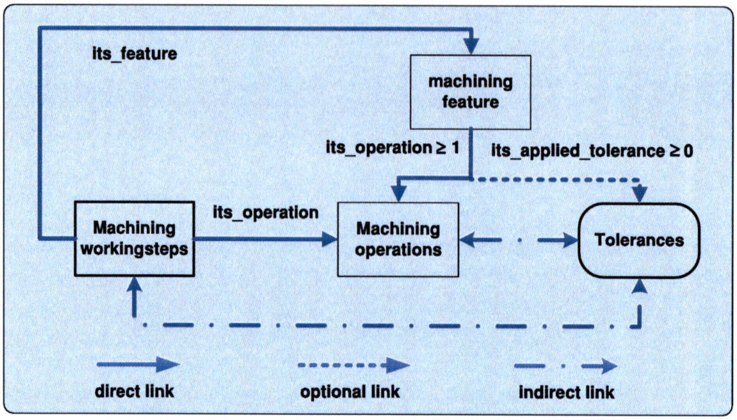

Fig. 4.17 Information linkage between machining, tolerance and features [77, 78]

measured features. Some of the entities relating to probing that were in AP 238 (stemming from ISO 14649 Part 10) had defects and have been removed. These entities include the workpiece_complete_probing entity and the touch_probing entity. A small hierarchy of physical resources for dimensional measurement has been defined in HIPP data model listed as following. More details (attributes, subtypes, meaning, etc.) are required.

- dm_resource is at the top of the hierarchy. It has subtypes dm_device_resource and dm_tool_resource.
- dm_device_resource has subtypes dm_machine and dm_gage.
- dm_tool_resource has subtype dm_probe.
- dm_probe has subtypes non_contact_probe and touch_probe. More details are required.
- dm_machine is defined as a stub for CMMs and other dimensional measuring machines that have controllers.
- dm_gage is defined as a stub for gages and other hand-operated dimensional measurement devices.

The measurement operation and physical resource information in HIPP are further developed in the integrated data model proposed by Zhao [77, 78] to include measurement technology and measurement result information. Some of these entities include: dm_result, touch_probing_strategy, dm_technology, and dm_machine_functions. However, the integrated data model was only tested with limited case studies in the research. In order to demonstrate the completeness and applicability of this data model to the industry, more sophisticated tests need to be done.

Some items related to high-level measurement process planning have been discussed in a series of meetings but have not yet been put into either the HIPP or the integrated data model. They are traceability, criticality weighting, assignable causes and reaction plan. The complete EXPRESS-G diagram of the integrated data model developed from the HIPP data model can be found in Appendix C.

4.3.4 Quality Measurement Plan Data Model

With rising industry demand for dimensional metrology information exchange standards, the International Metrology Interoperability Summit (IMIS) meeting was held at NIST, Gaithersburg, Maryland, USA in March of 2006. Attendees of this summit were 70 metrologists representing metrology supplier, end user, and standards organizations from Europe, Asia, and North America. The attendees assessed the worldwide status of metrology information exchange interoperability and developed a unified roadmap for future efforts. The element given the highest urgency was the "lack of implementations of non-proprietary data formats for CAD + PMI data downstream to inspection process planning." To address this high priority element, the HIPP (High-level Inspection Process Planning) meeting was held at NIST in April 2007 with 28 metrologists in attendance. Relevant action items of this group were to "complete the definition of requirements from the metrology community perspective that would define HIPP and relate to AP 203 and AP 238." After this meeting, NIST produced the HIPP data model and presented it in ISO meetings in Ibusuki Japan and Dallas USA in July and October 2007, respectively.

Subsequent to the proposal of the HIPP data model in 2007, the Dimensional Metrology Standards Consortium (DMSC) chose to establish a HIPP subcommittee to provide an organizational home for the HIPP effort. The DMSC HIPP committee determined that inspection planning needs in manufacturing went well beyond the CAD+PMI to a CMM scenario. Other metrology equipment and data acquisition systems needed to be considered. Quality requirements, such as Production Part Approval Process (PPAP) Advanced Product Quality Planning (APQP), Failure Mode and Effects Analysis (FMEA), and Control Charts of measurements necessary for quality control are also essential and required. Therefore, the scope of HIPP was expanded to include non-dimensional, attribute, and go/no-go measurements, as well as dimensional measurement. Subsequent to the DMSC HIPP committee meeting, two near future objectives were established:

1. define HIPP information in a "modeling language" such as UML or XML, since metrology software suppliers would then be more likely to implement HIPP, and
2. integrate HIPP information into AP238 (eventually), since AP238 is pursuing integrated machining and on-machine measurement, which is the future of metrology, namely, process-integrated metrology.

A UML version of HIPP data model was developed by NIST and presented to the HIPP committee at the DMSC meeting April 2008. Figure 4.18 shows the data classes in this UML data model of HIPP data model. At that meeting,

Fig. 4.18 UML support data classes of HIPP data model

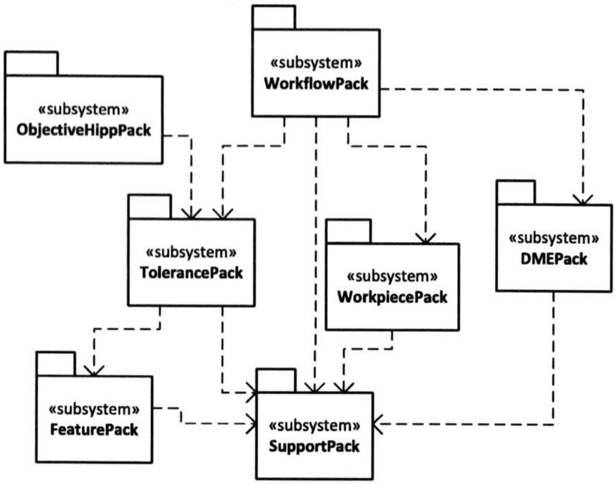

the committee/specification name was changed to the exchange of Quality Measurement Process Plans (eQuiPP). It was decided that eQuiPP would need to specify all the information required to generate a complete and precise measurement process plan for the execution of all types of quality measurements: dimensional, non-dimensional, attribute, and binary, both off-line and in-process. Therefore, eQuiPP defines information such as measurement features, characteristics, and tolerances, datum features, part geometry, part setups, part surface characteristics, measuring system uncertainties, and measurement priorities.

Since the effort of eQuiPP commenced, the committee had multiple casual meetings to develop the data dictionary for eQuiPP. Meanwhile, experts from standards efforts for other dimensional metrology areas, such as measurement result exchange, analysis, and reporting, recognized the need to consolidate the standards efforts in dimensional metrology systems. A proposal was developed by NIST in July 2009 and endorsed by all major dimensional metrology industries in North America. In the proposal, a Quality Information Framework (QIF) project was proposed under the DMSC to develop a set of four standards to address the major facets of manufacturing quality systems: Quality Measurement Plan (QMP), Measurement Resource Information (MRI), Measurement Execution Program (MEP), and Quality Measurement Results (QMR). The eQuiPP effort had then evolved into the QMP effort.

The QIF project is set to develop a common vocabulary and common data definitions for the entire range of quality measurement systems. The scope of QMP is defined as "the pre-requisite for the science-based downstream generation and execution of integrated measurement processes and for the fullest utilization of acquired measurement data". The purpose of the QMP Technical Working Group is to

1. consolidate existing standards and specifications that are related to the quality measurement process planning activity,
2. define the unique aspects of Quality Measurement Plans within the collaborative framework of the QIF Domain Working Group, and

3. ensure the flexibility and scalability of the QMP data model to support additional data genres.

In the International Manufacturing Technology Show (IMTS) held at Chicago in September 2010, a preliminary demonstration of the QIF data structure and the seamless information generation from dimensional measurement plan to downstream dimensional measurement plan execution and result collection was demonstrated [79] with the participation of Hexagon Metrology [80], Mitutoyo America [81], and Metrology Integrators [82]. The QIF data dictionary and QMP data model are still at an early stage. Interested readers can follow the development of QMP from the DMSC website. A more detailed discussion of the QIF project is in Chap. 9.

4.4 Summary

Design intent may be defined as the logically intended arrangements of features, dimensions and tolerances as the core determinants of the functional design of a discrete part. Design Intent may be encoded as an electronic data file from a 2D or 3D CAD system (i.e., ISO 10303 AP203), a printed paper drawing, a set of text-based instructions from a manufacturing execution system, or in some other medium and format. Whatever its source, design intent is the primary input to any discrete manufacturing process, which includes both production and measurement operations, where the latter is intrinsically necessary to efficient closed-loop production process validation and control.

Dimensional measurement process planning is the process of deciding what to inspect and how to inspect it (which equipment, how many points, etc.), given the design for a part. Measurement process planning may also involve computing estimated uncertainties for making specific measurements using different equipment and techniques. This chapter first discussed the functionalities of computer-aided inspection process planning activity. The most important functions include to extract or accept as input all the information necessary to generate a complete measurement process plan (called the high-level process plan) and then to generate a device-independent process plan containing the necessary information to execute the part measurement process. To generate the measurement process plan, information such as part material, machine accuracy, and measuring constraints need to be considered to support the decision making of what to measure, what features to measure, what are the measurands, the measurement purposes, how to handle outliers and filter measurement results, and recommended selection of measurement devices.

The CAIPP research in the past three decades was then reviewed. Research on CAIPP systems started from the early 1980s. Before the mid-1990s, most of the CAIPP systems research remained at the conceptual-level. These systems can be categorized into two groups: the tolerance-driven inspection process planning systems and geometry-based inspection process planning systems. The research in the former category focused on planning inspections for those features that have specific tolerance requirements. The research in the latter category focused on planning the inspection process to obtain a complete geometric description of a manufactured workpiece using the inspection data. Thus, comparison can be made with the design model for a complete geometry inspection. From the middle of the 1990s, research on CAIPP systems started to shift to one or some of the modules that a typical CAIPP system has, such as sampling strategies and probing path planning strategies. At the same time, non-CMM measurement, such as 3D optical scanner, attracted more and more attention. Therefore, CAIPP system research for non-CMM measurement methods has become another major characteristic of the research trend during this period.

The CAIPP systems can generally be divided into high-level and low-level process planning systems. The low-level process plan activity is closely associated with the chosen measurement devices; therefore, there is a large overlap between low-level measurement process planning and measurement plan execution activities. Often the measurement device manufacturers provide very limited low-level measurement process plan capabilities. It is the exchange of high-level measurement plan information that is opposing the interoperability barrier. The dimensional metrology industry needs a standard data model for the exchange of high-level measurement plans between different CAIPP systems. The information that should be included in the high-level measurement data model not only needs to include CAD design information, but also should include high-level production quality requirements information as well as information for integrated manufacturing and measurement operations (i.e., reaction plan).

In the dimensional metrology industry, commercial inspection software has achieved great advancement in the past 20 years. Although the commercial systems have many differences due to development strategy or target market focus they all have basic fundamental similarities. The five basic steps of generating a measurement plan using commercial CAIPP software systems were introduced. Even though great automation has been realized in CAIPP systems, primary information exchange obstacles still remain.

Standards organizations are aware of these problems and have made several efforts in developing a proper data model for the exchange of high-level measurement process plans. These data models include ISO 10303 AP 219, HIPP data model of AP 238, and the QMP data model. The remaining part of this chapter discussed the information defined in these data models. In summary, these standard efforts have made big achievements in defining core information for high-level measurement plans. Some of the information definitions are fairly complete such as the dimensional measurement feature definitions and GD&T definitions. However, throughout a series of meetings, the dimensional metrology society concluded the most pressing issues include developing non-proprietary data formats for CAD + PMI data downstream to inspection process planning and a proper data model to include quality requirements from the production point of view (i.e., PPAP, APQP, etc.). Therefore, information for non-dimensional, attribute, and binary measurements also need to be defined in this non-proprietary data model. The most recent standard effort is the QIF project, which was proposed to consolidate the standardization work in the entire quality measurements area and develop a neutral data library for the four main facets of quality measurement systems. The data model for high-level measurement plans is one of these four facets.

References

1. ElMaraghy HA, ElMaraghy WH (1994) Computer-aided inspection planning (CAIP). In: Shah JJ, Mäntylä M, Nau DS (eds) Advances in feature based manufacturing. Elsevier, Amsterdam, pp 85–89
2. Wong FSY, Chuah KB, Venuvinod PK (2005) Automated extraction of dimensional inspection features from part computer-aided design models. Int J Prod Res 43(12):2377–2396
3. Li Y, Gu P (2004) Free-form surface inspection techniques state of the art review. Comput Aided Des 36(13):1395–1417
4. ElMaraghy HA, Gu PH, Bollinger JG (1987) Expert system for inspection planning. CIRP Ann Manuf Technol 36(1):85–89
5. Helmy HA (1991) Feature recognition and CAD-directed inspection using solid geometric representation. Lehigh University, Bethlehem
6. ANSI (2004) Dimensional measuring interface standard, DMIS 5.0 standard, Part 1, ANSI/CAM-I 105.0-2004, Part 1
7. Joshi S, Chang TC (1988) Graph-based heuristics for recognition of machined features from a 3D solid model. Comput Aided Des 20(2):58–66
8. Hopp TH, Hocken RJ (1984) CAD-directed inspection. CIRP Ann Manuf Technol 33(1):357–361
9. IBM (1989) Valisys—for quality in the making
10. Audimess (1990) VW-GEDAS: Graphic-interactive programming system of CNC-coordinate measuring machines
11. Medland AJ, Singh R, Sittas E, Mullineux G (1990) Intelligent communication between CAD and manufacturing activities. In: Proceedings of the 28th MATADOR conference
12. Medland AJ (1992) The computer-based design process, 2nd edn. Chapman & Hall, New York
13. Medland AJ, Mullineux G (1992) Strategies for automatic path planning of coordinate measuring machines. In: 24th CIRP international seminar on manufacturing systems
14. Merat FL et al (1991) Automated inspection planning within the rapid design system. In: IEEE international conference on systems engineering
15. Yau HT, Menq CH (1991) Path planning for automated dimensional inspection using coordinate measuring machines. In: Proceedings—IEEE international conference on robotics and automation, pp 1934–1939
16. Yau HT, Menq CH (1992) Automated dimensional inspection environment for manufactured parts using coordinate measuring machines. Int J Prod Res 30(7):1517–1536
17. Menq CH, Yau HT, Lai GY (1992) Automated precision measurement of surface profile in CAD-directed inspection. IEEE Trans Robot Autom 8(2):268–278
18. Tannock JDT, Lee H, Williams JHS (1993) Intelligent inspection planning and computer aided inspection. Proc IME B J Eng Manuf 207(B2):99–104

19. Brown CW, Gyorog DA (1990) Generative inspection process planner for integrated production, In: American society of mechanical engineers, Production engineering division (Publication) PED, pp 151–162
20. Duffie N et al (1984) CAD Directed Inspection and Error Analysis Using Surface Patch Databases. CIRP Ann Manuf Technol 33(1):347–350
21. Menq CH, Yau H-T, Wong C-L (1992) Intelligent planning environment for automated dimensional inspection using coordinate measuring machines. J Eng Ind 114(2):222–230
22. Cho MW, Kim K (1995) New inspection planning strategy for sculptured surfaces using coordinate measuring machine. Int J Prod Res 33(2):427–444
23. Corrigall MJ, Bell R (1989) An inspection plan and code generation for coordinate measuring machines. In: Proceedings of 9th international conference of automated inspection and product control, pp 145–154
24. Corrigall MJ (1990) Inspection plan and code generation for coordinate measuring machines in a product modeling environment. Loughborough University of Technology, Loughborough
25. Sira (1992) Design to inspection project, Sira Ltd
26. Bogue R (2008) Car manufacturer uses novel laser scanner to reduce time to production. Assembl Autom 28(2):113–114
27. Vezzetti E (2007) Reverse engineering: a selective sampling acquisition approach. Int J Adv Manuf Technol 33(5–6):521–529
28. Minoni U, Cavalli F (2008) Surface quality control device for on-line applications. Measurement: J Int Meas Confed 41(7):774–782
29. Aguilar JJ et al (2004) Accuracy analysis of laser scanning probes used in coordinate measurement: simulation and experiments. In: VDI Berichte, pp 739–744, 797
30. Zhang SG et al (2000) Feature-based inspection process planning system for co-ordinate measuring machine (CMM). J Mater Process Technol 107(1–3):111–118
31. Vafaeesefat A, Elmaraghy HA (2000) Automated accessibility analysis and measurement clustering for CMMs. Int J Prod Res 38(10):2215–2231
32. Limaiem A, Elmaraghy HA (1999) CATIP: a computer-aided tactile inspection planning system. Int J Prod Res 37(2):447–465
33. Hwang CY, Tsai CY, Chang CA (2004) Efficient inspection planning for coordinate measuring machines. Int J Adv Manuf Technol 23(9–10):732–742
34. Elkott DF, ElMaraghy HA, ElMaraghy WH (2002) Automatic sampling for CMM inspection planning of free-form surfaces. Int J Prod Res 40(11):2653–2676
35. Menq CH et al (1990) Statistical evaluation of form tolerances using discrete measurement data. American society of mechanical engineers, Production engineering division (Publication) PED
36. Dowling MM et al (1997) Statistical issues in geometric feature inspection using coordinate measuring machines. Technometrics 39(1):3–17
37. Hwang I, Lee H, Ha S (2002) Hybrid neuro-fuzzy approach to the generation of measuring points for knowledge-based inspection planning. Int J Prod Res 40(11):2507–2520
38. Lee H, Cho MW, Yoon GS, Choi JH (2004) A computer-aided inspection planning system for on-machine measurement—part I: global inspection planning. KSME Int J 18(8):1349–1357
39. Cho MW et al (2004) A computer-aided inspection planning system for on-machine measurement—part II: local inspection planning. KSME Int J 18(8):1358–1367
40. Woo T, Liang R (1993) Optimal sampling for coordinate measurement: its definition and algorithm. Quality through engineering design, pp 333–346
41. Zhang YF et al (1996) A neural network approach to determining optimal inspection sampling size for CMM. Comput Integr Manuf Syst 9(3):161–169
42. Cho MW et al (2005) A feature-based inspection planning system for coordinate measuring machines. Int J Adv Manuf Technol 26(9–10):1078–1087
43. Jiang BC, Chiu SD (2002) Form tolerance-based measurement points determination with CMM. J Intell Manuf 13(2):101–108

44. Hocken RJ, Raja J, Babu U (1993) Sampling issues in coordinate metrology. Manuf Rev 6(4):282–294
45. Fan K-C, Leu MC (1998) Intelligent planning of CAD-directed inspection for coordinate measuring machines. Comput Integr Manuf Syst 11(1–2):43–51
46. Lee GL, Mou J (1996) Design the sampling strategy for dimensional measurement of geometric features using coordinate measuring machine. In: Proceedings of the Japan/USA symposium on flexible automation, pp 1193–1200
47. Lee G, Mou J, Shen Y (1997) Sampling strategy design for dimensional measurement of geometric features using coordinate measuring machine. Int J Mach Tools Manuf 37(7):917–934
48. Pahk HJ et al (1995) Integrated precision inspection system for manufacturing of moulds having CAD defined features. Int J Adv Manuf Technol 10(3):198–207
49. Orady E et al (2000) A fuzzy decision-making system for CMM measurements in quality control. In: The 2000 pacific conference on manufacturing
50. Kim WS, Raman S (2000) On the selection of flatness measurement points in coordinate measuring machine inspection. Int J Mach Tools Manuf 40(3):427–443
51. Fang KT, Wang SG, Wei G (2001) A stratified sampling model in spherical feature inspection using coordinate measuring machines. Stat Probab Lett 51(1):25–34
52. Kim D, Ozsoy T (1999) New sampling strategies for form evaluation of free form surfaces. In: Proceedings of the 1999 ASME design engineering technical conference
53. Edgeworth R, Wilhelm RG (1999) Adaptive sampling for coordinate metrology. Precis Eng 23(3):144–154
54. Albuquerque VA, Liou FW, Mitchell OR (2000) Inspection point placement and path planning algorithms for automatic CMM inspection. Int J Comput Integr Manuf 13(2):107–120
55. Ainsworth I, Ristic M, Brujic D (2000) CAD-based measurement path planning for free-form shapes using contact probes. Int J Adv Manuf Technol 16(1):23–31
56. Lin YJ, Murugappan P (2000) New algorithm for CAD-directed CMM dimensional inspection. Int J Adv Manuf Technol 16(2):107–112
57. Cho MW, Seo TI (2002) Machining error compensation using radial basis function network based on CAD/CAM/CAI integration concept. Int J Prod Res 40(9):2159–2174
58. Wong FSY, Chuah KB, Venuvinod PK (2006) Automated inspection process planning: algorithmic inspection feature recognition, and inspection case representation for CBR. Robot Comput Integr Manuf 22(1):56–68
59. Yuen CF, Wong SY, Venuvinod PK (2003) Development of a generic computer-aided process planning support system. J Mater Process Technol 139(1–3 SPEC):394–401
60. Chung SC (1999) CAD/CAM integration of on-the-machine measuring and inspection system for free-formed surfaces. Proc Am Soc Precis Eng 20:267–270
61. Kramer TR (1989) Automatic generation of NC-code for hole cutting with in-process metrology. In: Proceedings of the 1989 IEEE instrumentation and measurement technology conference, Washington, pp 45–52
62. Kramer TR et al (2001) A feature-based inspection and machining system. Comput Aided Des 33(9):653–669
63. Brecher C, Vitr M, Wolf J (2006) Closed-loop CAPP/CAM/CNC process chain based on STEP and STEP-NC inspection tasks. Int J Comput Integr Manuf 19(6):570–580
64. Suh SH et al (2002) Geometric error measurement of spiral bevel gears using a virtual gear model for STEP-NC. Int J Mach Tools Manuf 42(3):335–342
65. ISO(2004) ISO 14649-10: Industrial automation systems and integration—physical device control—data model for computerized numerical controllers—part 10: general process data
66. Ali L, Newman ST, Petzing J (2005) Development of a STEP-compliant inspection framework for discrete components. Proc IME B J Eng Manuf 219(7):557–563
67. ISO (2007) ISO 10303-219: Industrial automation systems and integration—product data representation and exchange—part 219: application protocol: dimensional inspection information exchange

68. ISO (2004) ISO/DIS 14649-16: Data model for computerized numerical controllers—part 16: data for touch probing based inspection
69. ISO (2004) ISO 10303-238: Industrial automation systems and integration—product data representation and exchange—part 238: application Protocols: application interpreted model for computerized numerical controllers
70. ISO (2002) ISO 14649-1: Data model for computerized numerical controllers: part 1 overview and fundamental principles
71. ISO (2004) ISO 14649-11: Industrial automation systems and integration—physical device control—data model for computerized numerical controllers—part 11: process data for milling
72. ISO (2005) ISO 14649-12: Industrial automation systems and integration—physical device control—data model for computerized numerical controllers—part 12: process data for turning
73. ISO (2004) ISO 14649-111: Data model for computerized numerical controllers—part 111: tools for milling machines
74. ISO(2003) ISO 14649-121: Data model for computerized numerical controllers—part 121: tools for turning machines
75. Hardwick M (2004) On STEP-NC and the complexities of product data integration. J Comput Inf Sci Eng 4(1):60–67
76. Honeywell (2007) FBMeas™ (cited 28 December 2010). http://www51.honeywell.com/aero/kcp/common/documents/FBMeas.pdf
77. Zhao YF (2009) An integrated process planning system for machining and inspection. Department of mechanical engineering, Ph.D. thesis, University of Auckland
78. Zhao YF, Xu X (2010) Enabling cognitive manufacturing through automated on-machine measurement planning and feedback. Adv Eng Inform 24(3):269–284
79. DMSC (2010) IMTS 2010 demonstration (cited 30 December 2010). http://www.dmisstandards.org/
80. Hexagon (2010) Hexagon metrology (cited 2010 December 30). http://www.hexagonmetrology.com/
81. Mitutoyo (2010) Mitutoyo America Corporation (cited 30 December 2010). http://www.mitutoyo.com/
82. Metrology integrators (2010) (cited 30 December 2010). http://www.hhisoftwaresolutions.com/

Chapter 5
Low-Level Dimensional Metrology Process Planning and Execution

For enterprise quality control, it is more sensible and pragmatic to generate a device-independent measurement plan (from the high-level measurement process planning activity) rather than a device-dependent one. The reason is because the workshops or factories may have different types of measurement devices with different measuring capabilities. The device-independent measurement plan will then give certain freedom to the shopfloor to choose available measurement devices. In industry, the dividing line between high-level and low-level measurement planning activity is vague. Generally speaking, the choice of measurement device for a particular measurement plan represents the starting point of low-level dimensional measurement planning activity. Also, measurement plan execution is tightly linked with the low-level measurement planning. Often when the low-level measurement plan is generated, it is executed on a measurement device straight away. This chapter will first discuss the activities involved with the low-level dimensional measurement plan generation. Then, a brief introduction of sensors used for measurement purposes in manufacturing systems is given. Sensor itself is a broad research area; the discussion about measuring sensors in this book is by no means able to cover the complete picture of measuring sensor instrumentation and technology. However, for readers who are not familiar with dimensional metrology, the discussion presented in this chapter is able to provide certain introductory knowledge.

Sensors are the tools of measurement. In the dimensional metrology industry, measurement execution systems include both sensors and the software to control the sensors. Therefore, a detailed discussion of today's measurement execution systems is provided in the following section. The three main groups of measurement execution systems are discussed. Due to the large variety of available measuring sensors and their control technologies, there are numerous types of data format for different types of measurement execution controllers. The remainder of this chapter provides the descriptions of proprietary and standard data models for dimensional measurement execution processes.

Y. Zhao et al., *Information Modeling for Interoperable Dimensional Metrology*,
DOI: 10.1007/978-1-4471-2167-1_5, © Springer-Verlag London Limited 2011

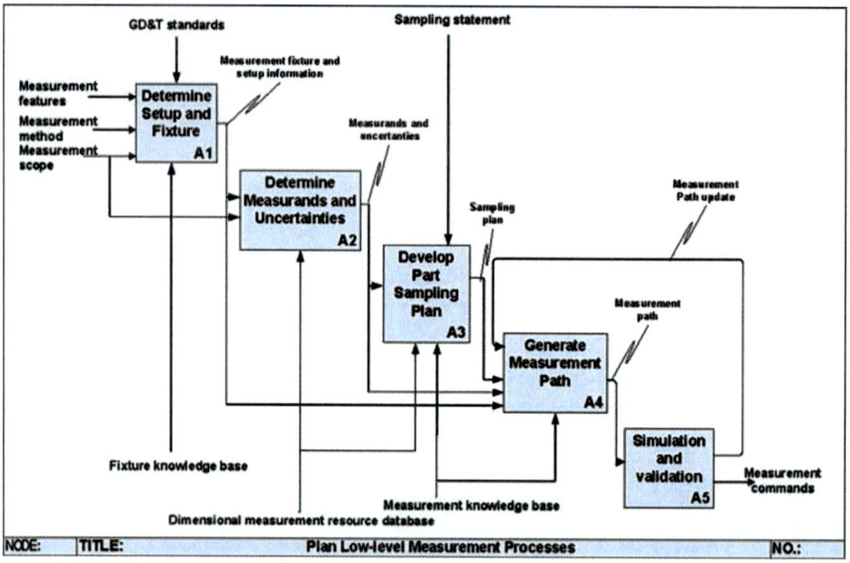

Fig. 5.1 IDEF0 activity model of low-level measurement process planning

5.1 Low-Level Dimensional Measurement Process Planning Activity

The main functionality of the low-level dimensional measurement process planning activity is to generate measurement commands for specific measurement devices. The measurement scope, method, and measurement feature information are determined through high-level measurement process planning activities. This information is passed on to the low-level measurement process planning activity, which consists of five sub-activities A1–A5 shown in Fig. 5.1.

In activity A1, the workpiece setup on a measurement machine with proper fixturing to enable measurement operations is determined. The information generated from this activity should include the detailed instruction of setups needed to inspect all the tolerances and dimensions within the measurement scope and all the information of the clamping devices. Then, measurands and uncertainties are determined based on the measurement device that is chosen for the operation in activity A2. For batch production, a part sampling plan is determined in the next step and measurement paths are generated in activities A3 and A4. The simulation and validation activity is then carried out in activity A5 for necessary update to the measurement path before real measurement is performed.

Once the low-level measurement plan is generated and simulated, the measurement program execution process begins. Activities within the measurement program execution include interfaces between executing low-level measurement tasks and executing high-level instructions. The measurement process execution activity needs to handle a huge number of types of measuring equipment and nearly limitless ways of measurement. The choice of measurement devices has significant effects on the low-level measurement process planning especially in generating the sampling plan and measurement paths (activities A3 and A4). For example, if a scanning probe is chosen for measurement, the measurement path should be a scanning path and measurement results will be a cloud of measured points. The typical benefit of this type of sampling is the large amount of data gathered through scanning. Whereas, if a touch trigger probe is chosen, the number of measurement points and their allocations must be determined first before the measurement path is generated. It is, therefore, necessary to introduce the major types of measurement sensors and devices commonly used in today's industry.

5.2 Measurement Sensors

The word sensor came from the Latin *sentire*, which means to perceive, and is defined as "*a device that detects a change in a physical stimulus and turns it into a signal which can be measured or recorded*" [1]. The development of measuring sensors began in about 1919 when a simple mechanical device was used for measurement by Tomlinson at the National Physics Laboratory (NPL), UK [2]. The first measurement instrument for engineering use is ascribed to German engineer Gustav Schmalz, who described an arrangement in which a stylus on the end of a pivoted arm traversed the surface and tilted a small mirror, as used in a reflecting galvanometer [3, 4].

An essential characteristic of the sensing process is the conversion of energy from one form to another. For measurement purposes, the following six types of signal are important: radiant, mechanical, thermal, electrical, magnetic, and chemical [5]. The signals obtained through the sensing process are converted into electrical signals suitable for processing by means of a transducer. A brief distinction between a transducer and a sensor is given here: a transducer is generally defined as a device that transmits energy from one system to another (often with a change in form of the energy); while a sensor is a device which first perceives an input signal and then converts that input signal or energy to another output signal or energy for further use.

Measurement sensors are playing an important role in manufacturing systems and processes. For example, sensors are used for monitoring large-scale systems and ultra-precision manufacturing processes. Sophisticated machine tools also need monitoring sensors to prevent machine failures. Heavy-duty machining with high cutting and grinding speeds depends on monitoring sensors to reduce human intervention from the safety point of view. Moreover, the rising environmental awareness in today's industry requires the monitoring of emissions from manufacturing processes. In general, sensors used in manufacturing processes are usually involved in four generic types of monitoring applications [6].

- *Production monitoring* sensors are utilized to determine the status of operations on the production floor. One frequent use for these sensors is to answer questions concerning the amount of material left which requires processing, the total number of parts produced, the number of good or bad parts, up-time, down-time, cycle time and so on.
- *Machining monitoring* sensors are used to gather data for the determination of whether or not a process is functioning properly. An early warning of the need for preventive maintenance or process adjustments is the objective of these measurements.
- *Environmental monitoring* sensors are used to provide information concerning the condition of an area. A common location for the installation of these censors is in the heating, ventilation and air conditioning system.
- *Machine control* sensors are used to control machine tool force, axes movement, and spindle speed. These sensors gather according information for closed-loop machine tool control.

As might be expected, each of these broad categories overlaps with the others to a certain extent. This is because one particular sensor may be applicable for use in multiple areas depending upon the manner in which it is employed. However, this classification still provides a useful guideline for discussion purposes.

Sensors used throughout the manufacturing process constitute a significant technology and help manufacturers to meet the challenges inherent in manufacturing a new generation of precision components. These sensors used at different manufacturing stages can address the tooling, process, workpiece, or machine status and accuracy. They allow manufacturers to improve the control over critical process variables. Figure 5.2 summarizes the level of precision that each type of sensor can achieve, and the parameters these sensors are used to control.

With regard to sensor systems for manufacturing process and workpiece monitoring, a distinction should be made between continuous and intermittent measuring as well as between direct and indirect measuring systems. The measured variable is available throughout a continuous measuring system but is only recorded during intervals in the machining process often known as pre-, inter-, or post-process measurements. Direct measuring systems employ the actual quantity of the measured variable (e.g., tool wear); whereas indirect measuring systems measure suitable auxiliary quantities such as the cutting force components and deduce the actual quantity via empirically determined correlations [5]. Generally, direct measuring possesses a higher degree of accuracy, whereas indirect methods are less complex and more suitable for practical application.

Continuous measurement permits the continuous detection of all changes to the measuring signal. It ensures that sudden and unexpected process disturbances, such as tool breakage, are detected in good time. Intermittent measurement normally interrupts the machining process or depends on certain measuring intervals. This generally entails time loss and subsequently high cost. However, intermittent measurement provides a continuous monitoring of workpiece dimensions and tolerances during manufacturing processes. Thus, it is suitable to be employed for

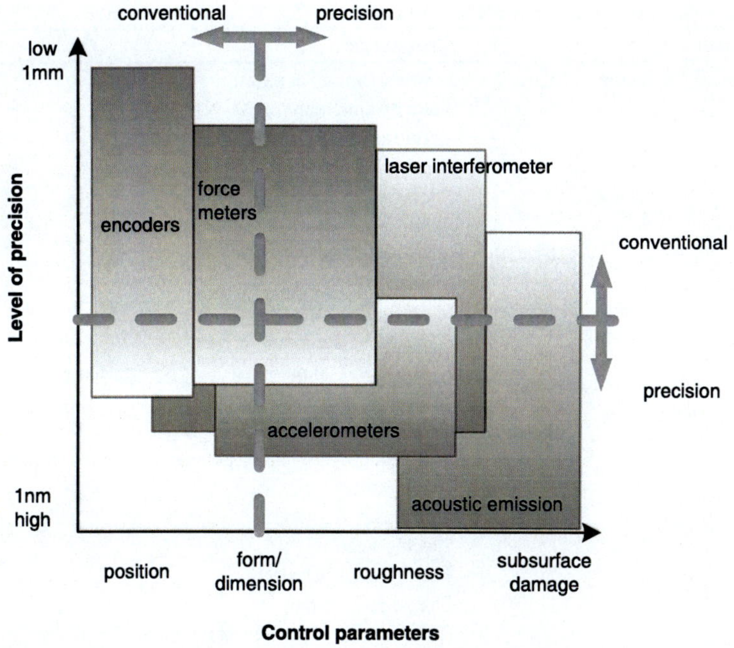

Fig. 5.2 Sensor application versus level of precision and error control parameters

expensive, one-of-a-kind products. It depends on the purpose of monitoring to decide which type of measurement is used in industry.

The philosophy of implementation of any sensing methodology for diagnostics or process monitoring can be divided into two simple approaches [5]. One approach is to use a sensing technique for which the output bears some relationship to the characteristics of the process. After determining the output and behavior of the sensor corresponding to "normal" machine operation or processing, one is able to observe the sensor signal for deviations that indicate a problem. Another approach is to determine a model linking the sensor output to the process mechanics and then use the model with information gathered by sensors to predict the behavior of the process. Both methods are useful in different circumstances. However, how to avoid misinterpretation of sensor signals is a key issue in sensor technology. Thus, signal process strategy is required. This book will not further discuss this topic. Interested readers can refer to related books for further study.

5.2.1 General Sensor Classification

There are many ways to classify sensors into categories. In this book we discuss the most common way for the classification of sensors—based on the signal output. There are six types of sensors listed as following. Each type of sensor is typically used for measuring certain types of measurands shown in Table 5.1.

Table 5.1 Sensors and measurands they can measure [5, 39]

Sensor output type	Measurands
Mechanical (including acoustic)	Position (linear, angular)
	Velocity (including wave velocity)
	Acceleration
	Force
	Stress and pressure
	Strain
	Mass and density
	Moment and torque
	Speed of flow and rate of mass transport
	Shape, roughness, and orientation
	Stiffness and compliance
	Viscosity
	Crystallinity and structural integrity
	Wave amplitude, phase, polarization, and spectrum
Thermal	Temperature
	Flux
	Specific heat
	Thermal conductivity
Electrical	Charge and current
	Potential and potential difference
	Electric field (amplitude, phase, polarization, spectrum)
	Conductivity
	Permittivity
Magnetic	Magnetic field (amplitude, phase, polarization, spectrum)
	Magnetic flux
	Permeability
Radiant (including optical)	Energy
	Intensity
	Emissivity
	Reflectivity
	Transmissivity
	Wave amplitude, phase, polarization, and spectrum
	Wave velocity
	Components (identities, concentrations, states)
Chemical (including biological)	Biomass (identities, concentrations, states)

- Mechanical;
- Thermal;
- Electrical;
- Magnetic;
- Radiant;
- Chemical.

Mechanical sensors are perhaps the largest and most diverse type of sensors because they have the largest set of potential measurands varying from position to mass to velocity (Table 5.1). They generally consist of the mechanism (i.e., piezoelectric crystals, strain gages, potentiometers, etc.) to convert the measurands into a signal.

Thermal sensors generally function by transforming thermal energy or the effect of thermal energy into a corresponding electrical quantity that can be further processed or transmitted. Electrical sensors are intended to measure charge, current potential, electric field, etc. The measurands that can be measured by electrical sensors overlap with some of those that can be measured by magnetic sensors.

Magnetic sensors convert a magnetic field into an electrical signal. They are normally applied directly as magnetometers. Radiant sensors convert the incident radiant signal energy into electrical signals as output. The radiant signals are electromagnetic, neutrons, fast neutrons, fast electrons, or heavy-charge particles [7].

Chemical sensors mostly rely on the interaction of chemical species at a semiconductor surface. Then, the change caused by the additional mass affecting the performance of the device is used as an indication of measurement value.

5.2.2 Sensors Used for Dimensional Metrology

Sensors used for dimensional metrology can generally be categorized into contact and non-contact measuring sensors. Contact type sensors measure objects precisely within wider ranges than non-contact types. Compensation of the ball radius of a touch probe and the time consuming process in measurement is one of the problems in contact sensors. On the other hand, the non-contact type technologies are currently deficient in precision, even if this is acceptable for soft objects and fast measurement.

The sensors are generally employed in five types of detection methods for dimensional measurement: mechanical, optical, pneumatic, ultrasonic, and electrical. Apart from the mechanical detection method, the rest of the five detection methods are non-contact [6, 8–13].

- *Mechanical method*—it is a method in which the measuring sensor operates in mechanical contact with the workpiece, although the actual signal may be electrical or pneumatic. There are three main types of sensors used in mechanical measurement methods:
  - *Caliper type*—the general features of caliper contact gages are: the wear of contact head relatively low gain, and low resolution.
  - *Friction roller type*—the friction-roller wheel contacts with the rotator surface along which it rotates by friction force. When the friction-roller wheel rotates, the encoder generates pulse signals. By counting the pulse signals in a revolution of the rotator, the diameter of the rotator is numerically evaluated. Both caliper and friction roller measurements are widely used for large-scale diameter workpieces, despite their low accuracy.

  - *Probe type*—the significant advantage of this type of sensor is that the same sensor can be employed to measure both the internal and outer diameter or workpiece lengths.

- *Optical method*—it is a method in which the sensor module produces and emits a light, which is collected and photoelectrically sensed through the object to be measured by a receiver module. This type of measurement method is highly sensitive and requires high accuracy of system alignment thus leading to certain limitations in the use of in-process measurement. There are two main sub-types of optical measurement methods: direct and indirect optical measurement:

  - *Direct optical measurement method*—the dimensions of a workpiece are generally measured by interrupting the light emitted from a sensor and by detecting this light electronically to obtain electronic signals. Then, the electronic signals are converted into dimensional readings. Scanning light beam, machine vision, and light gauging are the typical techniques that belong to direct optical measurement.
  - *Indirect optical measurement*—the dimensions of a workpiece are indirectly measured by means of features of the light reflected from the workpiece surfaces. In general, the light beam is projected onto the workpiece and reflected onto a photodetector or other electronic device. Any change in the dimension of the workpiece results in a change in the location of the image or a change in the location of the focusing point on the workpiece. Light focusing, light-spot detection, and light sectioning are the typical indirect optical measurement techniques.

- *Pneumatic method*—it is a method that measures the pressure drop in the gap between the measuring head and the workpiece. Then, the pressure drop is converted into an electrical signal for dimensional readings. This method is suitable to measure small distances that other types of sensor can not physically reach. The advantages of this kind of measurement method are: unaffected by the material, high speed gauging, and application over a wide temperature range.
- *Ultrasonic method*—it is a method that uses wave propagation principles and produces digital output. The signals employed are generally outside the frequency range of human hearing. Generally, an ultrasonic sensor uses the pulse-echo technique to generate a sharp acoustic pulse and change it into a sound wave form. This method is suitable for wet environments and its output is independent of target surface reflectivity. However, the accuracy of this type of measurement is not dependable.
- *Electrical method*—this type of method is non-optical and non-contact measurement methods using electrical field techniques, such as reluctance, capacitance and eddy current techniques. The main limitation of this type of measurement is the workpiece material, which must be electromagnetic for reluctance techniques and electrical for capacitance and eddy-current techniques.

The above sections of this chapter have discussed measurement sensors and sensors for dimensional measurements. However, before the sensor signal can be analyzed it must be transferred from the output of the device to the equipment that is used for the display and/or manipulation of this information. The data receiving system may be as simple as an analog or digital readout or as complex as a mainframe computer. The data transfer interface may be as simple as a set of wires connecting the output of one device to the input of another, or it may be as complex as a radio frequency transmitting system that sends analog data to an analog to digital (A/D) converter that is connected to a computer input port. In any event, the function of the sensor interface (data collection and data conditioning hardware/software) is to transfer the intelligence contained in the electrical output of the sensing system to the appropriate point in the data receiving system so that the data manipulation can be accomplished. There are a number of different sensor interfaces listed as the following:

- Analog interfaces
- Digital interface
- Signal multiplexers
- Sample and hold
- Parallel interface
- Serial interface
- Wireless interface
- AC input
- Input scaling
- Signal conditioning
- Pulse inputs
- A quad B
- Resolver/synchro interface
- Software interface.

A measuring sensor system consists of a number of modules, in which sensors and data processing for output reading are the essential ones. Three features need to be introduced here: the sensor picking up the information from a workpiece, the reference, and the means for comparison. The comparison normally needs signal amplification. While the sensor gathers information from the workpiece, the reference of the measurement device must not move in order to establish the loop for comparison. Figure 5.3 illustrates the schematic view of a measuring sensor system. The sensor is the device which picks up the information from the workpiece. Part of the sensor looks up a reference, the transducer establishes the difference between the two and converts the difference information into an electrical signal and thus produces a data reading. In today's industry, the interfaces between sensor signal receiver and data reading/output are hidden from the operators of measurement devices. When a measurement plan is generated, the measurement commands (i.e., measurement feature, measurement points and path information) are passed on to a measurement device, then the device is either manually guided to measure the workpiece or directed by a controller to carry out measurement operations. The sensors, sensor signal receiver, data reading, and interfaces within the sensor system are integrated into part of the dimensional measurement

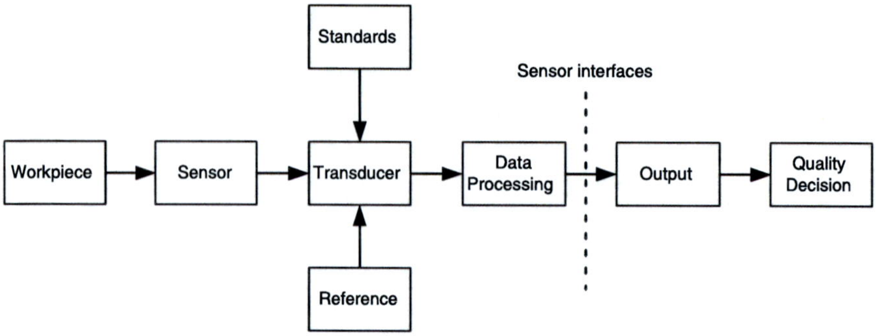

Fig. 5.3 A schematic view of dimensional measurement sensor systems

execution system. In the following section, the three main categories of dimensional measurement execution systems are discussed.

5.3 Dimensional Measurement Execution Systems

Modern state-of-the-art measurement execution systems can be categorized into three major types according to the three types of dimensional measurement employed in manufacturing processes introduced in Sect. 1.2: CMMs, portable measurement systems, and on-machine measurement systems. Most CMMs can generate low-level measurement plans and carry out measurement operations automatically once the high-level measurement plan is read into or created in the CMM control software. Most portable measurement systems, on the other hand, depend on the operators to guide the measuring arm onto the workpiece surface and manually collect measurement data. For on-machine measurement systems, the low-level measurement plans are typically manually generated by the operator (i.e., selecting surfaces to measure, deciding measurement sequence, etc.). The CNC machine controllers are then able to drive the measuring sensor to carry out measurements automatically. In this section, these measurement systems are discussed in detail.

5.3.1 CMM Systems

A CMM, as shown in Fig. 5.4, is a dimensional metrology instrument used to generate 3D points from the surface of a part for the purposes of establishing tolerance conformance and quality control. It digitizes portions of the surface of a part in three dimensions. In addition to three-dimensional inspection, a CMM is often used to make 2D measurements such as measuring the center and radius of a circle in a plane, or even one-dimensional measurements such as determining the

Fig. 5.4 A CMM

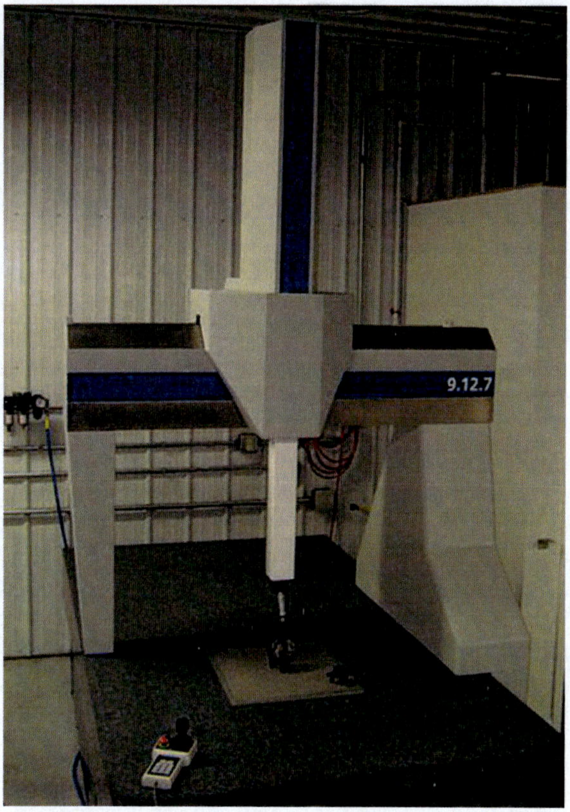

distance between two points. Typically, CMMs are configured to measure in Cartesian coordinates (X, Y, and Z). There are also CMMs that measure in cylindrical or polar coordinates. They can measure any part surface they can reach through bridge, arm, probe posture or stylus variation and movement [14].

A typical CMM consists of three main components: the main structure which includes three axes of motion, a probing system, and a data collection system. The data collection system includes the computer connected to the CMM, machine controller (embedded in the CMM) and the application software. Currently, computer-controlled CMMs are programmed by computer interface—pick points on screen or quasi-automatic simulation or through teaching techniques [15]. They can be programmed in both on-line and off-line mode.

The CMM is essentially a very precise Cartesian robot equipped with a tactile probe. The probe, under computer control, touches a sequence of points in the surface of a physical object to be measured, and the CMM produces a stream of x, y, z coordinates of the contact points. The coordinate stream is interpreted by algorithms that support applications such as reverse engineering, quality control, and process control [16].

In his paper "What Can CMMs Do?" Kurfess (2006) illustrated that "CMMs typically generate points in two ways: point-to-point mode, where the CMM taps or touches the part and generates a single point per tap; or scanning, where the CMM moves over a part, generating data as it moves. Scanning generates significantly more data than touch trigger, but is typically not as accurate. CMMs can be either manual or automatic depending on the mode. In manual mode, the CMM is moved by the user. An automatic CMM is typically actuated by electric drives (using ball screws or linear motors) driven by a CNC controller. Articulated-joint CMMs look very much like six-degree-of-freedom robots, and are almost always manually driven. Hybrid CMMs are a cross between articulated-arm systems and traditional CMMs and may have servo assist for making measurements [17]".

While the CMM hardware generates the coordinate data, the software bundled with the CMM (or in many instances sold separately) analyzes the data and presents the results to the user in a form that permits an understanding of part quality, and conformance to specified geometry.

According to Destefani "the most important advancement in CMM technology over the past several years is error mapping of the CMM. A machine is precisely measured and significant errors are corrected mathematically via software. As a result, looser tolerances can be used on the system hardware, and the resulting errors (as long as they are highly repeatable) are eliminated in software. This results in lower manufacturing costs while retaining or even improving the capabilities of the CMM. Other major design innovations in the past were linear air bearings and linear scales for improved repeatability and accuracy [18]".

There are many manufacturers of CMMs (i.e., Mitutoyo, Zeiss, Hexagon, Wenzel, etc.) that also develop user-friendly software that allows the CMM and probe to be accurately, quickly, and easily calibrated. This software has made the CMM more accurate and easier to use. Calibration is a key operation in CMM operation and is critical to achieving consistent maximum accuracy of measurement.

As indicated earlier, two types of probes dominate CMM operation: trigger probes and scanning probes. Trigger probes send a signal to the CMM when contact has been made with a surface. These probes operate in point-to-point modes, generating a single point of data every time contact with the part is made. Scanning probes are swept over the part surface, generating points as they move across the part and can be contacting as well as non-contacting (i.e., laser triangulation based probes, capacitance-based probes, and some probes based on laser intensity). Examples of geometries that are difficult to measure include very deep holes, where a probe must be inserted down the length of the hole. If the hole diameter is small, such as cooling holes on turbine blades, the task becomes even more formidable [17]. For this problem probe vendors (i.e., Renishaw) provide a wide variety of probes and stylus extensions.

Scanning CMMs already have dynamic compensation. However, scanning CMMs have difficulties at higher speeds, because inspecting a part faster increases vibration due to the higher acceleration and lower stiffness of the system. Much of this can be eliminated by "input shaping," which is the same technology that is used to reduce vibration on low-stiffness robot arms [17]. Another issue that must be overcome is probe error compensation due to non-normal incidence on the stylus tip that creates cosine error in calculating the contact point.

A controlled environment is important for efficient CMM operation. CMMs can operate well on the shop floor if they are equipped with thermal compensation capabilities that correct for temperature changes from standard temperature (20°C, 68°F). In most cases, the CMM needs to be kept in a relatively clean environment and located in a space that is isolated from vibration. Although CMMs are typically placed in an environmentally controlled Quality Control Lab, there is an increasing desire and movement to move CMMs out to the shop floor for "in-line" or "near-line" inspection. The challenge still exists for environmental stability for maximum measurement accuracy and a clean environment (e.g., dust-free, oil-free) to maximize the machine's physical performance.

Kurfess states "when considering that the accuracy of a stationary bridge-type CMM is usually better than that of a mobile articulated-arm CMM, it is recognized that recent advances in the articulated arm area, in particular related to error mapping, have yielded significant advances in the capabilities of the articulated arm. For many operations, the accuracy of articulated arm CMMs is sufficient for a variety of processes. The advantage of articulated-arm CMMs is that they generally have a larger work volume than bridge CMMs and at the same time are able to reach areas that are not easy to access with typical CMMs. Thus, if factory specified accuracies for articulated arm CMMs are sufficient for a particular application, it may be considered as a viable alternative to CNC CMMs. Also, articulated arm CMMs are more portable. Typically, they can be set up for measurement quickly [17]".

On the downside, articulated-arm CMMs are manually driven while gantry-type CMMs are both manual and servo-driven. Thus, articulated arm CMMs do not lend themselves as well to automation as servo-driven gantry-type CMMs. The size range of a CMM can span about four orders of magnitude with respect to part size. There are a variety of enormous CMMs that are used for measuring entire car bodies, the bodies of earth moving equipment, and even large aircraft elements (i.e., wings that are 10-m long). There are other CMMs that measure parts that have features on the order of 1 mm. This capability can offer significant advances in micro-manufacturing [12].

In quality and process control, the goal is to decide if a manufactured object meets its design specifications. This task is called dimensional inspection, and amounts to comparing the measurements obtained by a CMM with a solid model of the object. The model defines not only the solid's nominal or ideal geometry, but it can also provide the tolerances or acceptable deviations from the ideal [19]. The inspection results are used to accept or reject workpieces (quality control) and also to adjust the parameters of the manufacturing processes (process control).

Fig. 5.5 Scanning probe system

5.3.1.1 Touch-Trigger Probes Versus Scanning Probes

CMMs with touch-trigger probes have been common and effective measurement and inspection tools for decades. Touch-trigger probes function by contacting an individual point on the workpiece, then moving to measure the next point. However, CMMs that use contact scanning probes (shown in Fig. 5.5) provide much more information about part shape than touch-trigger probing, and the latest generation of scanning CMMs does so much more at an affordable cost than earlier models. Current scanning CMMs can read hundreds or even thousands of data points in the time it takes touch-trigger systems to register just a handful of touch trigger events. They can also do this in shop environments provided that the environment will support the system.

Automated 3D measurement of complex curved surfaces is easier with a CMM capable of scanning. Scanning is simply a method of collecting point cloud data that accurately defines the 3D shape of a workpiece. Scanning capability is no longer considered in the domain of only the highest of the high-tech manufacturers. Today, there are many vendors that offer sophisticated CMM solutions capable of scanning.

The ability to provide increased data density, along with improved measurement certainty is the key to scanning CMMs' improved accuracy in checking plastic injection molds, stamping dies, airfoils, engine blocks and other parts with complex contours. Contact scanning of contoured surfaces on a CMM can provide more information as a "point cloud" for software driven geometric calculation engines that, in turn, has the potential to yield higher measurement certainty than touch-trigger scanning, and achieves these results in a much shorter inspection cycle time than the traditional point-to-point touch-probe method [20].

However, in the real-world, CMMs must inspect less than perfectly formed features. This can be far more challenging than inspecting artifacts with perfect form, because on imperfectly formed features the result can vary based on where the CMM samples the feature. This leads to such challenges to "known path" versus "unknown path" scanning and the need for collision detection and avoidance within the CMM measurement program generating software.

Automated CMMs with scanning capability gather and analyze dimensional data that record information about contoured surfaces. This information reflects machine performance and quality of the workpiece. Contoured surfaces, shapes that curve in three directions, are being used more frequently by product designers who see the inherent benefits in them. According to Sheehan, "almost any surface that interacts with natural elements works better as a contour. Automobile bodies are more aerodynamic when they are streamlined, making cars more fuel efficient. Furniture is more comfortable, and equipment shaped for the body is more functional when designed with curved surfaces that match the shape of the body part. Artificial joints, such as replacement hips, are more functional when they closely resemble natural bone structure in form and size [21]."

Part form is not the only type of measurement that can benefit from high data density. Form deviation is present in all features that can be measured. When form is not controlled separately through the use of a modifier such as circularity (roundness) of a bore, the limits of size control the allowable form deviation. Therefore, when evaluating location and size, form must be understood to provide an accurate result. This means that the quality of size, location, and form information is directly related to the number of samples and location of each sampled point [20].

This type of data density is very valuable when a functional fit diameter needs to be determined for mating part analysis or for 'good part/bad part' determination. Simply put, the more data acquired, the more accurate the measurement, and therefore the more certainty. Also, less risk is achieved in determining whether any given part is in or out of specification.

For example, it has been suggested that a minimum of 300 data points are required just to approach the correct result for form such as circularity of a diameter in the 1–3″ (25–76 mm) range. At or above this level of data density measurement certainty is high enough to calculate a more correct form value. From this type of information we can begin to understand how much data may be required for size and location features as well, because you must measure the true form in order to know the functional size and location of the geometric and dimensional characteristics.

Technologies that enable scanning include the probe, the machine, and software for control, data acquisition, and analysis. Continuous-contact scanning probes are essentially small, accurate auxiliary measuring machines whose readings complement those of the CMM. During scanning, the probe stylus is in constant contact with the part surface. The controller of the CMM must maintain a consistent gaging force by detecting surface changes and adjusting the path in real-time. This force deflects the probe, and high-resolution electronic transducers track the displacement. Dimensional data are continuously read off the machine scales and scanning probe deflection sensor electronics, and sent to the software for analysis [20].

The CMM must also have mechanics and a control system, drives, and filtering functions suitable for scanning. For example, the mechanical system must provide rigidity for high repeatability. Accuracy depends on the linearity of the probe as it reacts to surface changes. The wider the linear range of the scanning system, the better the probe will handle dramatic surface changes while maintaining high speeds. According to industry experts, a linear scanning range of ± 1 mm is necessary for a scanning speed of 10 mm/s. If the linear range is smaller, the scanning speed must be reduced. The control system is critical since it links the mechanical system, the scanning probe, the drives, and computer analysis of the acquired data. Surface form changes must be identified quickly so that the contour path is precisely followed. The speed and accuracy of the adaptive mechanical system determine the throughput of the coordinate measuring machine [20].

Scanning software needs to have the filtering ability in order to detect subtle changes in the surface direction and other types of variability in a part's surface finish, such as a rough area on a turbine blade. Radius correction for the cosine error is considered an important function within scanning software. CMMs use probe center coordinates for measurement and the data generated by the machine is the location of the center point of the ball. Probe radius correction in scanning applications translates data by using the radius of the probe and a parallel curve function of the contact point to represent the real surface of the workpiece. During analysis, spline functions can be used to remove the mismatch between the scanning points and the nominal points so that deviations from nominal can be calculated [21].

Software must present measurement results in a way that helps operators easily identify possible part problems, find root causes, and take corrective actions on the process. This is typically accomplished through high level programming that can contain conditional branching for operator prompts or through SPC software that determines trends or other variation anomalies. State of the art systems support part program authoring and editing interactively and take advantage of three dimensional CAD views as a primary user display.

Many scanning software packages allow the import of 3D CAD part data, and automatically extract nominal values along with the correct vectors from the mathematical definition of the surface provided by the CAD model. Operators have the ability to select areas of the part to be scanned by clicking on the part's screen representation and gain significant efficiencies that ultimately reduce program generation and inspection cycle times.

From a design perspective, more and more manufacturers are also combining several parts into a single part to create lighter and less expensive components. This combination usually results in a contoured part, rather than a part that is assembled from many geometric pieces. Contoured shapes can be quite difficult to manufacture consistently. They present special machining challenges that have been largely overcome by the availability of computer numerically controlled (CNC) equipment and five-axis milling systems. As with any machined workpiece, the key to quality is the ability to control the machining process so that workpiece dimensions conform to the model or print. It is the job of the measuring system to capture these workpiece characteristics [21].

It is exactly in this area that new 5 axis scanning methods designed for measuring contoured surfaces have furthered the requirement of the acquisition of massive numbers of data points. Traditional methods of measuring the characteristics of shapes (e.g., the twist in a jet engine fan blade or the curvature in a mold cavity) using some form of a bench gage is quickly being replaced by CMM scanning techniques. The older time-consuming process that is subject to errors and inaccuracies is now being replaced with intelligent metrology. Having the ability to measure the shape and form of workpieces automatically, and use that information in conjunction with CAD systems, can make machining contoured surfaces and shapes a lot faster and a lot easier.

A contour-measuring CMM and a five-axis milling machine can be thought of as "thinking" the same way about contoured shapes. They describe an XYZ (point) and an IJK (surface normal vector). You can consider this by imagining a number of flag poles set up all over the earth's surface. The base of the flag pole represents the Cartesian point (XYZ) while the pole itself represents the vector (IJK). Each pole is perpendicular (normal) to the curve of the earth at its set point, but is not perpendicular to the earth at other points on the surface. Contour measurement systems represent deviations from nominal dimensions the same way, along a vector which is normal to the surface, in a very local sense [21].

Scanning techniques eliminate many problems. Fully compensated dimensional data from the workpiece is electronically overlaid on a graphic representation of the part to visually identify deviations of the measured surface from the nominal. A practical example is making dies for components such as under hood reinforcements for automobiles. To ensure a strong spot-welded attachment for automobile under hood reinforcements, a stamped metal strip must conform precisely to the flowing contours of the hood. The fit must be documented statistically prior to shipping the die. Simply comparing the part to a template is not accurate enough anymore and manual measurement of each part is far too time consuming. A typical stamped part may have 100 dimensions, with only 10 of them being key characteristics. In turn the production run off effort to validate a die usually requires about 500 pieces, and the significant characteristics of each piece have to be scanned. The use of a CMM reduces the inspection cycle time by several orders of magnitude [21].

CMMs excel in collecting large numbers of precise measurements quickly through scanning methods and provide the quality control personnel with high levels of accuracy, repeatability and meaningful conformance information, especially as it relates to product form specification.

5.3.1.2 Manual Versus CNC CMMs

Whereas CNC CMMs are programmed to be driven automatically for dimensional measurement, manual CMMs require human intervention to acquire inspection points. Manual CMM measurement sometimes uses a hard probe, rather than an electronic probe, and does not require a part program for probe movement around the workpiece. The drawback to this type of contour measuring is that the data is not organized in any particular way and is often difficult to evaluate correctly [21].

Many metrology vendors offer software that allows an inspector to follow prompted guided sequences against a visual display in order to make measurements against a workpiece. Several criteria are considered by a manufacturer when deciding to invest into a manual versus a CNC CMM. For example, increased inspector productivity, lowered scrap and rework costs, and elimination of wasted production time stemming from corrective work on defective parts in a multi-stage process are all issues when considering this decision.

The ability to use CMMs on the factory floor for real-time inspection has fueled a greater adoption of CNC CMM installations due to automation efficiencies, but there will always be applications where a manual CMM is a more practical choice, such as in job shops where every part being measured is different and the work envelope is relatively small.

Historically speaking a manual CMM could be purchased at nearly half the cost of a CNC driven machine and provided some economic incentive. However due to advancements in accuracy and ease of use the price gap between the two has closed in recent years. Manufacturers must consider the nature of their workload. Manual CMMs are often considered an entry level device for manufacturing startups. In more mature environments the repetitiveness of the dimensional metrology operation must be considered. For lengthy on-line production applications where measurements are repetitive, an automatic CMM is typically the better choice. Pre-programmed movements mean higher speeds without error, and minimal operator attention to the measuring process. On the other hand, in a small-lot production environment, the programming time can seldom be justified. Here, a manual CMM may be the better alternative.

Another consideration is the feature size and difficulty of access on the part being measured. If the features to be measured are small and inaccessible, an automatic CMM with better probe posturing will usually work better. This can save operator time and possible sequencing errors. On the other hand, a manual CMM makes more sense as the part characteristics get larger and are more accessible. In either case the pros and cons must be weighed when making an investment decision.

5.3.1.3 Multi-Sensor CMMs

Multi-sensor CMMs make it possible to test an extensive range of quantities such as geometric and dimensional, mechanical, electrical, optical and other material properties. To improve quality control, many metrology manufacturers are now combining different modern analytic technologies, tactile and non-tactile sensors. The integration of different sensor principles into one machine makes it possible to gather holistic information about all relevant attributes of modern parts, assemblies and products and to merge this information into one common coordinate system and measurement cycle [22]. Figure 5.6 illustrates a multi-sensor metrology system for holistic measurement.

For example, in order get a first overview of the workpiece to be measured, a measuring system is needed that helps to find details to be tested or to determine the position of different components in relationship to each other. The measuring range of these sensor types is rather big and the resolution low. In the actual measurement this overview system is used to navigate the other sensors integrated in the multi-sensor system. Optical systems based on image lend themselves naturally to this task. If an automatic navigation is needed, systems based on fringe projection, white light interferometry or video microscopy can be used. Often, the size of the workpiece is very large compared to the resolution needed to test fine structures on surfaces [22].

Data from fast scanning optical sensors are being used today and combined with measured data of higher resolution. In order to gain meaningful holistic information from data measured with different sensors, correct and exact integration into one coordinate system must be established. In addition to positioning or other high level information gathering (i.e., RFID or barcode reading) the variety of sensors applied in multi-sensor CMMs solves a myriad of measuring tasks. Form and positions of standard geometric features with a size of less than 1 mm, surface roughness in the μm or sub-μm domain, coating thickness are all considered in these types of hybrid systems. Depending on the kind of sensor many interior features and non-dimensional features can also be measured, if they are important for the performance or aesthetics of the part, consider color or missing hole identification as examples [22].

There are different setups for multi-sensor CMMs, 2.5D and 3D. Typically 2.5D systems are set up out of a plane scanning system and various 1D probing systems, similar to many conventional topography and surface roughness measuring systems. 3D systems on the other hand are mostly set up according to larger CMMs with modifications especially in the coordinate axes to improve resolution and to reduce measurement uncertainty [22].

Future challenges for the multi-sensor technology lie in the fusion of inhomogeneous data from different sensors. Unsolved problems include developing strategies for combining measurements taken with sensors with vastly different resolution. The automatic navigation of sensors in a multi-sensor CMM is most certainly a major part of today's research efforts in the field of metrology.

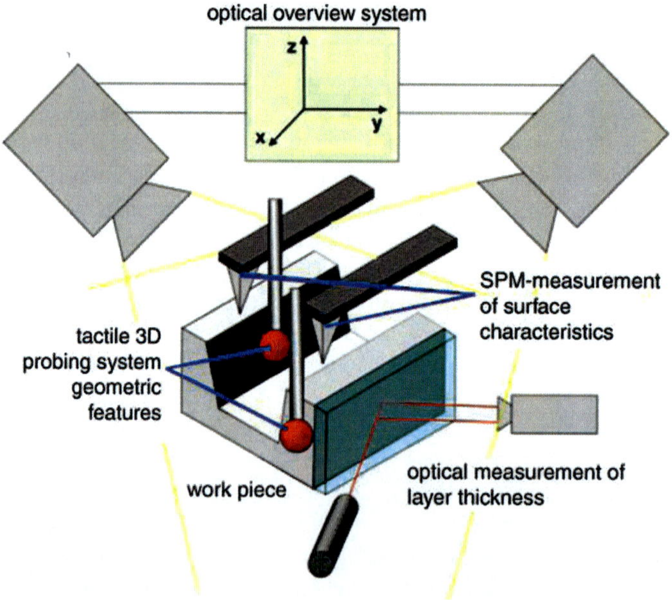

Fig. 5.6 Example of multi-sensor metrology system for holistic measurement

5.3.1.4 CMM Software

When considering the application software that drives the CNC machine we need to consider the following aspects:

- Probe calibration
- Part coordinate systems
- Feature constructions
- Tolerance analysis
- CNC movement of Sensor Probe and Head

Most CMM hardware vendors have proprietary methodologies and algorithms in which they calibrate the probe placement within the machine volume. When calibrating against a master ball (which is a spherical calibration artifact located within the machine metrology volume), the information is stored and recalled in order to maximize the accuracy of artifact inspection.

Part coordinate systems also depend on the hardware itself. A CMM program not only needs to consider the part coordinate system for dimensional metrology purposes, but it must also work within the native machine coordinate system (based on the proprietary encoding technology of the manufacturers scale system).

Each software vendor that writes CMM applications must also support feature construction and tolerance analysis. For example, feature construction may consider the way a slot is defined (perhaps as two circles connected by parallel tangential lines). Tolerance analysis, on the other hand, will be determined by specific point cloud processing algorithms that calculate dimensional characteristic actual observations that are then compared to nominal.

Lastly, in the case of CNC CMMs, the software must move the probe about the workpiece in order to acquire data. Part programs make calls to controllers via a hardware interface (Ethernet, USB, PCI, etc.). The controller then drives the machine through a series of commands that move the arm and the probe into contact (in the case of tactile probes) with the workpiece.

5.3.2 Portable Measurement Systems

Over the last several decades CMMs have revolutionized three dimensional measurements within the quality control discipline. They have evolved from slow and fragile machines that were unable to function reliably outside the protection of environmentally controlled rooms to become fast, robust, and reliable machines capable of providing accurate measurements in a wide variety of production environments.

Conventional CMMs, however, can be limited from the fact that workpieces must travel to them for measurement. Where workpieces are small, and transportation of inspection work to the CMM does not pose a problem, this is not an issue. As workpieces increase in size, moving them to CMMs becomes difficult. In many cases the workpiece is so large or is integrated into an even larger assembly, that it becomes impractical or physically impossible to bring the workpiece to the coordinate measuring machine [23].

Today factories that work with large manufactured parts can use the alternative portable CMMs (shown in Fig. 5.7) to bring the measuring capability to the workpiece. These portable units can be an ideal complement to conventional CMMs. They provide the ability to extend coordinate measuring capability into areas of the manufacturing plant that have traditionally been beyond reach. The most prominent feature of the portable CMM is its measuring arm, typically constructed from aluminum material along with precision bearings and rotary transducers at each of its joints. The base unit supports mobility through quick and easy mounting, and the freedom of movement of the probe sensor provides the inspector with a spherically accessible measurement envelope.

CNC, Manual and Articulating Arm CMMs can be used to monitor and identify process trends in production that in turn, support the elimination of out-of-tolerance parts. These systems also further detect variation in part dimensions caused by changes in process tooling, providing significant accurate data in order to establish optimal tool setup and subsequent tooling changes [24].

Non-contact laser scanning probes are also available from a number of metrology equipment vendors. Data from the arm is fed to a control that converts each recorded probe position to a precise location in three-dimensional space. Portable CMMs collect and record data one point at a time and/or in a continuous stream of points as the probe is moved along the surface of a measured workpiece.

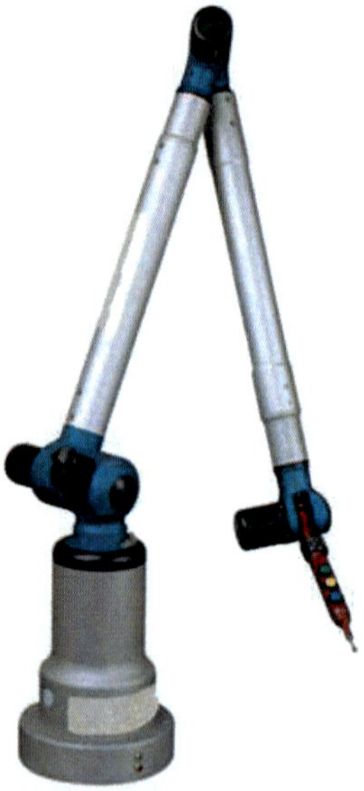

Fig. 5.7 A portable CMM

The control is typically supplied as user friendly system software capable of measurement tasks such as establishing coordinate systems and prompted guided sequencing for characteristic measurement for the operator to follow with manual movement of the probe head.

The portable CMM has certain flexibility advantages over conventional models. A portable unit has more measuring freedom, with an arm that is flexible enough to measure external features and then pass through a window, or other opening in the volume of the workpiece to measure internal features and surfaces in the same setup. This makes the portable CMM ideal for measuring difficult-to-access spaces. In addition to this scenario we must also consider the case of extending a measurement arm into the volume of a machining center workspace with a part still mounted to a fixture and ready for rework.

The portable CMM also has software flexibility. It is typically compatible with many CAD/CAM programs. Measurement data also can be exported to statistical process control programs to ensure that parts stay within tolerances and that the process remains under control. Portable CMMs can be purchased at a significantly lower cost than that of conventional CMMs. They bring coordinate measurement technology to companies that otherwise could not afford it [23].

For companies that presently operate conventional CMMs, using a portable CMM for less demanding measurement tasks provides a company with the opportunity to augment its measurement capacity at reasonable cost, and provides a means of relegating measurement tasks to the appropriate inspection machine. When companies cannot measure large components or assemblies simply because their conventional CMMs do not provide a large enough working envelope, many of them may consider securing larger CMMs. Quite often the solution may be to acquire a less costly solution and use a portable CMM to measure workpieces that exceed the working envelopes of the firm's conventional CMMs. Measurement accuracy requirements for such large workpieces may well be within the capabilities of the portable CMM.

Like the human arm, portable measuring machines have evolved. Articulated arms permitting six axes of movement, separated and linked by rigid supports are now available. Take a moment and lift your arm. Notice that there are two movements at the shoulder which provide coverage over a wide range, with no blind spots within the range. These same movements are repeated in the elbow and allow the arm to reach inside and around objects. Finally, the wrist provides directionality for the hand. The beauty and simplicity of this symmetry is duplicated in the portable measuring machine, from the base of the unit (body), to the probe (hand). The so-called portable CMM is sometimes referred to as a portable three dimensional measuring machine, designed to complement CMMs, not replace them. First introduced in 1988, these portable three dimensional measuring machines grew from the manufacturer's need to perform point-of-site inspections. Before assembly, most parts are already measured on a traditional granite table CMM, producing data that can be used as the global and final representation of accuracy and alignment [25].

In the automotive and aerospace industries, portable CMMs have become an integral part of the inspection process. In auto manufacturing, for example, it is possible to obtain spot measurements on the assembly line quickly and easily. As a result, considerable savings in time and manpower needed to obtain a sufficient number of unit samples needed to consider modifications in a part and/or changes in vehicle design have been realized. Before the portable CMM, it was perhaps only possible to check only three vehicles a day and it may have taken as many as 20 working days to obtain enough inspection data for an engineering department to statistically work with. Working with the portable CMM, as many as 15 vehicles a day can be inspected, which has introduced large efficiency gains in quality control programs.

In aerospace, the portable CMM is often used to complement and extend traditional measuring methods. Portable CMMs are sometimes replacing existing hard-tooled measurement fixture methodologies. By using the portable CMM instead of building expensive custom-tooled gages that have been traditionally used to measure large curved parts, engineers have realized significant cost savings in metrology tooling especially through equipment reuse when part designs change [25].

In another example, the portable CMM can also be used to shorten the feedback loop between production measurement activity and the design process for an installing an engine into the body assembly. Critical measurements are taken while engines are secure within the frame on the assembly line itself. In another scenario, portable CMMs are used to position and verify bracket placement in order to avoid misalignment during the installation of fuel lines, hydraulic tubing, wiring and other components. Sheet metal mechanics use the portable CMM as an integral part of the assembly process. A touch-sensitive probe locates and measures the coordinates of a point in space on the assembly and an accompanying software system will compare the measurements with the design engineering model. The result is instant verification of the accuracy of part placement within the larger assembly.

Two types of portable three dimensional measuring machines are currently available: the passive digitizing arm and the dynamic inspecting arm. The passive digitizing arm records the unstructured movement of the probe sensor across the surface of a part. The data is then transferred into a supporting CAD system for analysis. Once acquired the data is further processed into geometric elements and surfaces by the CAD system. The operator can then superimpose the data over the theoretical data to determine the accuracy of the recorded features. One of the advantages of this type of approach is that the geometric analysis of the data rests with design and manufacturing engineering specialists. It is also an economical method for reverse engineering to develop digital models of a part when the original engineering data doesn't exist.

The dynamic inspecting arm uses data originating from supporting CAD systems before or during measurement to guide the operator to the measurement location and to control the measuring sequence. It has the ability to offset for stylus radius and automatically construct the geometry needed to compare the measurements taken to theoretical part data within the associated software system. The chief advantage of the dynamic arm is that it provides the inspector with immediate quality conformance analytics. This places all of the necessary dimensional metrology resources in the hands of the manufacturing and quality engineering departments.

All of these systems provide the manufacturer with quality reports that can be generated quickly for compliance documentation of process trends, unnatural variation and quality characteristics. This information can be used to immediately determine defective parts and drive down scrap and rework costs. Many systems also provide the user with best fitting algorithms. Linear accuracies of better than [+ or −] 0.005″ are available with these types of articulating arm measurement systems.

When purchasing a portable 3D measuring machine, the buyer should keep in mind that this equipment is desired to enhance a company's existing inspection program and not to replace it. Prospective equipment should have the following features [25]:

- An articulated, six-axis measuring arm should mimic the structure and movements of the human arm, especially the elbow and shoulder.
- It should have the ability to develop geometric surfaces from measured points.
- Graphic and audible guidance to measuring location should be provided.
- Self-calibration methods and intuitive software must be part of the system in order to maintain measuring accuracy.
- The system should have the capacity to measure extended volumes through a single position and create new alignments to a global reference system.

- The portable CMM should have a method to compare geometry against CAD definitions for in- or out-of-tolerance results.
- The software must be able to provide documentation of results.
- The portable CMM should provide the use of non-contact, laser and other type sensor probes.

To summarize, manually operated, articulated-arm CMMs are used in many manufacturing operations because they can be moved quickly and easily to the shop floor to perform fast, accurate measurements. Articulated arm CMMs use an anchored, jointed arm with an attached probe tip at the moveable end [24].

Fixed CMMs, as the name implies, are located in one position. What fixed CMMs give up in mobility and portability, they return in accuracy and repeatability. The decision on which technology to use depends on the following factors:

- Size, configuration and transportability of the workpiece.
- Number of features that require inspection and accuracy required.
- Magnitude of the process control problem that must be corrected.
- Investment objectives.

Large workpieces such as complete auto bodies, auto body sheet metal panels and subassemblies, airframe components, molds, dies, and welding fixtures present special handling challenges. It's not only their size that makes these parts difficult to measure, but also the variety of different tolerances often found on them.

The number of workpieces, the number of features, and the accuracy required are all considerations that must be taken into consideration when deciding whether a portable-arm CMM or a fixed-position CMM is the best choice for a given dimensional metrology application. If there are more than one or two parts in a batch to be inspected, and if those parts have several features that need to be checked, it may be more cost-effective to inspect them using a fixed CNC CMM than a portable-arm CMM. The time savings and throughput efficiencies achievable through running automated measurement programs on a fixed CMM may more than compensate for the time required to move the part to the inspection station [11].

It is quite possible that the combination of the multiple types of CMMs may work best from a manufacturing perspective. The portable-arm CMM may be a good initial investigative tool that can quickly isolate process issues. Early in the manufacturing cycle, a portable-arm CMM may provide the level of accuracy necessary to determine if the process is in control, for example, if holes and slots are in the correct position. By adding this measure a manufacturer can improve metrology capacity and relieve the burden against fixed-position CMM assets, thereby improving the overall inspection efficiency. Later in the manufacturing cycle, such as finish-machining operations, it may be most appropriate, and faster, to inspect the finished part on a fixed-position CMM [24].

It is worth mentioning that equipment cost is always an important factor in a manufacturer's decision to invest in a CMM. For example, a portable arm CMM with a 15 ft (4.6-m) measuring volume carries a considerably lower price tag than a fixed-position CMM with the same measuring volume. However, total quality inspection operating costs must be factored in. These costs include inspection cycle-time. A CNC CMM may prove less costly to operate over the course of an entire production run, but in the long run, the final decision on which technology to use depends on careful examination and understanding of total system requirements.

5.3.3 On-Machine Measurement Systems

Manufacturers, in general, are reluctant to use measurement probes in machining centers even though they have been available for CNC machines for years. The biggest impediment to using them is the sometimes difficult and always time-consuming task of writing probing macros. In addition, the traditional mindset that "doing anything but cutting on a machine tool wastes precious time" and "you should never measure a part using the same machine that made it," also pose obstacles [26].

The views among academia are mixed with pros and cons on using on-machine measurement for in-process measurement. Using the machine tool as an inspection device eliminates the need for expensive inspection equipment, allowing the manufacturer to divert resources to other uses. There is no need for an inspection fixture either, because the machine tool part fixture serves as the inspection fixture [27]. The advantages of employing on-machine measurement for in-process measurements are summarized as follows [28–32].

- Cost and time saving through
  - reducing lead-time required for gages and fixtures,
  - minimizing need for design, fabrication, maintenance of hard gages, fixtures and equipment,
  - reducing inspection queue time and inspection time, and
  - eliminating rework of nonconforming product.
- Changing from "reactive" inspection to "proactive" control by
  - integrating quality control into the product realization process,
  - using characterized and qualified processes to increase product reliability,
  - focusing resources on prevention of defects instead of detection in the end (a post-mortem process),
  - utilizing real-time process knowledge and control, and part acceptance/disposition, and
  - enhancing small lot acceptance capability.
- Elimination of non-value added operations such as lot inspection, sampling plans, receiving inspection, design, fabrication and maintenance of hard gages, and reworking nonconforming parts;

- Agile machining by providing quick responses to product design changes. Since inspection operations are carried out on the same machining center, inspection gages and fixture changes are not required. New and existing technologies such as probing strategy, error compensation, data analysis software and fixture design technology can be integrated into the OMM system.

In comparison, the disadvantages of using on-machine measurement for in-process measurement include the validity of dimensional measurement on the same machine that makes the part. It is often known that measurements performed by a cutting machine are subject to some of the same error producing factors as the cutting progress, the errors that are most difficult to eliminate through machine maintenance and certification can easily be detected and accounted for with in-process measurement. For example, machine flexing, tool wear, and vibration will all be absent during measurement. Additional error compensation techniques such as laser measurement, ball-screw compensation, and measuring pre-cut proofing parts for future reference should also be applied to compensate for other machine inaccuracies [33]. Apart from the concern of measurement accuracy, a traditional objection to on-machine measurement is that it diverts machine time away from actual machining. However, this notion can be overcome by measuring productivity in terms of total in-process time rather than machining cycle time. The view that on-machine measurement steals machining time overlooks the fact that checking a part off-line, a step that on-machine measurement seeks to replace, can impose the need for additional part handling and another setup; this adds to in-process time, as well as introducing the potential of fixture errors [28].

With new software introduced by several independent software vendors, the primary barriers to on-machine measurements are now being overcome. Some vendors today provide interactive graphic tools for automating the task of creating probe macros on a workstation offline from the machine tool. Through point and click activities against a CAD model, programmers can now create probing routines in a fraction of the time. Once these dimensional metrology programs are generated, operators can launch these routines as they are needed or cutting programs can call them automatically.

By measuring a few key features at setup, machine operators can detect a range of errors before they become costly mistakes. Probing makes it easy to provide a machine operator with the means to detect several problems related to improper tool and work offsets, erroneous feeds and speeds, buggy NC programs, and many other machining related issues. Unlike conventional inspection, it is not necessary or even desirable to measure all features of the part. Most critical errors can be found by checking a select group of key features or characteristics. As always, the trick is to identify the measurements most likely to highlight problems.

Tool wear is another major contributor to scrap and rework. By periodically measuring select features while machining, users can monitor the degree of tool wear and make informed decisions about the need to replace a tool, adjust a cutting program, or leave things alone. As a result, today manufacturers can use actual process data, rather than some other type of empirical schedule to decide when to replace their tools that may be dull or worn.

In today's manufacturing environment, operators spend an inordinate amount of time aligning fixtures and parts during setup. For short runs, for example, more time can be spent on setup than on actually removing metal. By using automated, on-machine measurement, the operator can significantly reduce the time required to set up fixtures and parts through pre-process measurement. In addition, he or she can quickly calculate and cut production cycle time by more efficiently entering tool and work offset information in order to bring the dimensional key characteristics under statistical control and machined closer to tolerance during production runs than through post-process measurement.

In most machining situations, manual, on-machine first piece inspection is another major setup bottleneck. First the machine is idle waiting for an inspector to arrive. Then additional time is lost during the inspection process and while results are being calculated and entered. With automated on-machine inspection the machinist only needs to load and launch previously developed dimensional measurement routines. If adjustments are needed, some in-process metrology software, can, in many cases, make necessary program adjustments automatically. With on machine probing systems, the operator can easily interpret the results and make appropriate corrections quickly. Once production is under way, automated inspection, integrated with part cutting, has the ability to alert machine operators of actual and potential problems before they result in scrap or rework costs.

When parts, particularly large ones, have to be removed from the CNC machine for inspection, productivity has the potential to be reduced significantly. The machine may stand idle for hours (sometimes days) while the part is removed, transported to the inspection machine, set up, measured, evaluated, removed, transported back, set up again, and adjusted. Using in-process metrology software to automatically measure the part on the machine puts an end to this convoluted process and dramatically reduces the potential for errors.

In addition to giving immediate feedback, the in-process metrology package can direct its output to external databases and software systems to perform such functions as, SPC, machine and process capabilities studies, and many types of graphical reports for easy information interpretation. Furthermore, there is industry movement toward calculating operational efficiency in real-time in order to provide manufacturing engineering better insight into their processes. Through a powerful new set of tools for automating CNC probing to be used in machining centers, users are increasingly applying creative solutions to complex and otherwise insolvable problems through the use of dimensional metrology through pre-process, in-process and post-process on machine metrology.

5.4 Information Modeling for Low-Level Dimensional Measurement Process Plan and Execution

Having discussed the sensors and low-level measurement process plan generation and execution systems in the above sections, it is obvious that there are a huge number of sensors that can be used for dimensional measurements. Some of these sensor systems require computerized controllers to drive the sensor for out-of-tolerance detection. Most dimensional measurement execution system vendors provide limited choices of sensors with their software. When it comes to the integration of sensors and measurement software systems from different vendors, compatibility and interoperability problems occur.

The interoperability issue in low-level dimensional measurement process plan generation and execution is more important in large, enterprise-level corporations, where a single-vendor solution is impractical if not impossible. An equipment-independent data format for representing both high-level (measurement feature level information) and low-level (point to point level information) measurement process execution plan is necessary and critical for big corporations. However, there is no such standardization in industry.

Low-level dimensional measurement process plans are embodied in programs that may be executed by the controller of a CMM (or other numerically controllable piece of dimensional measurement equipment). There is only one standard language for such programs. That is the Dimensional Measuring Interface Standard, DMIS (commonly pronounced DEE-miss). The semantics of DMIS and the syntax for programs are given in DMIS Part 1 [34]. The most recent version of DMIS is 5.2. That version became an ANSI standard in 2009 and an ISO standard in 2010.

There is also a DMIS Part 2 standard. This puts the semantics of DMIS part 1 into a collection of object interfaces that provide interoperability between DMIS client applications, a DMIS server, a DMIS mathematics module and a DMIS equipment module.

In all known implementations of CMM programming languages (which include many proprietary languages as well as DMIS), there is a distinct interface between the controller that executes the program and the controller that controls machine motion and collects raw data. The two controllers generally run on different computers and often run software built by different companies. This interface usually consists of messages that are generated dynamically, sent back and forth over a communications system, and not saved in a program for reuse. There are two publicly available specifications for this interface. One is the DMIS equipment module of DMIS Part 2 (DMIS Part 1 does not deal with this interface).

The other is the I++DME Interface Specification [35] which was developed by several European automakers and measuring equipment vendors. There is a great deal to be gained by using a standard CMM programming language level. If two different CMM program execution systems execute the same language, it may be possible to run a given program on either one. This allows the CMM system buyer to buy whichever he or she prefers and allows the CMM system user who has multiple CMMs to use the same program on different machines. If the standard is widely used, as DMIS is, it also gives CMM execution system developers the ability to reach a large market with a single product.

There is also a great deal to be gained by using a standard messaging specification between a CMM program execution system and the equipment controller. If a standard is used, different program execution systems can be plugged in the same CMM hardware, and different CMM hardware can be controlled by the same program execution system. This can be a winner for all three parties: CMM users, CMM hardware vendors, and CMM program execution system vendors.

This section describes the DMIS programming language and the two equipment level interfaces (DMIS equipment module and I++DME Interface Specification).

5.4.1 DMIS Data Model

DMIS was originally developed under the auspices of Computer-Aided Manufacturing-International (CAM-I). Version 2.1 became an ANSI standard in 1991. CAM-I stopped supporting DMIS in 2005, so members of the development committee formed the Dimensional Metrology Standards Consortium (DMSC) to ensure its continued life. It is still maintained by the DMSC.

A formal system is in place for receiving DMIS "standard improvement requests" (SIRs) and acting on them. From time to time, usually after a number of SIRs have been dealt with, a major or minor release of the standard is produced. The current version is numbered 5.2 and was approved in 2009.

DMIS is a large, statement-based language. The DMIS 5.2 specification is over 700 pages long. A bit over half of those pages are used for 226 sections, all but one of which (intrinsic functions) describe an entire statement or a variant of a statement. Moreover, the implementation of many statements is extremely complex. A full implementation of DMIS would require about ten times as much source code as a full implementation of a typical language for programming a machining center.

The DMIS specification actually describes both a language for writing programs to execute and a language for writing output reports about what was done when a program was executed and what the results were. The programming language is described in Sect. 5.4.1.1. The output language is described in Sect. 5.4.1.2, and the DMIS Part 2 equipment module in Sect. 5.4.1.3.

5.4.1.1 DMIS Programming Language

The specification itself divides statements into 18 types. Here, we will compress those into seven types: program, geometry, metrology, equipment, motion, output, and miscellaneous. The following six sections describe all of them except output, which is covered below.

1. Program

A DMIS program consists primarily of one-line DMIS statements, each of which tells the executing system to do something. A statement always includes a "major word" indicating the nature of what to do. If the statement defines or sets something (a feature or variable, for example), the major word is preceded by the name (and possibly type) of the thing to define or set followed by an equal sign. What a statement does may be modified by "minor words", and arguments. For example in the following statement that defines a circle

```
F(circle1) = FEAT/CIRCLE, OUTER, CART, 2, 3, 4, 0, 0, 1, 7
```

- The major word is FEAT
- CIRCLE, OUTER, and CART are minor words
- All the numbers are parameters (2, 3, 4 is the center, 7 is the diameter)
- F indicates that circle1 is a type of feature
- circle1 is the name of the feature

Programs are, by default, executed in the order in which the statements occur in the program file, but several statements change the flow of execution. These include, for example JUMPTO, IF, SELECT, DO, and ITERAT.

JUMPTO gives the name of a jump label indicating where to continue execution. IF and SELECT provide for testing a condition in order to determine which of alternative sets of statements to execute. DO and ITERAT provide for looping.

Programs have structures called "blocks" which consist of sets of statements. Each type of block has a statement that starts the block and a statement that ends it. For example, an IF block starts with an IF statement and ends with an ENDIF statement. An entire DMIS program may be considered to be a block starting with a DMISMN statement and ending with an ENDFIL statement. Blocks may be nested.

Programs may declare and use variables. Any parameter to any DMIS statement may be replaced by a variable of an appropriate type. All variables must be declared and set before they are used. Programs may also use expressions, such as (1 + SIN(x))/y. Expressions may include constants, variables, operators, and functions. Any parameter to any DMIS statement may be replaced by an expression of an appropriate type.

The example just given returns an arithmetic value, but other types of expressions may return strings, booleans, or vectors. DMIS defines over 60 "intrinsic functions", such as the SIN(x) in the example above.

Programs may be constructed by combining several files using the INCLUDE statement. The MACRO statement provides a limited method of defining a function (that does not return anything) consisting of DMIS statements. A MACRO works by text substitution, in the same manner as a macro may be written in C or C++ using #define. Via the CALL statement, a main DMIS program may execute a subordinate DMIS program, a DMIS MACRO, or even a non-DMIS program. The main DMIS program resumes executing statements following the CALL statement.

2. Geometry

DMIS deals with geometry in terms of features. Features are defined as ideal forms; these are called "nominal" features. Once a nominal feature has been defined, it may be measured (with MEAS or RMEAS) or constructed (with CONST) using data from features that were measured or constructed. A measured or constructed feature is called an "actual" feature. To distinguish between a nominal feature and the corresponding actual feature, the type identifier F is used for nominal, while FA is used for actual. For example F(circle1) and FA(circle1) refer to the nominal and actual versions of circle1.

All of the DMIS features represent points, curves (including straight lines), or surfaces in three dimensions. Curves and surfaces may be bounded intrinsically (for example, a sphere) or not (for example, a plane). All curves and surfaces (intrinsically bounded or not) may be bounded using a BOUND statement.

The following types of feature from classical geometry may be defined with the FEAT statement.

- circle (FEAT/CIRCLE)
- cone (FEAT/CONE)
- cylinder (FEAT/CYLNDR)
- ellipse (FEAT/ELLIPS)
- line (FEAT/LINE)
- plane (FEAT/PLANE)
- point (FEAT/POINT)
- sphere (FEAT/SPHERE)
- torus (FEAT/TORUS)

The FEAT statement may also be used to define portions of the following classical elementary features.

- circle (FEAT/ARC)
- cone (FEAT/CONRADSEGMNT)
- cylinder (FEAT/CYLRADSEGMNT)
- sphere (FEAT/SPHRADSEGMNT)
- torus (FEAT/TORRADSEGMNT)

The FEAT statement also provides for defining 13 other types of geometric entity in 1, 2, and 3 dimensions. In general, features may be enclosed by material (the cylindrical surface inside a coffee mug, for example) or may enclose material (the outside of the mug). For many features, this is indicated by the minor words INNER and OUTER. Where INNER and OUTER are not appropriate (for points, lines and planes, for example), a vector pointing away from material is used.

Features may be defined in Cartesian coordinates (indicated by CART) or cylindrical coordinates (indicated by POL).

3. Metrology

Metrology statements in DMIS provide for handling: tolerances (TOL), datums and coordinate systems (DATDEF, DATSET, ROTATE, TRANS, SAVE, RECALL, etc.), measurement uncertainty (UNCERTALG and UNCERTSET), simultaneous requirements (SIMREQT and ENDSIMREQT) and key characteristics (KEYCHAR). Of course all of DMIS deals directly or indirectly with metrology, so many other statements might be considered to be metrology statements.

Tolerances in DMIS follow the ASME Y14.5 specification. The TOL statement has 29 variants that cover all of the Y14.5 tolerances.

4. Equipment

The bread and butter of DMIS is programming a single-armed coordinate measuring machine. This includes articulated arm machines as well as machines with motions along three orthogonal axes.

DMIS is most well-developed for using a sensor that is a touch trigger probe or a scanning probe. The SENSOR and SNSDEF statement may be used for describing sensors of those types as well as sensor components such as extensions (EXTENS, WRIST) and multiple styli. Other types of sensor are supported, but not as comprehensively. These include video with lighting, laser, capacitance, and infrared.

Scanning has been a major focus of improvements to DMIS in the recent versions. It is implemented using PAMEAS, PATH, SCNMOD, and SCNSET. DMIS also provides statements for using rotary tables (ROTDEF, ROTAB, and ROTSET) and multiple arms (CRGDEF, CRMODE, CRSLCT, CZONE, and CZSLCT). Tool holders that hold sensors when they are not in use may be defined with THLDEF.

5. Motion

Motion in free space (moving without measuring) may be commanded with GOTO and GOHOME statements. Multiple free space moves may be grouped in a GOTARG-ENDGO block. Rounding corners at the intersection of consecutive free space straight line moves may be controlled with FLY.

Motion for measuring is done inside a MEAS or RMEAS block, either of which may be used for measuring a feature. The motion statements inside the block are for (1) free space motion, or (2) measuring specific points (PTMEAS) or points lying on a scanning path (PAMEAS).

DMIS provides three modes of motion controlled by the MODE statement: AUTO (autonomous), PROG (program), and MAN (manual). In AUTO mode, the controller may ignore GOTO and PTMEAS statements inside a measurement block and pick points where it pleases on the feature being measured. In PROG mode the controller does what the program says to do. In MAN mode, motion statements are carried out by an operator using a joystick or other control device.

The point that is controlled by motion statements is usually the center of the ball at the tip of a stylus mounted on a sensor. The alignment of the stylus and sensor axes (if there are any) may be controlled (at the same time the tip location is controlled) by using the appropriate minor words and parameters.

Speed and acceleration of motion may be controlled with FEDRAT (feed rate), ACLRAT (acceleration rate) and RAPID.

6. Miscellaneous

Interaction with the operator may be done using the TEXT and PROMPT statements. TEXT simply sends a message. PROMPT sends a message asking for the operator to respond and gets the operator's response. About 20 statements are available for in-process verification and quality information systems.

5.4.1.2 DMIS Output

The DMIS output consists of several levels of information control. Some of the measurement output information is defined in DMIS Part 2. This section discusses the output information in DMIS Part 1.

1. Output control

There are several levels of control on DMIS output. The first level is exercised by the DISPLY statement, which selects one or more types of destination for output (terminal, printer, file, or communications port) and the language (DMIS output language or vendor output language) to use for sending output to each destination.

If DMIS output language is chosen for a destination, output to that destination begins when a FILNAM statement is executed, and the first line of the output is the FILNAM statement. Additional destinations to which to send output may be defined with the DEVICE statement, which assigns a name and one of the four destination types listed above. These output devices may be opened with OPEN and later closed with CLOSE anywhere in a program. Output to all destinations ends and all open devices are closed when the ENDFIL statement at the end of a program is executed. The format used for DMIS output language files is very similar to the format of the input language. The description of each DMIS statement in the standard indicates whether there is an output format (all statements have input formats). Many statements (all the program flow statements, for example) have no output format. Many other statements have output formats identical to the input (the 15 varieties of CONST, for example).

Using the REPORT statement, a DMIS program may specify that information beyond that specified in the output format descriptions be included in the output whenever output is produced. If the minor word OUTFIL is used with the TEXT statement, the specified text will be written into the output.

2. Feature and tolerance output

The most important output, of course, is the results of taking measurements. This applies particularly to features and tolerances. Tolerances are generic, in that they are defined with no connection to features. A tolerance is associated with a feature only when an OUTPUT, EVAL, or KEYCHAR statement is executed. The OUTPUT and KEYCHAR statements cause internal feature fitting and tolerance calculation (if not already done) as well as writing output. The EVAL statement causes only calculation (which may be necessary so that parameters for an actual feature are available for use in defining or constructing other features).

Actual feature data may be output as a feature description in the same format as the input, except with FA to indicate an actual feature where the nominal feature has F. Feature data may also be output as (1) a collection of uncompensated Cartesian measured points or (2) as individual compensated or uncompensated measured points. If an OUTPUT statement has associated one or more tolerances with a feature, the actual tolerances will be output immediately after the feature.

Output of nominal features and tolerances and actual sensors and gauges may also be produced, as specified in OUTPUT statements.

5.4.1.3 Equipment Module of DMIS Part 2

DMIS Part 2 is largely a repackaging of DMIS Part 1 as an object model written using the CORBA Interface Definition Language (IDL). The most recent version, "ANSI/CAM-I 104.0-2003, Part 2", is numbered 1.0 and dated June 2003. It implements version 4.0 of DMIS. DMIS Part 2 was conceived and built by one person, Dietmar May.

Section 6.3 of DMIS Part 2 describes a DMIS Equipment Module (dmisEquip), which has no counterpart in DMIS Part 1. As mentioned earlier, dmisEquip provides commands, responses, and a communication protocol that may be used between a system executing a DMIS program and a system controlling equipment.

The communication protocol is not described in the Part 2 document in any detail. The document specifies that messages be handled by an object request broker (ORB) implementation. The communication protocol is therefore whatever the ORB uses.

The Part 2 document describes enumerations, structures, interfaces, and exceptions to be used by the ORB. The document describes these for two other modules (DMIS Server Module and DMIS Mathematics module) in addition to the DMIS Equipment Module. The other two modules are not discussed further here.

The dmisEquip Module includes four principal interfaces: carriages, sensors, rotary tables, and general equipment. The carriages interface has commands that:

- create a new carriage object
- get the current position
- move the controlled point to a new position without measuring
- measure a feature under automatic DME control

- measure a single point manually or under automatic DME control
- scan along a path and collect point data
- scan along a path using a rotary table
- select a sensor
- set measurement parameters (speed, acceleration, search distance, retract distance)
- set free space motion parameters (speed, acceleration)
- set video parameters (filter, focus, window, lighting, scale)
- turn fine positioning on or off
- turn probe compensation on or off
- update the active sensor algorithms
- update the sensor mount information for the carriage

The sensors interface has commands that:

- obtain the type of the sensor
- obtain a reference to the actual sensor or the nominal sensor
- create a new probe sensor object of any of the following types: probe, video, laser, infra-red, non-contact, X-ray
- initiate a sensor calibration algorithm within the DME
- initiate a sensor calibration using a specified routine
- create or recall an actual sensor
- obtain calibration data in a DME-specific format
- get actual probe data including tip diameter and offset

The rotary table interface has commands that:

- create a new rotary table object
- initiate calibration of the rotary table
- provide the DME with the results of calibration
- initiate a calibration algorithm
- obtain calibration data associated with the rotary table
- provide previously obtained calibration data to the DME
- cause the table to rotate
- set the current position to a new value (without moving the table)
- get the current position

The equipment interface has commands that:

- initialize, halt, reset, or shut down the DME
- destroy objects
- initiate DME diagnostics
- get DME status
- enumerate carriages
- execute a DME-specific command
- get environmental data (temperature, humidity)
- set temperature compensation
- convert DME-specific raw data into Cartesian format.

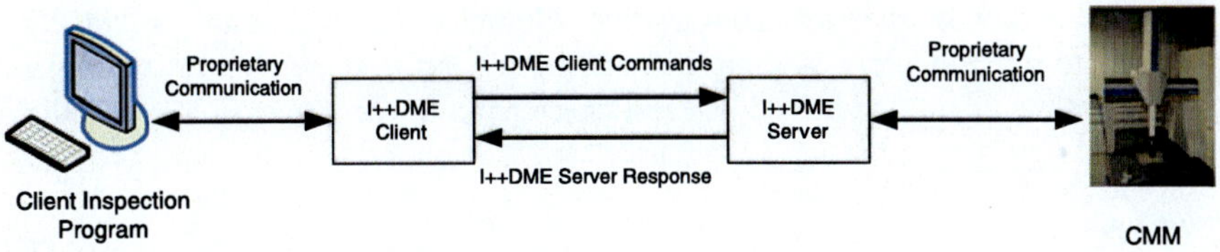

Fig. 5.8 I++DME activity model

5.4.2 I++DME Data Model

Traditionally, CMM vendors have sold a tightly-coupled software-hardware system for programming and controlling the inspection process. The last 15 years have seen large manufacturers acquire CMMs from many different vendors and endure the overhead of supporting multiple software applications. Furthermore, third party software vendors have been offering high quality products that often cannot be used because they are incompatible with some CMMs. Automakers are major users of measurement equipment, and suffer from the cost and time to work around these incompatibilities [36].

In order to solve the dimensional measurement equipment interoperability problem, major European automakers supported the development of I++ Dimensional Measuring Equipment Interface specification (I++DME) [35]. The goal of I++DME is to allow automakers, and any other manufacturers, to select the best software and equipment for their purposes and budgets and ensure that they work together seamlessly out of the box. The I++ committee is comprised of measurement equipment end users primarily from the automobile manufacturing sector. The I++DME specification was written by I++ members and targeted toward equipment and software vendors. Dimensional measurement devices covered by I++DME include the following types:

- 3D CMMs including multiple carriage mode
- Form testers
- Camshaft, crankshaft measuring machines

I++DME is a messaging protocol between measurement plan executors and measurement equipment. It uses TCP/IP sockets as the communication mechanism, and defines a message set and a client–server architecture. Clients are measurement plan executors, and servers are the equipment that carries out the measurements. For example, a client could read DMIS measurement plans produced by some upstream application, interpret the DMIS statements, send I++DME messages to the measuring equipment, accumulate the measurement results that return as I++DME messages from the server, and output a DMIS or DML [37] measurement report. Figure 5.8 illustrates the activity model of I++DME.

I++DME uses UML as the information modeling language to model descriptions of the messages, accompanied by natural language (English) that describes the semantics. Production rules in Backus-Naur Form (BNF) are provided that define the syntax of message composition. Numerous examples are provided as guidance to implementers. A sample I++DME session is shown below, with messages from the client underlined and responses from the server in Bold and Italic.

```
00002 StartSession()
00002 &
00002 %
00003 GetDMEVersion()
00003 &
00003 # DMEVersion(1.4.2)
00003%
00027 ChangeTool(''ProbeB'')
00027 &
00027%
00078 SetProp(Tool.GoToPar.Speed(25.0))
00078 &
00078 %
00079 GoTo(X(2.626), Y(-4.656), Z(-4.100))
00079 &
00079 %
00094 PtMeas(X(2.47), Y(-4.13), Z(-5.10), IJK(-0.01,-0.99,-0.00))
00094 &
00094 # X(2.44), Y(-4.64), Z(-5.99), IJK(-0.019,-0.997,0.074)
00094 %
```

The command lines from I++DME client to server always start with a tag, which is represented by the first 5 characters of each command line. There are two types of tags: CommandTag and EventTag. A CommandTag is a 5 digit decimal number with leading zeros present. The number must be between 00001 and 99999. The client must make sure that command tags sent to the server are unique while the server processes the commands related to the tags. The easiest way to accomplish this is to increment the tag number each time a new command is sent. For example, in the above I++DME program example, line 00002 `StartSession()` represents a command from client to server with CommandTag 00002 indicating start a measuring session. For a command line, the first 5 characters of a response line represent a tag (ResponseTag). During normal command processing by the server it will use the tag received from the client as the ResponseTag so the client can use this tag to relate the response line to a command line. In addition the server can send a response line using ResponseTag E0000 for reporting unsolicited events to the client. The first 5 characters in each response line represent the ResponseTag. The 6th character of a response line must be a space and the 7th character must be one of: *& (represents Ack), % (represents Transaction complete), # (represents Data),* and *! (represents Error).*

An EventTag is a 4 digit decimal number that is preceded by the character E (ASCII code = 69). The number must be between 0001 and 9999. The client also needs to make sure the event tags sent to the server are unique while the server

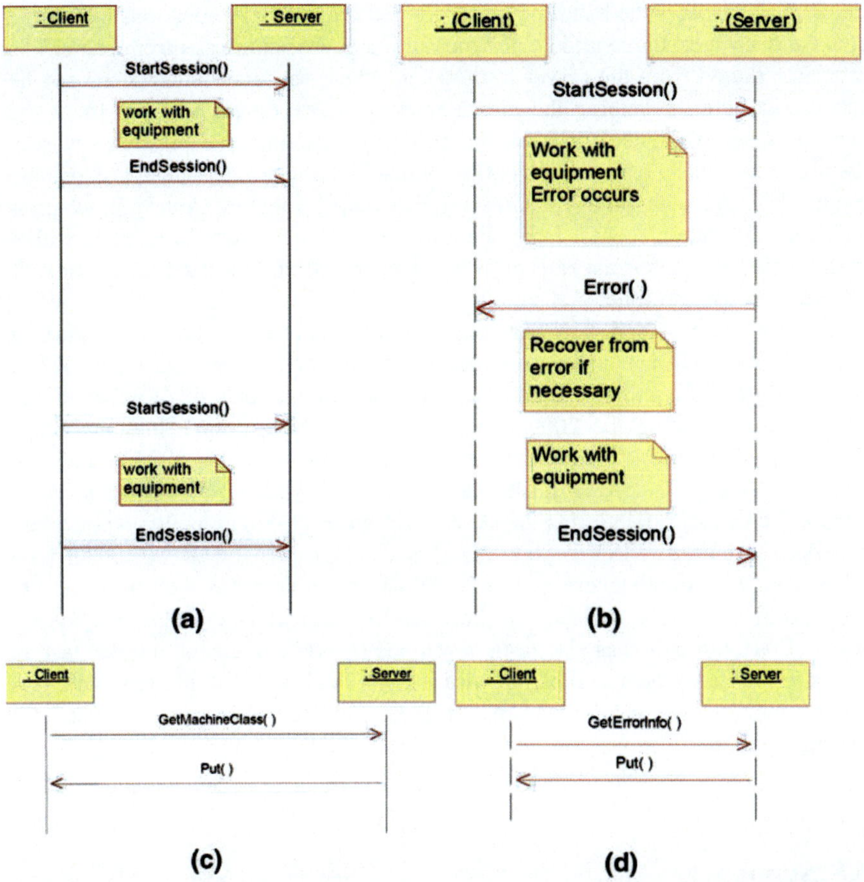

Fig. 5.9 Basic use cases of I++DME. **a** Sequence diagram, **b** Handling of unsolicited errors, **c** Standard queue diagram, **d** Event and fast queue diagram

processes the commands related to the event tags. At any point during a measurement operation, the server may notify that something happened by sending an event to the client. If the event is triggered by a transaction, the tag used is that of the transaction. The server must first send an Ack before it can send the response with the EventTag. This Response can then be sent before or after the Transaction complete. An example of an EventTag in the above I ++DME example data is `00094 # X(2.44), Y(−4.64), Z(−5.99), IJK(−0.019, −0.997, 0.074)`. It represents an event indicating the measured coordinates of XYZ and IJK of a 3D point.

The basic use cases of I ++DME are summarized in Fig. 5.9. Figure 5.9a shows a typical sequence of I++DME messages. Each session starts with a `StartSession()` command and ends with a `EndSession ()` command.

Figure 5.9b displays the handling of unsolicited errors communications between a client and a server. In the middle of a measurement procedure, an error occurs. The error is reported from the server to the client, then, necessary corrections can be carried out before resuming the measurement process. Figure 5.9c illustrates the normal queue. The client calls for measurement machine type information from the server and gets a response when the command in the normal queue is executed. Figure 5.9d represents an event and fast queue communication. An event, which is an error information request from client to server in the figure, is handled with a higher priority in a special queue. This causes bypassing of standard commands and shorter reaction time.

The use cases listed here are only some examples of the capabilities of I++DME. I++DME has undergone testing in a series of demonstrations involving real software and equipment at several important international quality technology expositions, including the 2004 International Manufacturing Technology Show (IMTS), the 2005 Quality Expo in Chicago, USA, and the 2005–2007 International Trade for Quality Assurance in Germany. These multivendor demonstrations have included combinatorial testing of several software packages with several measurement machines. NIST in the USA has conducted I++DME validation tests to validate the completeness of the I++DME specification as well as to ensure the specification has defined the information correctly and unambiguously. An I++DME test suite has also been developed by NIST to enable internal testing of conformance to the specification with a given I++DME data file [38]. In a real implementation, I++DME files are not used. I++DME files are only used for testing purposes.

5.5 Summary

Low-level dimensional measurement plan generation and execution activity is closely related to the chosen measurement device. Touch-trigger probes require probing points information for measuring a feature, while scanning probes need probing path information. In low-level dimensional measurement process planning activity, the workpiece setup and fixturing is firstly determined, after which the measurands and uncertainty information is generated based on the chosen measurement device. Then, the sampling plan and measurement paths are determined. The measurement paths are normally simulated and validated before the real measurement is performed.

In dimensional metrology industry, the low-level measurement process planning is normally integrated with measuring sensors as part of the execution system. There is a huge variety of measuring sensors used throughout manufacturing processes. Based on the signal output they can be categorized into six types: mechanical, thermal, electrical, magnetic, radiant, and chemical. Each type of these sensors is able to measure a range of measurands and can be used to serve different monitoring functions. For dimensional metrology, mechanical sensors (including acoustic sensors), electrical sensors, and radiant sensors (including optical sensors) are commonly used. The detection methods can be divided into five types: mechanical, optical, pneumatic, ultrasonic, and electrical. Also, the detection can be carried out in direct and indirect manner.

Commercial dimensional measurement execution system vendors normally provide a complete system including software, controller, sensors etc. The three main types of commercial measurement systems are fixed CMM, portable measurement system (including portable CMM), and on-machine measurement system, among which fixed CMM is the most commonly used measurement system in manufacturing industry. Most of the commercial measurement execution system vendors provide limited choices of sensors with their software. When it comes to the integration of sensors and measurement software systems from different vendors, compatibility and interoperability problems occur. The interoperability issue in low-level dimensional measurement process plan generation and execution is more important in large, enterprise-level corporations, where a single-vendor solution is impractical if not impossible. An equipment-independent data format for the exchange of low-level measurement process plans is necessary and critical for big corporations.

There is one standard specification for the interface between high-level process plan execution and low-level process plan execution: the DMIS Part 1 data model and the I++DME specification. DMIS is a large, statement-based language. The implementation of many statements is extremely complex. A full implementation of DMIS would require about ten times as much source code as a full implementation of a typical language for programming a machining center. The DMIS specification actually describes both a language for writing programs to execute and a language for writing output reports about what was done when a program was executed and what the results were.

I++DME Interface Specification was developed for the exchange of dimensional measuring equipment information between CMM internal controllers and external connected computers that are used for CMM programming. This specification was supported by the European automakers and measuring equipment vendors. I++DME is a messaging protocol between measurement plan executors and measurement equipment. It uses TCP/IP sockets as the communication mechanism, and defines a message set and a client–server architecture. It uses UML as the data modeling language. A number of implementations and demonstrations of I++DME have been carried out in international technology shows and exhibits. The Equip module of DMIS Part 2 is an alternative to I++DME for the same interface.

References

1. Usher MJ (1985) Sensors and transducers. Macmillan, London
2. Whitehouse DJ (1994) Handbook of surface metrology. IOP, Bristol
3. Schmalz G (1929) Zeitschrift. VDI. 73:144–161
4. Whitehouse DJ et al (1994) Gloss and surface topography. CIRP Ann—Manuf Technol 43(2):541–549
5. Tönshoff HK, Inasaki I (2001) Sensors in manufacturing sensors applications. Wiley-VCH Verlag GmbH, Weinheim
6. Barkman WE (1989) In-process quality control for manufacturing. M. Dekker, New York
7. Sze SM (1994) Semiconductor sensors. Wiley, New York
8. Shiraishi M (1988) Scope of in-process measurement, monitoring and control techniques in machining processes-Part 1: In-process techniques for tools. Precis Eng 10(4):179–189
9. Shiraishi M (1989) Scope of in-process measurement, monitoring and control techniques in machining processes-Part 3: In-process techniques for cutting processes and machine tools. Precis Eng 11(1):39–47
10. Shiraishi M (1989) Scope of in-process measurement, monitoring and control techniques in machining processes-Part 2: In-process techniques for workpieces. Precis Eng 11(1):27–37
11. Yandayan T, Burdekin M (1997) In-process dimensional measurement and control of workpiece accuracy. Int J Mach Tools Manuf 37(10):1423–1439
12. Vacharanukul K, Mekid S (2005) In-process dimensional inspection sensors. Measurement. J Int Meas Confed 38(3):204–218
13. Barkman WE (2000) In-process measurement. In: Richard ELH, Shell L (eds) Handbook of industrial automation. Marcel Dekker, New York
14. Gu P, Chan K (1996) Generative inspection process and probe path planning for coordinate measuring machines. J Manuf Syst 15(4):240–255
15. Medeiros DJ et al (1994) Off-line programming of coordinate measuring machines using a hand-held stylus. J Manuf Syst 13(6):401–411
16. Schaffer GH (1982) Taking the measure of CMMs. Am Mach 126:145–160
17. Kurfess TR (2006) What can CMMs do? Manuf Eng 136(3):173
18. Spitz SN (1999) Dimensional inspection planning for coordinate measuring machines. Department of Computer Science, University of Southern California
19. ANSI (1994) ASME Y14.5 M-1994: Dimensioning and Tolerancing
20. Destefani JD (2001) CMMs make contact. Manuf Eng 127(3):100–105
21. Sheehan K (1993) Measurement verifies machine performance. Mod Mac Shop 66(5):76
22. Weckenmann A, Kraemer P, Hoffmann J (2007) Manufacturing metrology-state of the art and prospects. In: 9th international symposium on measurement and quality control, ISMQC-Proceedings
23. Rakowski LR (1996) Portable CMMs: affordable accuracy. Foundry Manag Technol 124(2):29
24. Dove J, Schueneman D (2005) Portable or fixed-position CMM? Manuf Eng 135(1):53–55
25. Eaton H (1994) How to buy a portable CMM. Tool Prod. 60(3)
26. Logee S (2004) On-machine measurement. Tool Prod 70(10):40–43
27. Kim KD, Chung SC (2001) Synthesis of the measurement system on the machine tool. Int J Prod Res 39(11):2475–2497
28. Kamath R (2000) On-Machine Inspection and Acceptance (OMIA). University of Missouri-Rolla, USA
29. Cho MW et al (2004) A computer-aided inspection planning system for on-machine measurement-Part II: Local inspection planning. KSME Int J 18(8):1358–1367
30. Lee H, Cho MW, Yoon GS, Choi JH (2004) A computer-aided inspection planning system for on-machine measurement-Part I: Global inspection planning. KSME Int J 18(8):1349–1357

31. Chung SC (1999) CAD/CAM integration of on-the-machine measuring and inspection system for free-formed surfaces. Proc Am Soc Precis Eng 20:267–270
32. Renishaw (2000) Renishaw technical specifications: probing systems for CNC machine tools
33. Davis TA et al (2006) Flexible in-process inspection through direct control. Measurement. J Int Meas Confed 39(1):57–72
34. ANSI (2004) Dimensional Measuring Interface Standard, DMIS 5.0 Standard, Part 1, ANSI/CAM-I 105.0-2004, Part 1
35. I ++DME (2005) Dimensional measurement equipment interface, the International Association of Coordinate Measuring Machine Manufacturers
36. Proctor F et al (2007) Interoperability testing for shop floor measurement. Performance Metrics for Intelligent Systems (PerMIS) Workshop
37. Dimensional Markup Language (DML) (2009) http://www.aiag.org/
38. Horst J et al (2004) User's manual for version 3.0 of the NIST DME interface test suite for facilitating implementations of version 1.4 of the I++DME interface specification. National Institute of Standards and Technology
39. White RM (1987) Sensor Classification Scheme. IEEE Trans Ultrason Ferroelectr Freq Control UFFC 34(2):125–127

Chapter 6
Quality Data Analysis and Reporting

Once measurements are carried out, the measured data are collected and analyzed. The quality data analysis and reporting activity is an important element of dimensional metrology. The functionality of this activity is to receive input from measurement process execution and product definition activities, to analyze the part measurement data in terms of production definition requirements, to perform a statistical analysis of the measurement results and present them in the form of a statistical process control report, and to archive whatever measurement values and derived statistics are necessary.

Today, most Coordinate Measuring Systems (CMSs) typically obtain discrete 3-D measurement points from measurement processes (i.e., they generate Cartesian (x-y-z) coordinates of a measured part's surface). Though there are limitations on the number of measurement points and the accessibility of probes, the surface profile of an arbitrary item of 3-D geometry can be obtained by coordinate measurement. The capability of a CMS in measuring a part surface profile is essential for the verification of form and profile tolerances. Since the measurement results are presented as discrete coordinate points, the points must be associated with the design geometry in order to examine the actual part deviation or a tolerance conformance. Typically the data are fit to target geometry or a design model via different types of data fitting methods. In order to correctly perform the data fitting process, proper mathematical representation of geometry elements must be used together with appropriate data fitting algorithms. This chapter first gives an inductive discussion of the quality data analysis and reporting activity followed with a detailed introduction of data fitting theories and algorithms. The remainder of this chapter discusses existing data models for quality data analysis and reporting.

Y. Zhao et al., *Information Modeling for Interoperable Dimensional Metrology*, DOI: 10.1007/978-1-4471-2167-1_6,

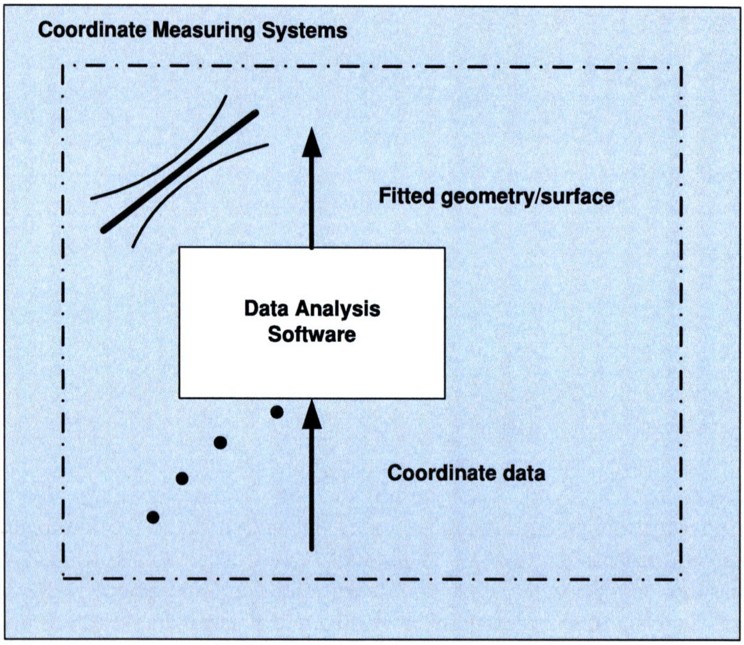

Fig. 6.1 Data analysis software

6.1 Quality Data Analysis and Reporting Activity

Most of the commonly used dimensional measurement devices, such as vision systems, theodolites, photogrammetry, and CMMs, can be categorized as CMSs. They assess length-based characteristics of mechanical parts by measuring points on the part surface and analyzing the point data. Since tolerances are defined as geometric bounds, and the measured data are sets of points, a supplemental numerical algorithm must be provided in order to make a pass/fail decision. Data fitting algorithms are used to locate the measured points on the design model, and then the tolerance is verified by comparing the deviation of measured points to the tolerance limits. A critical question that must be answered is "which fitting method yields the most accurate and reliable results?'' For the same measured data, different verification methods may yield different verification results [1]. In dimensional metrology industry, the data fitting process is encapsulated in the data analysis software systems (shown in Fig. 6.1).

The data analysis function with which we are concerned is geometric fitting. This function lies at the core of most inspection tasks. The role of geometric fitting is to reduce measured point coordinates to curve and surface parameters. The resulting curves and surfaces are called the *substitute geometry* for the feature. In further processing, these parameters are compared to the tolerance limits for the part. Most CMSs are characterized in terms of how accurately point measurement can be made. However, it is the uncertainty of the computed substitute geometry that determines the quality of a measurement. Data analysis software can contribute significantly to the total measurement error of a CMS. Factors affecting software performance include the choice of analysis method, the quality of the software, and characteristics of the specific measurement task [2]. The research that centers on measurement data analysis and computation is often referred to as *computational metrology.*

The phrase *computational metrology* refers to "*the study of the effects of data analysis computations on the performance of measurement system*" [2, 3]. Computational metrology involves the application of core concepts of metrology to the computational components of a measurement system [4]. Not everyone shares the view that computational metrology is a significant area of study. For instance, an ASME standard on measurement uncertainty [5] states: "*Computations on raw data are done to produce output (data) in engineering units. Typical errors in this process stem from curve fits and computational resolution. These errors are often negligible*". The standard does deal with the propagation of errors through computations, but the above is the only mention of computations as a source of errors. During the mid-1980s to early 1990s, however, much evidence has been discovered that data analysis can be a significant source of errors. In the mid-1980s, Germany began a program [6–8] of testing coordinate measuring machine software, with the express purpose of improving what was perceived as low quality of commercial fitting algorithms. In 1988, Walker [52] issued an advisory in which he reported the results of experiments with commercial inspection systems: "*Certain algorithms ..... are capable of stating that the measurement is worse than the actual data gathered up to an error of 37% and that the measurement is better than the actual data gathered up to an error of 50%*".

In 1989, Estler analyzed a measurement device for inspecting the casings of the solid rocket boosters for the NASA space shuttle. He reported that the data analysis software was the single largest source of error in the entire system [3]. Since then, considerable ongoing research has been carried out on CMS algorithms. There is also growing interest by government standards laboratories in testing the performance of CMS software, particularly in Great Britain [9] and Germany [6]. Both countries offer services to test CMS software by comparing results for test data sets to results obtained from reference software. In the U.S., NIST has developed a similar service [2]. These test services focus on examining how well the software fits a set of measured points into the geometry on a part surface. Fitting is at the core of most measurements made by CMSs (shown in Fig. 6.2). From a metrological point of view, two factors determine fitting software performance: the choice of fitting objective and the quality of the software implementing that objective. Theories of fitting objective and fitting algorithms are explained in detail in the following sections. How well the fitting objective is implemented in commercial software systems depends on their proprietary implementation technique, thus it is not discussed in this book. The government test services mentioned above provide a way to measure the performance of commercial software.

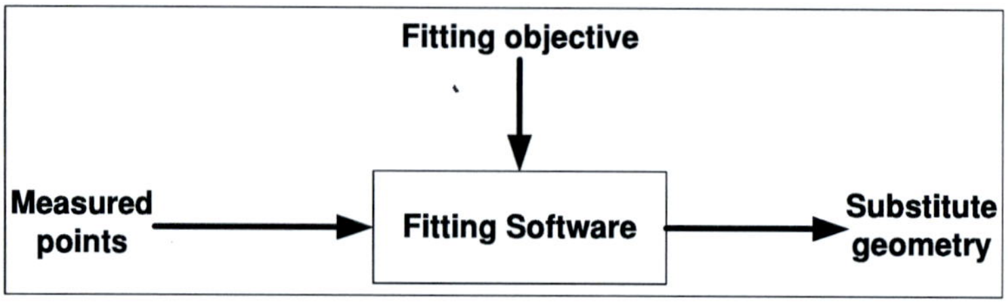

Fig. 6.2 Ideal model of fitting

Once the substitute geometry is obtained through fitting process, it is compared with the design geometry for out-of-tolerance check, which is the final step of quality data analysis activity. The results gathered through this activity are often reported to manufacturing process and product quality control systems. In industry, SPC systems are most commonly used. SPC is the application of statistical methods in monitoring a process to ensure that it operates at its full potential to produce conforming product. While SPC has been applied most frequently to controlling manufacturing lines, it applies equally well to any process with a measurable output. Key tools in SPC are control charts. Much of the power of SPC lies in the ability to examine a process and the sources of variation in that process using tools that give weight to objective analysis over subjective opinions and that allow the strength of each source to be determined numerically. Variations in the process that may affect the quality of the end product or service can be detected and corrected, thus reducing waste as well as the likelihood that problems will be passed on to the customer. With its emphasis on early detection and prevention of problems, SPC has a distinct advantage over inspection for detecting and correcting problems after they have occurred [10]. The applications of SPC and quality data in manufacturing processes are discussed in Chap. 8. From the interoperable dimensional metrology point of view, it is critical to have a standardized data format for analysis and report of quality data to various SPC systems. The theories for the development of such a standard data model are discussed in this chapter.

6.2 Data Fitting Theories and Computational Metrology

Fitting can be viewed as an optimization problem: find the parameters of substitute geometry that optimize a particular fitting objective for a set of points. Data fitting in dimensional metrology is the major part of computational metrology. The past decade has seen the emergence of computational metrology as a separate discipline in CAD and manufacturing. It deals with fitting and filtering discrete geometric data that are obtained by measurements made on manufactured parts [11]. In manufacturing, measurements are made on the surface of the part and the measured data are then reduced to a few numbers or attributes by increasingly

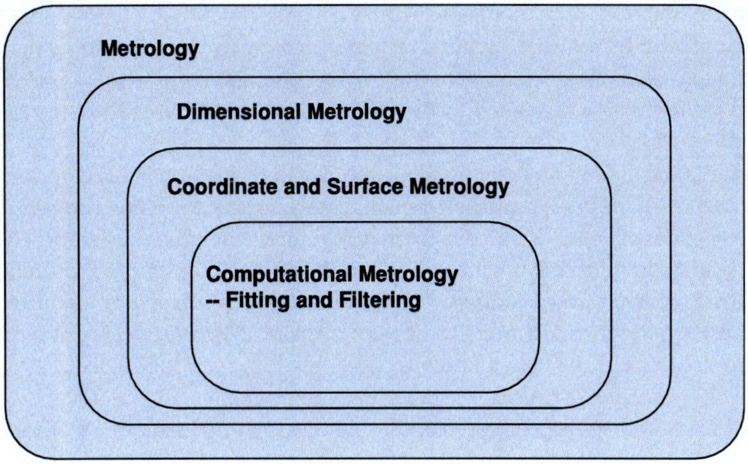

Fig. 6.3 Relation between computational metrology and other metrology areas

sophisticated computational techniques. Computational metrology, defined here as fitting and filtering of discrete geometric data, is a subset of coordinate and surface metrology, which is a further subset of dimensional and geometric metrology practiced extensively in industry (shown in Fig. 6.3). It is in this sense the phrase "computational metrology" was first coined by Srinivasan in the early 1990s [12, 13]. In this section, we first discuss the functionalities of computational metrology, namely its fitting and filtering processes, followed with detailed discussion of data fitting algorithms.

6.2.1 Introduction to Computational Metrology

It is known that manufacturing imprecision and measurement uncertainty are two basic facts existing in all manufacturing and measurement industry. All manufacturing processes are inherently imprecise and produce parts that vary. In fact, it is the primal fact that no man-made artifact has Platonic ideal form. There is increasing experimental evidence that the geometry of a manufactured surface behaves more like a fractal in the range of dimensional scale that is of interest in engineering. Therefore, at least conceptually, a manufactured surface can be modeled as a fractal set and the only information we can obtain about it comes from a discrete set of points sampled on that surface.

As for the measurement, no measurement can be absolutely accurate and with every measurement there is some finite uncertainty about the measured attribute or measured value [14]. If we take the discrete set of measurements as input to computations, it is good to remember that these input values can never be taken as absolutely accurate. It is equally important to remember that results of the computations should be accompanied by statements about their uncertainty, which is partly inherited from the input uncertainty and partly attributable to the computational scheme itself. Manufacturing imprecision and measurement uncertainly are two preliminaries that need to be addressed before discussing fitting and filtering processes.

Fitting is the task of associating ideal geometric forms to non ideal forms, such as discrete set of points sampled on a manufactured surface. Fitting is usually used for two purposes: datum establishment and deviation assessment. Historically such fitting was accomplished by the use of surface-plates, collates, mandrels, and specialized measurement fixtures. More recently, manufacturing industry has started using modern measurement devices such as CMMs and optical scanners. This has accelerated the use of fitting by computation. Its initial success is placing increasingly complex demands on our ability to compute.

The computational scheme used for fitting is one of optimization. For example, it may be of interest to fit a plane to a set of points in space such that the sum of the squares of the distances of the points from the plane is minimized. This can be recognized as a least-square fitting problem. Such problems have been studied in science for over two centuries. However, there are also other seemingly simple fitting problems that can tax our computational skills. For example, an engineer may want to find the smallest cylinder (that is, a cylinder with the smallest diameter) that encloses a set of points in space because it gives him some quantitative information about how a part will fit in an assembly. This can be easily posed as a minimization problem, but computational methods to solve this problem are not simple. The choice of computational methods is based on the fitting objective. Part tolerances are generally interpreted in terms of extremal fits. That is, the fitting objective is to find the geometry that fits the extremes of the data: the largest inscribed, smallest circumscribed, or minimum separation geometry. Also, simulation of functional gages can be interpreted as finding the maximum clearance or minimum interference solid model fit to the data. It is customary to divide the fitting problems broadly into the following two categories on the basis of the objective function that is optimized [3, 11, 15–17].

- *Least-square fitting* the objective is to find an ideal geometric object (a smooth curve or surface) that minimizes the sum of squared deviations of data points from this object. It includes linear least squares, total least squares, and nonlinear least squares techniques.
- *Chebyshev fitting* the objective is to minimize the maximum deviation. Some of these fitting problems have been studied by discrete and computational geometers in the last 20 years. This has added some valuable insight in designing algorithms to solve such problems.

National and international standards groups are actively working on standardized definitions for the objective functions and constraints for the fitting problems.

Filtering is the task of obtaining scale-dependent information from measured data. At a more mundane level, filtering can be used to remove noise and other unwanted information from the measured data. In the context of engineering metrology, engineers are interested in filtering mainly for the following two reasons.

- *Surface roughness*: many engineering functions depend on how rough or smooth a piece of surface is. Designers define bounds on certain roughness parameters obtained by observation on rather small scale to ensure functionality of parts. These small-scale variations are subtracted from the surface measurement data before form and other deviations are assessed.
- *Manufacturing process*: *diagnosis* manufacturing processes leave tool marks on surfaces. By measuring surfaces at fine scale, it is possible to detect tool erosion and its effect on the surface quality.

Historically, filtering techniques were pioneered by communication theorists. Developments in analog and digital signal processing strongly influenced how filtering was carried out in surface metrology. More recently, developments in digital image processing have been influencing computational surface metrology.

The computational scheme used for filtering is one of convolution. Engineers use the following two types of convolutions [11, 15–17].

- *Convolution of functions*: filtering is often implemented as discrete convolution of functions. In the most popular version, the measured data is convolved with the Gaussian function (Eq. 6.1). It has a smoothing effect on the surface data.

$$y(x) = \int_{-\infty}^{+\infty} z(s)K(x-s)ds \qquad (6.1)$$

  where *z(s)* is the unfiltered input profile, *y(x)* is the filtered profile, and *K(x−s)* is the kernel function.
- *Convolution of sets*: morphological filters are implemented using Minkowski sums (Eq. 6.2). These can be regarded as convolutions where the input set is convolved with a circular or flat structuring element.

$$A \oplus B = \{x + y : x \in A, y \in B\} \qquad (6.2)$$

  where A and B are two sets of data in R^n.

It is important to note that computational metrology is a crucial part of the dimensional measurement analysis process. However, in order to obtain good data for this computational process, some procedures must be taken beforehand. One of the earliest standard efforts for standardizing the dimensional measurement assessment process is BS 7172:1989 [18]. It recommended that the dimensional measurements assessment process should be carried out in four stages:

1. apply an appropriate measurement procedure, i.e., a strategy for obtaining a representative set of measurements on the workpiece;

2. (optionally) pre-process the data, i.e., replace the measured data by modified values in order, for example, to smooth the data, to remove inappropriate points or to compensate for environmental effects;
3. compute the reference (e.g. an approximating circle in terms of its centre coordinates and radius), to give position and size;
4. assess, in terms of the reference, the departure from nominal form.

Research on the first stage—appropriate measurement procedure—has been reviewed in Sect. 4.2. If the gathered data is considered of sufficiently high quality for purposes of the assessment it should be left unaltered. Alternatively, if it contains random or systematic errors that, it is judged, would adversely affect the results of the assessment, the data should be pre-processed. Pre-processing can be used to remove outliers, to reduce data errors by smoothing, to operate on data according to the functional requirements of the workpiece under tests, to account for flexing of the probe, and to make corrections for the effect of temperature, humidity and vibration. Commercial dimensional measurement systems require particular environment conditions. The dimensional measurement result analysis software normally has the functionality to detect outlier data and pre-process the data. Stages three and four belong to computational metrology.

6.2.2 Mathematical Representation of Geometric Elements

In order to obtain a reliable assessment of a geometric form in any particular case, the corresponding geometric element should first be represented, i.e., parameterized, by a set of measured data points in a Cartesian coordinate system in a mathematically sound way. It is possible to parameterize each of the geometric elements in more than one way. The parameterizations given in BS 7172:1989 [18] are recommended as being generally applicable. They have the property that small changes in the geometric element usually result in correspondingly small changes in the parameter values. Certain other parameterizations may be equally sound, although it should be noted that the use of some parameterizations can yield unreliable results. The following mathematical representations of geometric elements are listed in this book.

6.2.2.1 Lines

The parameterization of cylinder or cone requires the specification of an axis.

1. One point and the direction cosines
   A line L, related to a set of data points, should be specified by:

   (i) a point (x_0, y_0, z_0) on L; and

(ii) its direction cosines (a, b, c).
The point (x_0, y_0, z_0) should be taken at or near G, the point on L closest to the centroid of the data points. Any point (x, y, z) on L satisfies the equation:

$$(\mathrm{x},\ \mathrm{y},\ \mathrm{z}) = (x_0, y_0, z_0) + t(\mathrm{a},\ \mathrm{b},\ \mathrm{c}) \tag{6.3}$$

for some value of t.

2. Two points on the line
   A line L, related to a set of data points, should be specified by two points (x_1, y_1, z_1), (x_2, y_2, z_2) that:

   (i) lie on L; and
   (ii) are such that all data points lie between the two planes perpendicular to L passing through (x_1, y_1, z_1) and (x_2, y_2, z_2), respectively; and
   (iii) are as close together as (reasonably) possible.
   Any point (x, y, z) on L satisfies the equation:

$$(\mathrm{x},\ \mathrm{y},\ \mathrm{z}) = (1 - t)\,(x_1, y_1, z_1) + \mathrm{t}\,(x_2, y_2, z_2) \tag{6.4}$$

   for some value of t.

6.2.2.2 Planes

A plane should be specified by a point on the plane and either of the following two items.

1. The direction cosines of the normal to the plane
   A plane P, related to a set of data points, should be specified by:

   (i) a point (x_0, y_0, z_0) on P; and
   (ii) the direction cosines (a, b, c) of the normal to P.
   Point (x_0, y_0, z_0) should be taken at or near G, the point on P closest to the centroid of the data points. Any point $(x,\ y,\ z)$ on P satisfies the equation:

$$\mathrm{a}(x - x_0) + \mathrm{b}(y - y_0) + \mathrm{c}(z - z_0) = 0 \tag{6.5}$$

2. A point on the normal to the plane passing through the first point.
   A plane P, related to a set of data points, should be specified by:

   (i) a point (x_0, y_0, z_0) on P; and
   (ii) a point (x_1, y_1, z_1) on the normal to P at (x_0, y_0, z_0).
   Point (x_0, y_0, z_0) should be taken at or near G, the point on P closest to the centroid of the data points. Point (x_1, y_1, z_1) should be determined such that its

distance from P is comparable with the span of the data points. Any point (x, y, z) on P satisfies the equation:

$$(x_1 - x_0)(x - x_0) + (y_1 - y_0)(y - y_0) + (z_1 - z_0)(z - z_0) = 0 \tag{6.6}$$

6.2.2.3 Circles

A circle in three dimensions should be specified by its centre and radius, and the plane in which it lies. Since the centre of the circle lies in the plane, this point should be used in specifying the plane.

A circle C should be specified by:

1. its centre (x_0, y_0, z_0); and
2. its radius r; and either
   - (i) the direction cosines (a, b, c) of the normal to the plane containing C; or
   - (ii) a point (x_1, y_1, z_1) on the normal at the centre of C to the plane containing C.

   The point (x_1, y_1, z_1) should be chosen such that its distance from the centre is comparable to the radius.

6.2.2.4 Cylinder

A cylinder C, related to a set of data points, should be specified by:

1. the axis of C, and
2. its radius r.

   If the axis is specified by a point (x_0, y_0, z_0) on a line L; and its direction cosines (a, b, c), the point (x_0, y_0, z_0) should be taken close to the midpoint of the part of the axis that is enclosed by the data. Any point (x, y, z) on L satisfies the equation:

$$(x, y, z) = (x_0, y_0, z_0) + t(\mathrm{a}, \mathrm{b}, \mathrm{c}) \tag{6.7}$$

   for some value of t, where t is a parameter proportional to distance.

6.2.2.5 Cones

A cone should be specified by its axis, angle, and information about where on the axis the cone is situated. The use of the vertex is not recommended in general.

A cone C, related to a set of data points, should be specified by

1. the axis of C, specified according to Sect. 6.2.2.1; and
2. the apex angle ψ of the cone; and
3. the distance s to the surface of C from a point (x_0, y_0, z_0) on the cone axis. If the axis is specified according to Sect. 6.2.2.1 (1), the point (x_0, y_0, z_0) should be taken close to the midpoint of the part of the axis that is enclosed by the data. Numerical inaccuracies are likely to arise in the use of this parameterization for a cylinder and cone related to a set of data points that, when orthogonally projected onto a plane perpendicular to the cylinder axis, lie on or near an arc whose length is much smaller than the cylinder radius.

6.2.2.6 Spheres

A sphere S should be specified by its centre (x_0, y_0, z_0) and its radius r.

Any point (x, y, z) on S satisfies the equation:

$$(x - x_0)^2 + (y - y_0)^2 + (z - z_0)^2 = r^2 \tag{6.8}$$

Numerical inaccuracies are likely to arise in the use of this parameterization for a sphere related to a set of data points that span a region whose area is small compared with the surface area of the sphere.

6.2.3 Geometry Data Fitting Criteria

The computed reference should give the position and size of the geometric element and can be used in the assessment of the workpiece. For example, the reference for circularity should normally be the centre coordinates and the radius of a computed circle. For instance, this circle may be the smallest circle enclosing the data points.

The reference is defined by the parameters of the corresponding geometric element that best fits the measured points. The fit is represented by the values of its parameters, e.g. radius and centre coordinates of a circle. Many different criteria for specifying the best fit are possible. In general, the criterion should be to make some combination of the residuals as small as possible. In mathematical terms, the reference is obtained by optimizing the chosen combination of the residuals with respect to the parameters. Examples of criteria for specifying best fit are:

- least squares: $\min \sum_i \text{res}_i^2$; and
- minimax: $\min(\max_i |\text{res}_i|)$

Here, the residual, res_i is a measure of the departure of the ith point from the fit. The residual is conventionally defined as the distance of the point from the reference. However, when calculating a reference circle by least squares, a particularly simple algorithm can be obtained if the residual is taken to be the difference between the squared distance of the point from the circle centre and the squared radius of the circle. Not all criteria are of this general form. Frequently used criteria are least squares, minimax, maximum inscribed and minimum circumscribed.

The purpose of geometry data fitting is to apply an appropriate algorithm to fit a perfect geometric form (e.g. line, plane, circle, ellipse, cylinder, sphere, cone) to sampled data points obtained from the inspection of a manufactured part. The perfect form approximation obtained through fitting is called a substitute feature. The substitute feature is represented by a shape vector b. The exact nature of b varies with the geometric form being fitted. The substitute feature is a one-dimensional curve or a two-dimensional surface that we designate as a function $f(u;b)$ of a parameter vector u. The values of f are points in space (or on a surface, if fitting is being done in two dimensions). As u varies, f moves along the geometry represented by b; as b varies, the surface changes shape and location. A particular geometry need not have a single representation. In fact, much of the research on fitting techniques is based on developing clever representations for curves and surfaces.

The fitting problem, generally stated, is to minimize some objective function with respect to b. For some kinds of fitting (to be described below) the minimization may be subject to certain constraints. The most frequently used fitting algorithms are based on the L_p norm:

$$L_p \doteq \left[\frac{1}{N} \sum_{i=1}^{N} |e_i|^p \right]^{1/p} \tag{6.9}$$

where $0<p<\infty$, N is the total number of data points, and e_i is the shortest distance between p_i, the ith data point, and the considered feature. The best fit feature is the feature that minimizes the L_p norm. Since N is a constant, 1/N is usually omitted from the equation in most surface fitting applications. Similarly, since p is fixed, the exponent $1/p$ is often omitted. The resulting objective function is

$$S_p \doteq \sum_{i=1}^{N} |e_i|^p \tag{6.10}$$

S_p represents the same problem as the L_p norm. The value of b that minimizes L_p(and S_p) is called the L_p estimator of the feature. When $p = 1$, the fitting problem is least-sum-of-distances fitting, a generalization of finding the median of a data set. When $p = 2$, it is total-least-squares fitting, also called orthogonal distance regression. L_p fitting can be extended to $p = \infty$ by noting that $\lim_{p\to\infty} L_p = \max_i |e_i|$. Minimizing L_∞ is called the two-sided minimax problem, because the solution minimizes the maximum e_i on both sides of the feature. One-sided minimax fitting is a constrained two-sided minimax fitting. The objective functions for smallest circumscribed features and the largest inscribed features are somewhat different than one-sided minimax objective functions, but the fitting results often appear similar. Feng and Hopp [19] from NIST have made a thorough review of data fitting algorithms. Some of their findings are summarized here.

6.2.3.1 Least-Sum-of-Distances Fitting

This fitting problem is also known as the median-polish fit. The sum of distances can be formulated as:

$$S_1 = \sum_{i=1}^{N} |e_i| \tag{6.11}$$

The objective of this fitting is to minimize the sum of absolute distances, S_1. The result of the least sum of distance fitting passes through the median of the distribution of e_i's. This fitting is less sensitive to the data outliers than total least squares fitting.

6.2.3.2 Total-Least-Squares Fitting

Total-least-squares fitting is by far the most widely used approach in CMM data analysis. The sum of squared distances, S_2, can be formulated as:

$$S_2 = \sum_{i=1}^{N} |e_i|^2 = \sum_{i=1}^{N} e_i^2 \tag{6.12}$$

Each e_i is a function of the data point p_i and the point $f(u_i; b)$ on the substitute feature closest to p_i. Therefore, the least squares fitting can be expressed as:

$$\min_b \left(\sum_{i=1}^{N} |p_i - f(u_i; b)|^2 \right) \tag{6.13}$$

If each e_i is a nonlinear function of b, then the problem is a nonlinear, total least squares fitting, and is usually solved using an iterative process. The commonly used Gauss–Newton and Levenberg–Marquardt iteration algorithms can be applied in finding the minimal S_2. One problem with these methods is that they can easily find only a local minimum. Therefore, it is important, although sometimes difficult, to find a good starting value for the iterative process.

If the substitute feature is a line or a plane, then e_i can be expressed as a linear function of b, and S_2 can be then formulated as:

$$\begin{aligned} S_2 &= \sum_{i=1}^{N} |e_i|^2 \\ &= \sum_{i=1}^{N} \left|p_i - x^T b\right|^2 \\ &= |Ab|^2 \\ &= b^T A^T A b \end{aligned} \tag{6.14}$$

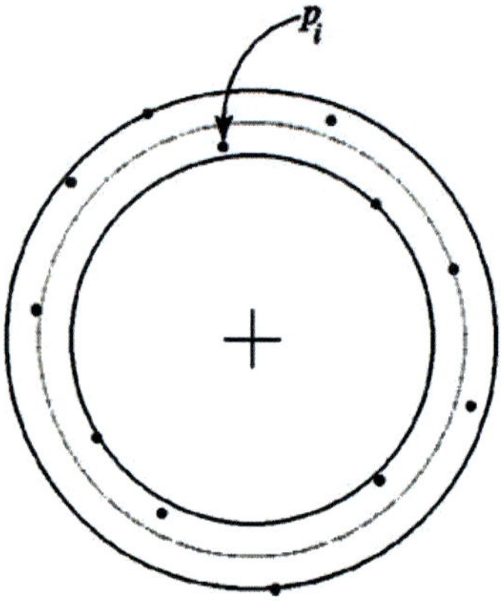

Fig. 6.4 Circles by two-sided fit

where A is a matrix that is not dependent on b. For the linear total least squares problem, the estimator b can be obtained by using Gaussian elimination [20]. The Gauss–Markoff theorem shows that the total least squares estimation is the best linear unbiased estimation. However, this theorem refers only to the class of linear estimations. It does not considered nonlinear alternatives. The total least squares fitting passes through the sample mean of a normal distribution of distances e_i's. The total-least-squares fitting is more sensitive to data outliers than the least sum of distances fitting.

6.2.3.3 Two-Sided Minimax Fitting

When p in the L_p-norm formula approaches infinity, the norm becomes the maximum absolute distance. The estimator b is then called the two-sided *minimax* fit for the feature. Minimax fitting minimizes the maximum distance between all the sampled data points and the ideal form. The problem can be formulated as:

$$\min_{b}\left(\max_{1\le i\le N}|e_i|\right) \tag{6.15}$$

This fitting problem is known as the L_∞-norm estimation problem. A roundness tolerance zone, as a result of two-sided fitting, is shown in Fig. 6.4.

The resulting fit is strongly affected by data outliers. The algorithm for calculating the two-sided-minimax fit for circles is presented by [21].

6.2.3.4 One-Sided Minimax Fitting

Two-sided minimax fitting is useful for estimating roundness, cylindricity, and other form deviation. When estimating the size of a feature, however, one is usually interested in having the substitute feature lie entirely outside the material of the part. This can be accomplished using one-sided minimax fitting. The formulation is based on representing the errors e_i so that they are positive on one side of the feature and negative on the other side. So, for instance, one side of a line or the inside of a circle in a plane, or one side of a plane or the inside of a cylinder, can be chosen as positive side and the other the negative. By properly choosing the representation of the substitute feature, the e_i can always be expressed in this way. One-sided minimax fitting can then be formulated as a constrained optimization problem:

$$\min_b \left(\max_{1 \leq i \leq N} |e_i| \right) \tag{6.16}$$

Subject to $e_i \leq 0$, $i = 1,\ldots., N$ or

$$\min_b \left(\max_{1 \leq i \leq N} |e_i| \right) \tag{6.17}$$

Subject to $e_i \geq 0$, $i = 1,\ldots, N$.

The choice of formulation depends on where the material of the part is with respect to the positive side of the curve or surface.

6.2.3.5 Smallest Circumscribed Fitting and Largest Inscribed Fitting

For features of size such as circles, cylinders, and spheres, a common objective is to find the largest inscribed substitute feature or the smallest circumscribed substitute feature. The problem can be formulated as:

$$\min_c R \tag{6.18}$$

Subject to $e_i \leq 0$, $i = 1,\ldots.,N$ and

$$\max_c R \tag{6.19}$$

Subject to $e_i \geq 0$, $i = 1,\ldots, N$

$$\begin{gathered} 0 = \sum_{i=1}^{N} \lambda_i (p_i - c) \\ 1 = \sum_{i=1}^{N} \lambda_i \\ \lambda_i \leq 0,\ i = 1,\ \ldots,\ N \end{gathered} \tag{6.20}$$

where c is the parameter vector which can be the centre of a circle or a sphere, or the axis of a cylinder and R is the radius of the circle, sphere, or cylinder.

Equation 6.18 can be used to find the smallest circumscribed circle, while Eq. 6.19 can be used to find the largest inscribed circle. The different between the

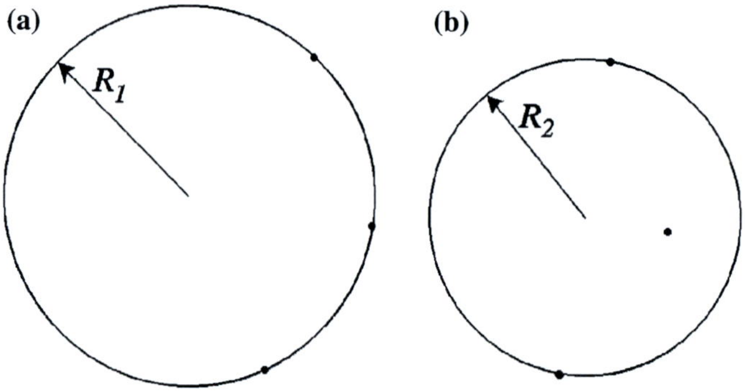

Fig. 6.5 Circles of one-sided minimax; (**a**) and smallest circumscribed; (**b**) fits

substitute circle obtained from Eq. 6.16 and the smallest circumscribed circle using Eq. 6.18 is shown in Fig. 6.5. The figure shows how the radius of the one-sided, circumscribed minimax circle is larger than the radius of the smallest circumscribed circle ($R_1 > R_2$). A similar difference can be seen in Fig. 6.6 for a one-sided, inscribed minimax circle, obtained from Eq. 6.17, and the largest inscribed circle obtained from Eq. 6.19. The radius of the one-sided minimax inscribed circle is smaller than the radius of the largest inscribed circle ($R_1 < R_2$, $d_1 = d_2 < d_3$).

Research carried out by Etesami and Qiao [21] presented efficient algorithms in finding these circles. The algorithm is to search the smallest circumscribed circle on the farthest-point Voronoi diagram generated from the convex hull of measured points on a plane. (A *Voronoi diagram* of a set of points is a partition of space into regions. Each region corresponds to a point of the set in that all points of the region are closer to or father away from the corresponding point than from any other point in the set.) Similarly, the largest inscribed circle can be searched on the closest-point Voronoi diagram. As with two-sided minimax fitting, one-sided minimax fitting is very sensitive to data outliers.

Once the reference has been determined, the deviation of the measured workpiece from nominal form can be assessed. The departure from nominal form is defined as the spread of the measured data about the reference. First, the deviation, e_i, of a single measured point from the reference should be taken as the distance from the point to the reference, where appropriate, the distance is given a sign according to the side of the reference on which the data point lies. The spread of the deviations is then computed from these e_i. Since the departure from nominal form is derived from the reference and the data points, points that have been modified or deleted in the pre-processing stage of the assessment process should be included in assessing departure.

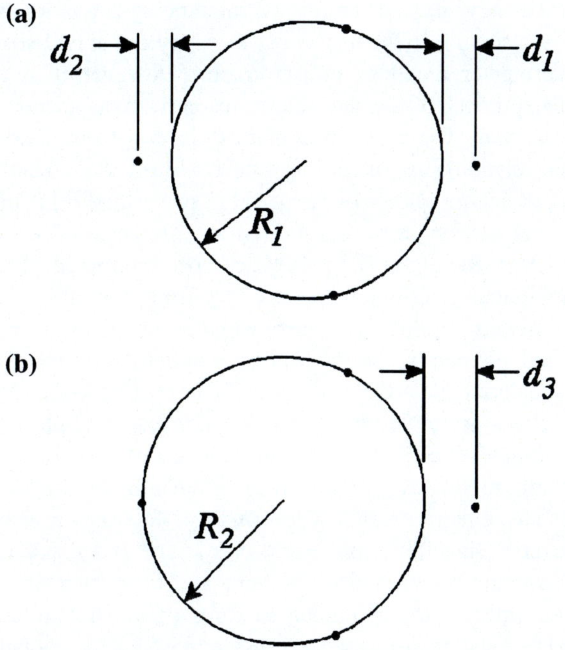

Fig. 6.6 Circles of one-sided minimax; (**a**) and largest inscribed; (**b**) fits

6.2.4 Algorithms for Minimum Tolerance Zone Calculation

If a measured point set is regarded as a complete replication of an unknown actual surface, and if the tolerance is interpreted as geometric constraints on the point set, then the verification of the tolerance can be formulated as a geometric problem. The minmax fit returns the solution for the geometric problem. If the measured points are interpreted as sampled data from the unknown actual surface, the least-squares-fit is more appropriate because even though the sampled points may satisfy the given tolerance by the geometric solution, the unknown actual surface may violate the tolerance. Thus, when dimensional measurement is accomplished by fitting results of the dimensional measurement data, both the interpretation of the measured data and the criteria for tolerance conformance verification must be carefully defined, examined and understood.

In this section, a brief introduction of traditional and modern tolerance theory and a review of minimum tolerance zone algorithms are given. Traditional plus/minus tolerancing provides a basis for defining the limit of size used in dimensioning mechanical parts. The size tolerance indicates the quantity of the allowable variation of a dimension, either linear or angular. A dimensioning theory developed by Hillyard [22] furnishes a scheme to specify sizes of interrelated features and to check whether a feature is over, under, or exactly defined by a set of specified dimensions on a part. It laid foundation for traditional tolerancing and geometry variation computing.

The advantage of traditional tolerancing is that it is simple for designers to use. It is also simple for inspectors to verify the actual size variation of parts using a micrometer, a caliper, or a protractor. However, there are several shortcomings in this approach. Only size tolerances and simple forms of positional tolerances are supported. There is no specification for form tolerances or complex feature interrelationships (including true position). As a result, assembly and alignment requirements cannot be represented or verified. Plus/minus tolerancing also lacks the abstraction power in representing tolerances of mechanical parts in CAD/CAM systems. Requicha [23] discusses how traditional plus/minus tolerancing can be ambiguous in how dimensions vary from nominal.

Modern tolerancing theory was developed to overcome shortcomings in traditional tolerancing theory. Modern geometric tolerancing methods are based on two major principles: the Maximum Material Condition (MMC) principle, also called Taylor's principle; and the Independence principle [24]. The MMC requires an envelope which is the boundary surface of a similar perfect form of the nominal feature in the design. The envelope must totally contain the feature and must meet the shape requirements. The similar perfect feature is the feature at the maximum material size limit (the worst case). The Independence principle makes a clear distinction between size tolerance and form tolerance. It requires tolerancing for size without any reference to form or location tolerances. The latter must be defined separately, when necessary. ANSI Y14.5 is based on the MMC principle. As stated in ANSI Y14.5 [25], a tolerance zone is a virtual region formed around the true feature. It can be interpreted as regulating the movement of a dial indicator.

Requicha [23] proposed mathematical formulations for tolerance zones. In his theory, a tolerance zone is a region bounded by similar perfect geometry, offset from the nominal feature surface. This research is one of the earliest research efforts in the mathematical modeling and computing of modern tolerance zones. Several important techniques have since been developed for computing offset surfaces in the early to mid 1990s. Approaches developed by Lin [26], Rossignac [27], and Yu [28] are well-suited for constructive solid modeling, while the method of Rogers [29] is based on boundary representations. Etesami [30] proposed a method for testing the conformity of actual manufactured parts to a tolerance zone. His approach is to construct tolerance zones for a feature and verify whether part boundaries lie entirely within the constructed tolerance zone. The approach uses a boundary representation technique in solid modeling to generate offsets of curves and surfaces, called constructors. The constructors are equivalent to Requicha's tolerance zones [19].

Both of the above theories are aimed at defining part conformance. A different approach was developed by Hoffmann [31]. Hoffmann proposed a set of mathematical models of manufacturing process errors. Traditional plus/minus tolerancing was used to formulate error models that included machining errors (tool wear and machine errors), part setup errors, and alignment errors. The tolerance of a shape dimension was specified by a set of inequality equations for the geometric parameters of the shape. These equations are closely related to the manufacturing error models. The resulting theory is well-suited to developing feedback from inspection to the manufacturing process.

A brief review of algorithms for calculating minimum-tolerance-zone is given here. Most minimum-tolerance-zone [32] algorithms have been developed for two dimensions. These algorithms include straightness, flatness, and roundness.

1. *Straightness and Flatness tolerance zones*. Algorithms used in the calculation of actual straightness and flatness tolerance zones applying convex hulls were initially developed by Traband et al. [33] and by Cavalier and Joshi [34]. A convex hull needs to be established first; algorithms for constructing three-dimensional convex hull are in [35, 36]. Then, the minimum zone is found by searching the maximum distance between vertex and edges (line or surface) of the established convex hull.

   The medial axis transformation method can be used to approximate the median axis (real axis) of a cylindrical feature (hole or shaft). ANSI Y14.5 allows the application of a straightness tolerance to the axis of a feature. This tolerance controls the deviation of the median axis from a straight line. Algorithms for computing the medial axis transformation in two and three dimensions can be found in [37–39] and [40, 41], respectively. The medial transformation method can also be used to calculate the median axis of a conical feature.
2. *Circularity tolerance zone*. Circularity (roundness) is a tolerance zone bounded by two concentric circles with minimum radial separation within which all measurements should lie. The methods proposed by Lai [42] and Etesami [21] are first, to construct both the near-point Voronoi diagram and the farthest-point Voronoi diagram and then to find intersections of the two diagrams. Each intersection is the center of two concentric circles that form an annular zone within which all the inspection points lie. A search method is then used to find a pair of concentric circles that has the minimum radial separation. This pair of circles forms the actual minimum circularity zone. An interactive process was developed by Chetwynd [43] as an alternative to linear programming for calculating circularity.
3. *Cylindricity tolerance zone*. An in-process measurement of cylindricity was first developed by Kakino and Kitazawa [44]. This method used an extended principle of three-point roundness measurement. A special cylindricity measuring instrument was created for this purpose. A comparison study of algorithms for cylindricity tolerance zone calculation was carried out by Murthy [32, 45, 46]. One method Murthy describes used Fourier series and orthogonal polynomials for representing the actual profile of a cylinder. Then the normal least squares method and simplex search method were applied for searching the minimum cylindricity tolerance zone. The main research efforts in evaluating cylindricity error can be found in [47–49].

Three-dimensional tolerance zone calculation is still on-going research. Choi and Kurfess proposed a zone fitting algorithm, which fits a set of measured points into a specified tolerance zone [1, 50]. The literature cited here includes mostly the foundational research works in tolerance zone calculation. It should be noted that tolerance zone calculation is closely related to tolerance modeling. Therefore, interested readers should first understand basic tolerance modeling (refer to Sect. 3.2.2) theories before further exploring the recent research in tolerance zone calculation.

6.3 Information Modeling for Quality Data Analysis and Reporting

The above sections have discussed the basic quality data analysis algorithms including data fitting, data filtering and tolerance zone calculation. In dimensional metrology industry, most of the quality data analysis processes are encapsulated in commercial software systems. The software generates measurement feature pass/fail results. These results are then reported to various quality control departments.

The quality data analysis and reporting activity differs from other dimensional metrology activities in the way that this activity is connected to both enterprise level and shop-floor process level for quality control. Quality control is the practice in which the quality of factors involved in a production is reviewed. The factors include product quality conformance, process accuracy, product defects, etc. The analyzed quality data from data fitting software (i.e., pass/fail results) are further analyzed through statistical software for quality control processes. Obviously, different dimensional measurement systems may have different formats for quality data. Much has been said about the need for standardization and the open exchange of information. Perhaps nowhere is the need more compelling than in the realm of quality measurement. Quality is the lynchpin for success in every enterprise, and the absolute prerequisite for quality is measurement. Achieving interoperability for quality data analysis and reporting means to have a neutral data format for the exchange of analyzed quality data for different types of industry departments and enterprise systems. In this section, the commercial and proprietary data models are first discussed. It is followed with detailed introduction of standardized data formats including QMD, DMIS output data, and DML data formats. The DMIS output data module has been introduced in Sect. 5.4.1.2 together with the major part of the DMIS standard data model. The other two data modules are introduced in this chapter.

6.3.1 Commercial and Proprietary Data Models

Quality data analysis and reporting requirements differ between industries; they differ between manufacturers within each industry and many times differ from department to department within a single enterprise. In many cases the motivations for analytical reporting drive these differences. Some manufacturers produce documentation as proof of lot conformance in the supply chain, some have quality improvement programs in which these tools are vital and some do both. This section describes in a broad brush the techniques and tools used by the market today.

6.3.1.1 Tally Sheets and Spreadsheets

Before the proliferation of personal computers, most shop-floor data collection occurred on paper with pencil. Surprisingly enough, these same techniques are still in widespread use today. An operator may measure a workpiece with a mechanical hand tool and then write down the value on a tally sheet. Some operations may plot the point on a line chart in order to determine trends. This method of quality data analysis, although simple to implement, is fraught with the potential for human error when capturing or plotting the data.

More often, we see wide use of spreadsheet applications such as Microsoft Excel, to enter data from quality sampling. In many cases this data may come directly from a digital gage through a computer USB port acting as a keyboard wedge. Although this method does increase operator efficiency and accuracy, it does have several drawbacks to industry best practices.

6.3.1.2 Computer-Based Systems

Since the advent of the personal computer in the 1980s, manufacturing quality control has benefitted from the development of both proprietary and commercial data management and analysis applications. From humble beginnings in the DOS environment to today's powerful Microsoft Windows based systems, we have seen the evolution of various packages that serve all aspects of quality control throughout all of industry.

Real-time data acquisition became synonymous with advanced and mature quality control programs. Regardless of the inspection device, the operator could simply press a button or a footswitch and immediately see the results on the computer screen. Not only the numeric observation value is tabulated, but the control chart limits are calculated in real-time. The histogram is maintained in real-time. All of this with the additional benefits to capture traceability information at the point of data collection. In addition, many of today's SPC software vendors provide for real-time alarms for out of control or out of tolerance conditions. They have the ability to present the machine operator or inspector with a pick list of Assignable Causes of variation and corrective actions for each one.

Email notifications can be made to the maintenance department when a tool breaks and other process alert signals can even shut down an entire production line automatically. Pattern recognition on control charts can determine Western Electric data test failures to indicate the need for a process adjustment. Taking the real-time data collection one step further, we have seen a growing increase in the

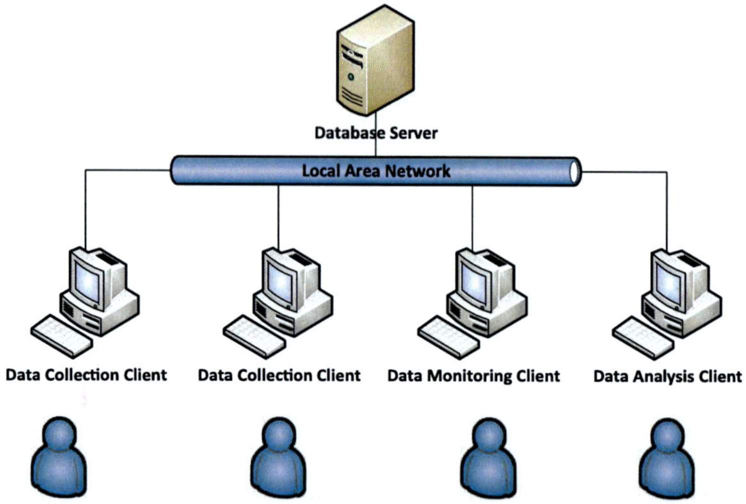

Fig. 6.7 Simple client server topology on Local Area Network (LAN)

number of integrated solutions that provide feedback to the machining process itself in order to establish better capability through adaptive tool wear compensation.

Network and Client Server applications as shown in Fig. 6.7 were the next major step in commercial quality control offerings. Moving beyond the standalone system, we saw the rise of Ethernet that provided the ability to connect many personal computers on a local area network. Combined with the power of Relational Databases, a new paradigm was realized: Enterprise Quality. Many systems today provide powerful data acquisition, management and analysis solutions. All data maintained in a centralized database offers significant advantages over the antiquated island approach to quality.

Database management systems, such as Oracle and Microsoft SQL Server, provide standard tools to the industry to build powerful, wide scale solutions. Combined with report writers such as Crystal Reports, we see much flexibility in how data can be represented for decision support systems. Relational databases provide the ability to easily store and retrieve large amounts of data in client/server based topologies allowing information sharing between the shop-floor and quality engineering offices. These databases are designed with table structures that have relationships in order to increase storage capacity through the reduction of duplicated information and provide excellent referential integrity and efficient query processing through the use of primary and foreign key relationships. Figure 6.8 shows a sample relational table structure with key relationships.

It can be seen that CollectSummaryInfo element consists a list of quality control related information such as CollectDataID, NumberofParts, etc. CollectDataID belongs to CollectedInfo element, which contains the collected data of a single measured workpiece. Therefore, when a number of workpieces are measured,

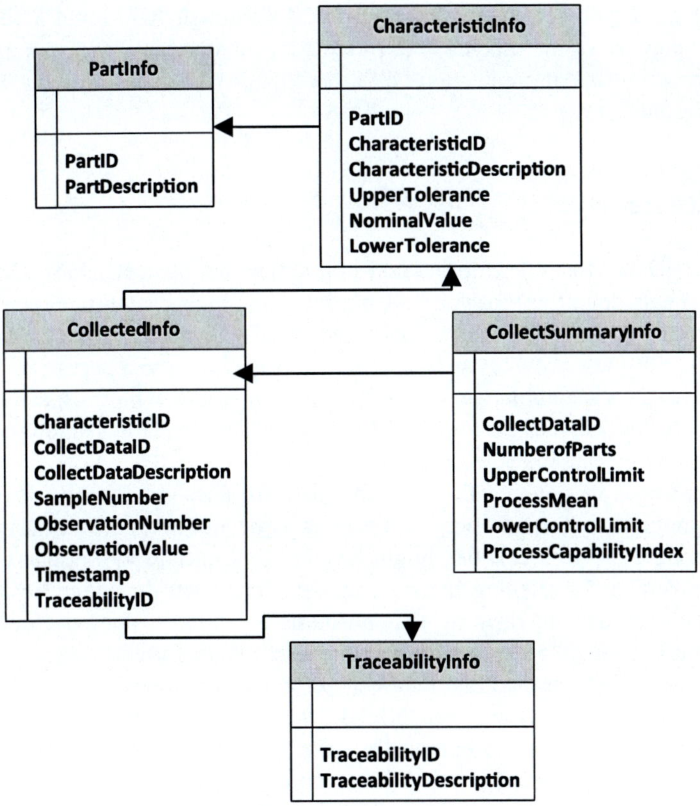

Fig. 6.8 Sample relational table structure with key relationships

CollectSummaryInfo element is able to store all the collected quality data and these quality data are traceable to each workpiece. If the characteristic information needs to be recalled, it (CharacteristicID) can be traced from the CharacteristicInfo element which is associated with workpiece information (PartID). This example is only one relational table structure. Each enterprise or industry department may have its own proprietary relational table structure. When the quality measurement data information from one department needs to be passed to another department, the interoperability issue occurs. Thus, a standardized data model for exchanging quality measurement data is desired.

6.3.2 Quality Measurement Data Model

This section deals with the formation of the QMD XML Schema [51], along with rules and associated conformance classes. This specification was created by several gage manufacturers, statistical data acquisition and analysis software vendors and Original Equipment Manufacturers (OEMs) through AIAG and MEPT. The intended purpose of the schema is to provide a data structure for the exchange of data between different applications that serve quality control efforts in the manufacturing industry.

6.3.2.1 Overview

The QMD Data Model describes a non-proprietary and open standard XML XSD (XML schema definition) for variable, attribute, and binary quality measurements. It is directed to anyone who is concerned with both quality measurement and interoperability and provides an inherently simple solution to a pervasive problem (i.e., the need for a common language for quality measurement, irrespective of the source or the intended target for that information). The standard is unidirectional—it defines the measurement export only.

Perhaps equally significant is what this standard does not describe. It does not offer a prescription for methods or processes associated with acquiring measurements, nor for the protocols that might be employed in transporting that data, nor any prescription for queuing or inserting data much less to define how people might use that data in their in core business. In short, it does not define any process. It defines only the quality measurement export format itself.

The XML XSD uses two familiar concepts in Quality Measurement. The first is that of Conformance Classes, which lend certain predictability and fixed format, with each class suited to progressively more complex reporting tasks, from the most basic to the most sophisticated. The second is the concept of Catalogs, which is familiar to anyone who has ever selected anything from a predefined pick list. The QMD implementation of catalogs is somewhat unique, however, because it also embraces the concept of Strong Parent elements. These elements make this standard completely extensible and offer the users completely customizable content within a common framework, with the flexibility to create any number of their own supplementary element names. These, in turn, point at any of the predefined (and human-readable) Catalog XML element names explicitly defined in the standard.

The standard also encompasses a Data Dictionary of elements as are assigned to the various Conformance Classes and Catalogs. Again, the standard does not limit itself to only these elements, but it does provide fixed reference definitions for the terms most commonly used in industry, along with use cases.

6.3.2.2 Business Case for a QMD Schema

The compelling need for creating a standard quality measurement data model should be clear to anyone that has suffered the pains associated with lack of application interoperability. As is the case with any business, it is the benefits that really speak to the business case for embracing such a standard information model.

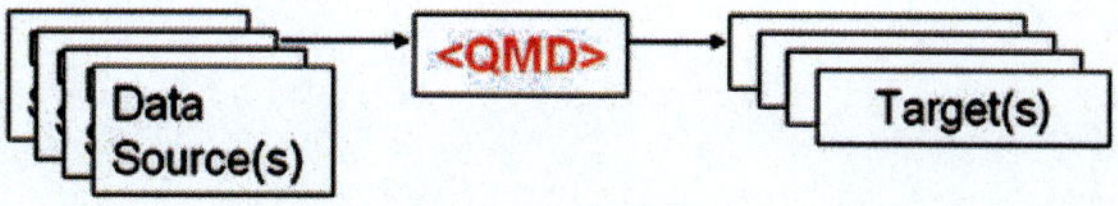

Fig. 6.9 Use case scenario of QMD

In a classic case of 20% of the effort yielding 80% of the benefit, it is the export format for quality measurement that presents the greatest business opportunity. The actual cost resulting from the lack of standardization is enormous and, in terms of the impediment it places on innovation, this lack of standardization is not even quantifiable. At the very least, it consumes a significant portion of IT resources in most enterprises. Here is a summary of the benefits of a standardized information data model:

- Eliminates wasted resources, money, and time in data integration tasks.
- Redirects these savings to value-added activities, enhancements, etc.
- Allows Solutions Providers and Gage manufacturers to redirect more energy to new development.
- Enables Gages to communicate with more reporting tools, making gages more useful.
- Enables Reporting tools to accept data from more sources, making reporting tools more useful.
- Permits customers to focus more on core business.
- Maps to virtually any legacy database schema.
- Uses standard identifiable tags, yet with provision for familiar user-defined names.
- Moves away from Gage dependencies and proprietary schemas that require separate technical support.

6.3.2.3 QMD Scope

The scope of the schema is limited to an XML quality measurement export format (XSD extensible schema definition) for variable, attribute, and binary quality measurements. Figure 6.9 illustrates this at a high level, with the QMD XML export standard situated between source and target.

To expand upon this simplistic view, the accompanying UML model shown in Fig. 6.10 illustrates one possible use case scenario for data acquisition, export, and reporting processes surrounding the creation of an XML quality measurement file. However, we emphasize that the full scope of the QMD standard actually begins and ends with defining the format for the XML file.

The QMD data model was developed for the following main use cases in industry:

1. Inspect part or process
2. Write file
3. Publish file
4. Read file

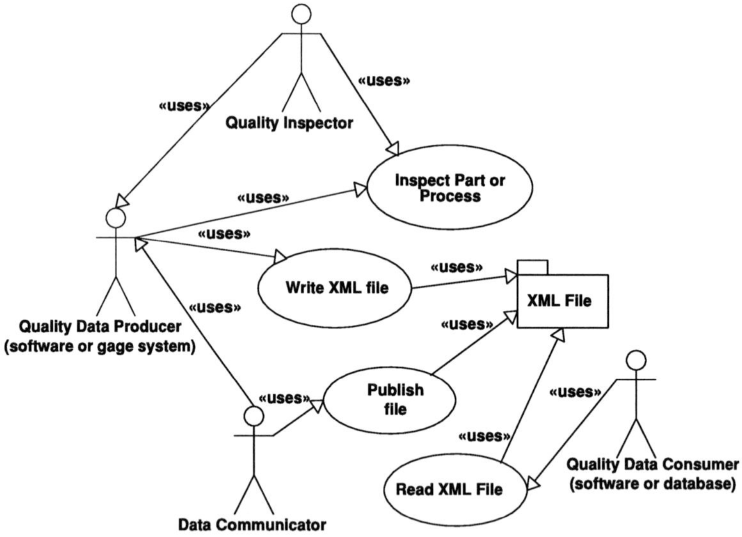

Fig. 6.10 UML model

For each use case, the specification defined related information including actors, description, pre-condition, post condition, begins when, scenario for main flow and sub flow, alternative flows, and ends when. The detailed QMD use case information can be found in Appendix D.

6.3.2.4 QMD XML Schema

The QMD XSD describes the architectural constructs for the XSD that define the set of QMD "Conformance Classes" and "Catalogs" and also provides some Implementation Guidelines for the integrator.

1. *Architectural Considerations*. XML was chosen for the standard because XML is itself completely database and schema non-specific. Therefore, it can be mapped to any legacy schema at any level as table header, body, or trailer information. XML also provides human-readable tags to facilitate understanding among integrators. Further, XML is increasingly used in industry and on the Web for structuring and transmitting data of all kinds. The W3C naming conventions for XML have been rigorously applied. The primitive types for all elements default to "String" to accommodate any case that might be encountered among countless legacy data sources, with recommended options for "Integer" and "Double" primitive types in cases when the values will most probably contain only values associated with those primitive types.

   The basic construct for every element in the XSD's includes a predefined XML "tag" together with an Element Name (for which users can provide their own descriptions) and a value. Consideration was also given to the application of so-called "Free Text" that appears in the Conformance Class XSD, but is also an optional attribute for every element in every catalog. This provides enormous flexibility in associating remarks to any element. Finally, it is important to note that the elements populating the various Conformance Classes and Catalogs are not mutually exclusive to either. Some elements belong to Conformance Classes, some to Catalogs, and some to both. In fact, every conformance Class element will also belong to a Catalog.

2. *Data Block Packaging in XML Payload.* The QMD XML specification and schema are designed to maximize interoperability between all data producers and data consumers. From this perspective, the data structure is designed so that each quality measurement is treated as a complete data block, containing all necessary information to extract meaningful quality data. This design lends itself well to "streaming" data flow. However, from a batch processing perspective (consider large data sets), this design has the potential to carry redundant data.

   The QMD XML specification does not support a "header/body/footer" approach but employs an "encapsulated data block" approach in order to ensure complete processing for the consuming application. Although this data model may result in larger XML payloads than a more "normalized" approach, it does offer certain additional simplicities and possibilities for implementation.
3. *XML Element Hierarchy and Use of Attributes.* The QMD XML specification uses a Parent–Child node approach for its data structure. Although the use of element attributes does provide a flatter XML structure with certain advantages, the use of an explicit hierarchy for all information maximizes flexibility of navigation through the data using currently available tool sets.
4. *Conformance Class Construction.* The QMD Conformance Class schema is comprised of five hierarchical Conformance Classes, each of which contains one or more elements that append to those already designated in the preceding classes. The intent of separating the classes is to provide increased levels of functionality with each successive class. Table 6.1 enumerates the Conformance Classes along with their intended level of functionality.

   Any data source said to be conformant with the QMD Conformance Class must be capable of exporting all of the elements that define that class. In this sense, all of the elements are mandatory for the conformance class, even if the values for some of them in the measurement export are null.
5. *Catalogue Construction.* Six "Catalogs" complement the Conformance Classes defined in this quality measurement specification. The following is the list of QMD catalogs:

   - Setup
   - Measurement
   - Traceability
   - Derived Value
   - Gage
   - MSA

Table 6.1 QMD conformance classes

Conformance class	Intended functionality
One	Basic Measurement: Measurement at its most rudimentary level, with facility to capture measurements from "dumb" gages. PLCs or simple lists of measurements.
Two	Basic Quality: Measurement with quality study capability. Includes all of the preceding elements along with some essential SPC elements for sub groupings, time stamps, etc.
Three	Quality with Traceability: SPC/Quality capability, with basic Traceability. Includes all of the above, plus some commonly used traceability elements. (Note that Traceability is considered to be related to environmental factors, discernible at the point of data acquisition. Expanded Traceability is provided for through the association of Traceability Catalog elements, to be covered later in this document)
Four	Advanced Quality: Basic Traceability as above with provision for Attribute and Binary data collection. (Note that Binary data is treated as distinct from Attribute data in the QMD dictionary because (like variable data) they are associated with a "right" or a "wrong" answer and thus processed differently in quality reporting.)
Five	Extended: Multi Lingual provisions and embedded Files. This Conformance Class contains elements typically associated with more sophisticated data collection and reporting activities.

Any of these catalogs can associate with any of the Conformance Classes, and each (catalog, class) includes at least one element.

6.3.2.5 QMD XSD Rules

This section describes the basic rules provided by the XML Schemas that are at the core of the QMD XML Standard.

1. *Basic Rule #1—Measurement Root Element.* "Measurements" is the root element, holding "n" numbers of measurements. It is helpful to highlight some of the basic tenets of the QMD XSD to integrators everywhere who are tasked with integrating data from their own schemas. The basic precept for QMD is that it provides a wrapper for quality measurements, nothing else. As such, the root of the XSD is aptly labeled "Measurements". All QMD XML files require the root node to be named "Measurements" as in the following example.

```
<?xml version = ''1.0'' encoding = ''UTF-8''?>
<Measurements xmlns = ''urn:aiag:meqm:qmd''>
  <MeasurementData uniqueID = ''1''>
   <ClassOne>
    <PartID>
     <StringValue>252</StringValue>
```

```
    </PartID>
     <CharacteristicID>
     <StringValue>Distance Between Hinges</StringValue>
    </CharacteristicID>
     <MeasuredValue>
     <StringValue>139.76</StringValue>
    </MeasuredValue>
      <DateTime>
      <StringValue>8/15/2009 2:24:44 PM</StringValue>
    </DateTime>
    </ClassOne>
   </MeasurementData>
 </Measurements>
```

2. *Basic Rule #2—Measurement Data Blocks.* Each discrete measurement is clearly bounded within MeasurementData blocks. One of the most attractive aspects of the QMD XSD is the ease with which one can view the measurements. In any QMD export, each measurement is separated from the next as shown in the above XML example. The above example has one MeasurementData block, which carries a uniqueID attribute whose value is a long primitive data type.
3. *Basic Rule #3—Conformance Classes.* Every measurement must comply with one of the conformance classes; Conformance classes are defined as ClassOne; ClassTwo; ClassThree; ClassFour and ClassFive. Each class extends the features of the previous class. ClassOne is the base conformance class. This means that a QMD XML file with a ClassTwo MeasurementData block implicitly carries ClassOne information; a ClassThree MeasurementData block implicitly carries ClassOne and ClassTwo information, and so on.

ClassOne Elements. ClassOne is the most basic data structure designed to carry measurement information. The following data are carried in a ClassOne QMD XML file:

- PartID
- CharacteristicID
- MeasuredValue
- DateTime

The following represents an example of a ClassOne MeasurementData block.

```
<MeasurementData uniqueID = ''1''>
    <ClassOne>
         <PartID>
           <StringValue>252</StringValue>
         </PartID>
             <CharacteristicID>
           <StringValue>Distance Between Hinges</StringValue>
         </CharacteristicID>
           <MeasuredValue>
```

```
      <StringValue>140.24</StringValue>
    </MeasuredValue>
      <DateTime>
      <StringValue>8/15/2009 3:36:32 PM</StringValue>
    </DateTime>
  </ClassOne>
</MeasurementData>
```

ClassTwo Elements. A ClassTwo MeasurementData block supplements a ClassOne data block by carrying additional information required for Basic Quality measurement information. The following additional information is carried in a ClassTwo QMD XML file:

- NominalValue
- LowerToleranceLimit
- UpperToleranceLimit
- SubgroupSize
- ObservationID
- SubgroupID
- SampleID

ClassThree Elements. A ClassThree MeasurementData block supplements a ClassTwo data block by carrying additional information required for Quality measurement with traceability information. The additional information carried in a ClassThree QMD XML file includes:

- ReasonForTest
- CharacteristicClass
- Unit
- FreeText
- Operator
- MachineID
- GageID

ClassFour Elements. A ClassFour MeasurementData block supplements a ClassThree data block by carrying additional information required for Advanced Quality measurement with traceability information. The additional information carried in a ClassFour QMD XML file includes:

- CharacteristicType
- Defect
- Event

ClassFive Elements. A ClassFive MeasurementData block supplements a ClassFour data block by carrying additional information that extends Advanced Quality measurement information. The additional information carried in a Class-Five QMD XML file includes:

- Language
- File Attachment

4. *QMD Implementation Guideline*. The QMD specification describes best practices for several scenarios in order to maximize consistency in implementation for producing QMD files. The purpose is to reduce the level of effort required for an integrator to develop an appropriate and uniform parsing of the QMD information for the consuming application. For example, all QMD files should have the *.xml extension in order to maximize interoperability. Although the XSD defaults to the use of the string primitive data type for all tag elements, it is recommended that the use of specific primitive data types be used where possible. It is highly recommended that integrators use the QMD "suggested" primitive data types for all element values. The QMD specification data dictionary section provides the list of recommended primitive data types.

6.3.3 *Dimensional Markup Language Data Model*

DML is an XML format definition tailored to the needs of dimensional results for discrete manufacturing. This specification was also developed under AIAG's MEPT team. Information defined in DML overlaps with some of the information defined in QMD. The purpose of DML is to haul the results between applications that generate or use dimensional information. A typical scenario is where an inspection device collects dimensional data and sends the information to an SPC package for process analysis or a database for long term storage. Compared with QMD, DML carries more information on dimensional measurement resource and devices, measurement cloud points, and raw data.

The DML data file contains three basic components: header information to identify the data, tolerance specification and feature specification. A simplified DML dataset is shown in the following:

```
<?xml version = ''1.0'' encoding = ''UTF-8'' ?>
<! -- DATE : Thr March 31 10:52:33 2001---->
<! -- UTC DATE: 2011-03-31T17:52:33Z---->
<dimensional_inspection_results version = ''1.04''
id = ''RUN1''>
    <results_header>
    <transform_list>
    <datum_definition_list>
    <tolerance_nominal_list>
    <feature_list>
      <feature_analysis_modes_default>
      <tolerance_analysis_modes_default
      <analysis_dofs_default>
```

```
<feature name = ''basic_hole_hole_center_lower_TPRH701''>
      <applied_tolerances>
      <cone_feature type = ''INNER''>
        <cone_feature_nominal>
        <cone_feature_actual>
      </cone_feature>
      <tolerance_actual_list>
      <point_list>
    </feature>
  </feature_list>
</dimensional_inspection_results>
```

The top-level element in the example is dimensional_inspection_results. There are five level-two elements in a DML file:

1. The *results_header* section is used to store implementation specific information like company name, location, etc. This is a required element for a DML processor to validate.
2. The *transform_list* is for storing transformations associated to reporting the data out in different coordinate systems. One of the fundamental rules of the DML is that all data is stored in a common coordinate system. It is the job of the reporting tool to display the information in the proper space. This is an optional element for a DML processor to validate.
3. The *datum_definition_list* is used to store the datum definitions used for the tolerances. In the tolerances either the datum definition or the feature ID can be used. This is an optional element for a DML processor to validate.
4. The *tolerance_nominal_list* is used to store the nominal information for the tolerances. The measured information is stored as part of the feature information. This is an optional element for a DML processor to validate.
5. The *feature_list* is the most important top level element because it stores the bulk of the inspection results. It includes the nominal, measured and raw data for the feature. As well as the measured instance of any tolerance applied to it. This is a required element for a DML processor to validate.

The feature_list element has five elements:

1. feature_analysis_modes_default (optional),
2. tolerance_analysis_modes_default (optional),
3. analysis_dofs_default (optional),
4. feature, and
5. scan_result_list (optional).

The feature element is required in a DML file. It stores the bulk of information and is subdivided into five elements:

1. feature_analysis_modes (optional). It allows a particular feature to be calculated using analysis modes that differ from the defaults.

2. applied_tolerances (optional). It associates tolerances with features.
3. feature specific element,
4. tolerance_actual_list (optional), and
5. point_list (optional). It stores raw data from measurements.

It is important to note that the features and tolerances defined in DML have not been fully validated and they overlap with those defined in DMIS and STEP AP 219. DML is the first standard effort in standardizing dimensional measurement result data. However, since the publication of DML Version 1, a number of problems have been found by industrial users. Further development and validation of DML is needed.

6.4 Commercial Application of Quality Data Analysis and Reporting

As mentioned in the above section, information generated from quality data analysis and reporting activity is widely used by many enterprise production planning and resourcing departments for quality control, statistical analysis, and process control. Having introduced the mathematical algorithms and information models used in quality data analysis and reporting systems, it is thus necessary to discuss how the quality data is used commercially in industry. This section introduces the two main areas of quality data applications. The first one is business intelligence, which is the newest trend in quality data analysis. The other area is quality and production engineering.

6.4.1 Business Intelligence

The most current trend in data analysis is BI or Business Intelligence. The term business intelligence (BI) refers to technologies, applications and practices for the collection, integration, analysis, and presentation of business information and also sometimes to the information itself. Its purpose is simple, improve business decision making by using new software applications and analyzing the organization of raw data. This is accomplished through the manipulation and presentation of key data by leveraging knowledge management and data mining capabilities leading to cost cuts and the identification of new business opportunities.

The evolution of BI began decades ago when the analysis of large volumes of data became more practical with the advent of the electronic computing age. Consider for example that in 1880 it took nine years for the United States to compile and analyze the population census. In 1945 the first general purpose computer, ENIAC (Electronic and Numerical Integrator and Computer) was designed to calculate ballistic firing tables for the United States Army. However, it would take until the 1980s for powerful computer server systems and personal computers to transform the way we process data. From punch cards to spreadsheets and on to data warehouses we now move into the next generation of information modeling.

With this new technology, information technology departments can create n-dimensional online analytical processing (OLAP) cubes of data (as shown in Fig. 6.11) that are in turn bound to web based dashboards that provide management with new and fast ways to analyze data.

An OLAP cube is a data structure that allows fast analysis of data. For example an OLAP cube should provide fast results—90% of queries back should be returned under 10 s and no query takes longer than 30 s. OLAP cubes must provide analysis capability with drill down, multiple aggregation techniques, and structures that support sophisticated graphics. An OLAP cube is multi-dimensional and may be considered an Excel pivot table on steroids by providing the ability to have any multiple dimensions of information on each axis of a cross-tab with other dimensions being used to further filter the results returned. Take dimensional metrology system for example; Fig. 6.12 depicts an OLAP cube consisting of measurement results data such as date, time, characteristic, traceability, part, etc.

The ability to slice and dice data among various measures or dimensions and drill down to meaningful data that is summarized from a fact table continues to take root in advanced data analysis environments. Consider what it would take to calculate the capability index (CpK) across all manufactured parts within a department on a weekly basis yet separated by shift, then be able to drill down to a specific machine to see if there is a correlating factor to some type of performance problem. The simple cube structure shown in Fig. 6.12 will be extremely useful.

Business Intelligence relies on the ability to perform data mining from a data warehouse. This means sorting through large amounts of data and picking out relevant information. It allows users to analyze data from many different dimensions or angles, categorize it, and summarize the relationships identified. The infrastructure of business intelligence is displayed in Fig. 6.13.

Within the discipline of BI a primary factor is the identification of Key Performance Indicators (KPI). KPIs are defined as the financial and non-financial metrics used to help an organization define and measure progress toward organizational goals. In order to develop KPIs they must be SMART.

- Specific
- Measurable
- Achievable
- Result-oriented or relevant
- Time-bound

KPIs can be categorized. For example, KPIs may be quantitative indicators which can be presented as a number, practical indicators that interface with existing company processes, directional indicators specifying whether an organization is getting better or actionable indicators that are sufficiently in an organization's control to effect change.

Fig. 6.11 Representation of an n-dimensional OLAP cube

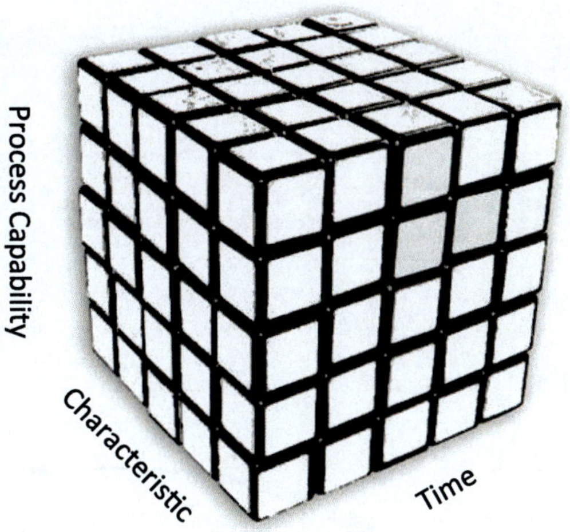

Fig. 6.12 Simple dimensional metrology cube structure

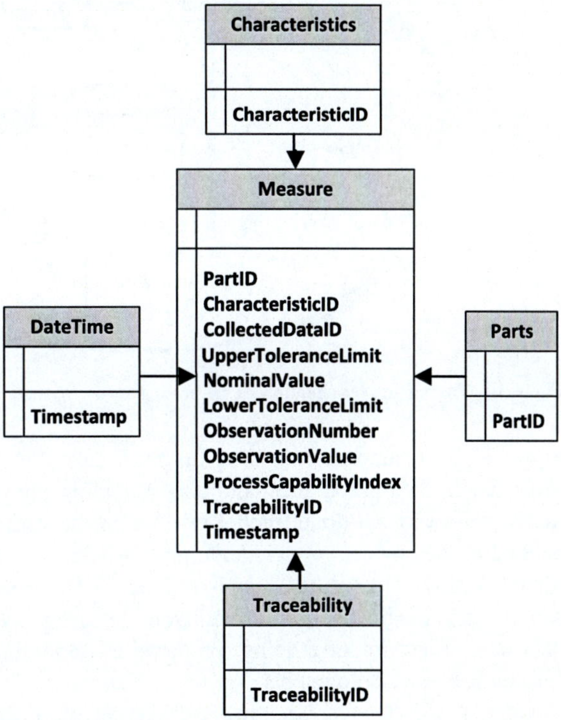

Once KPIs are identified for a BI system, the implementation may commence. Figure 6.14 displays the six phases of BI implementation. The first two phases are business and data understanding. They correlate with each other. Business understanding identifies the high-level information that needs to be included

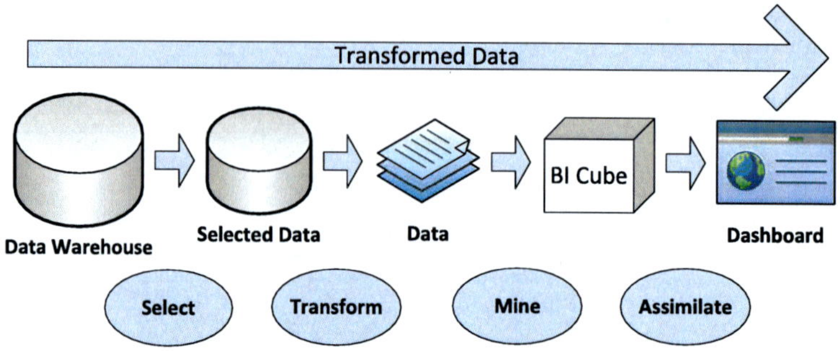

Fig. 6.13 Infrastructure of business intelligence

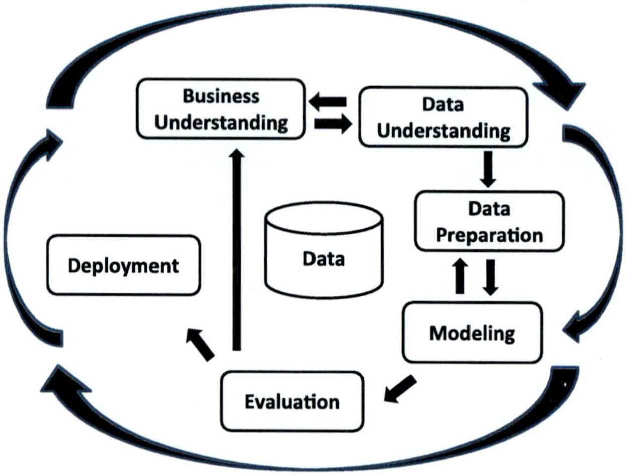

Fig. 6.14 Phased implementation of BI strategy

in the BI, and data understanding further breaks down the information at the data level. Then, data preparation and data modeling phases produce a data model for the BI followed with data model evaluation (according to the business use cases defined in the business understanding phase) and validation. The final stage is data model deployment for business intelligence. These six phases of implementation can also be categorized into three layers: business layer (business understanding), data layer (data understanding, preparation, modeling), and representation layer (evaluation and deployment).

Once the data layer and business layers of business intelligence have been established through cube creation with the fact and measure tables the presentation layer is displayed through development of dashboards. Based on the metaphor of the instrument panel in a car, the computer, or "digital" version of a dashboard provides a business manager with the input necessary to "drive" the business. Graphical elements such as red/amber/green lights, alerts, drill-downs, summaries, graphics such as bar charts, pie charts, bullet graphs, spark lines and gauges are usually set in a browser based portal environment that is often role-driven and customizable. A BI dashboard example is shown in Fig. 6.15.

6.4.2 Quality and Production Engineering

Compared with business intelligence, the more traditional applications of quality data are within the quality and production engineering discipline. This section briefly introduces the three major applications of quality data: first article inspection (FAI), production part approval process, and statistical process control.

6.4.2.1 First Article Inspection

FAI is one of the primary methods for the inspection and testing of vendor components. The testing of a pre-production sample is considered essential in the process of approving an order or contract; the FAI should determine if the product meets acceptance requirements and quality control requirements.

The purpose of the FAI is to give objective evidence that all engineering, design and specification requirements are correctly understood, accounted for, verified, and recorded. FAI is able to provide a consistent documentation requirement for aerospace components FAI. In general this is the aerospace equivalent of the automotive Production Part Approval Process (PPAP). PPAP requires a larger quantity of components than would be typically manufactured for aircraft components.

With the use of modern computers in the manufacturing environment, FAIs are no longer being used with the traditional three form layout on paper but instead recorded digitally and stored on servers for easy access and organization. Recording the first article digitally eliminates errors with the help of software that keeps tracks of the FAIs and generates reports immediately after successful completion of an FAI.

6.4.2.2 PPAP

The goal and deliverable of AIAG's Production Part Approval Process (PPAP) is a series of documents gathered in one specific location (a binder or electronically) called the "PPAP Package" and ultimately provided downstream to the product consumer in the automotive industry supply chain. The PPAP package contains information which needs a formal approval by both the supplier and customer. The forms that comprise a PPAP exercise are summarized in a package called the Part Submission Warrant (PSW). The approval of the PSW indicates that the supplier responsible person (usually the Quality Engineer or Manager) has reviewed this

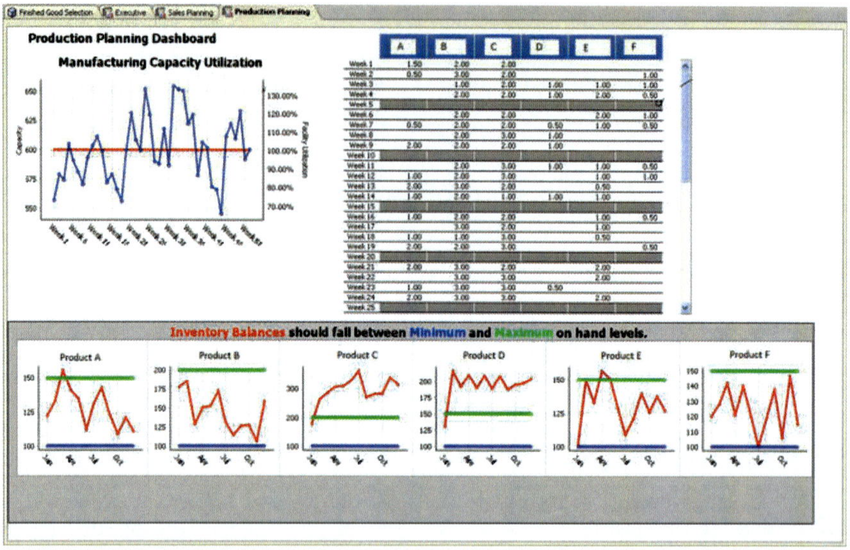

Fig. 6.15 Sample dashboard

package and that the customer has not identified any issues that would indicate the potential for supplier quality problems.

The documentation on the PPAP package is very closely related to the AIAG's Advanced Product Quality Planning process (APQP) that is used during the design and development of new vehicles and component systems to reduce the risk of unexpected failure due to errors in design and/or manufacturing. The PPAP manual is published by AIAG, and specifies generic requirements for obtaining PPAP approvals. Additional customer specific requirements may be imposed by particular customers (e.g. vehicle manufacturers) depending on the need for other types of conditional constraints and are typically incorporated in the purchasing contracts.

Suppliers are required to obtain PPAP approval from the vehicle manufacturers whenever a new or modified component is introduced to production, or when the manufacturing process has changed. Obtaining approval requires the supplier to provide sample parts and documentary evidence representing that:

1. The customer requirements have been understood.
2. The product supplied meets those requirements.
3. The process (including any external supply dependency) is capable of producing product that conforms to specification.
4. The production control plan and the supporting quality management system have the ability to prevent any non-conforming product to be received by the customer and compromising the safety, reliability and performance of final product.

PPAP may be required for all components and materials incorporated in a finished product. It may also be required if components are processed by external sub-contractors. Below is the list the PPAP elements along with a brief description of each:

1. *Design Records.* A copy of the part drawing. If the customer is design responsible this is a copy of the customer drawing that is sent together with the Purchase Order. If the supplier is design responsible this is a released drawing in the supplier's release system.
2. *Authorized Engineering Change Documents.* A document that shows the detailed description of the change. Usually this document is called the "Engineering Change Notice" or ECN, but it may be covered through the customer purchase order, or any other engineering authorization.
3. *Engineering Approval.* This approval is usually contains information from engineering trials during the manufacturing pilot phase with production parts. A "temporary deviation" may be required where parts are sent to customer before PPAP.
4. *Design failure mode and effect analysis (DFMEA).* These procedures are reviewed and signed-off on by both parties. In certain situations the customer may not share this document with the supplier. However, the list of all critical product key characteristics is shared with the supplier, so they can be addressed within the quality control plan.
5. *Process Flow Diagram.* This is the design that indicates all steps and operations in the manufacturing process, including incoming components and sub assembles.
6. *Process Failure Mode and Effect Analysis (PFMEA).* This procedures are reviewed and signed-off on by both supplier and customer. The PFMEA follows the steps in a workflow, and identifies "what could go wrong" during manufacturing and assembly of each component workpiece.
7. *Control Plan.* Control plan procedures are reviewed and must be signed-off by both supplier and customer. The Control Plan follows the PFMEA steps, and provides additional details as to how "potential issues" are checked for incoming quality, the manufacturing or assembly process, or during inspections of finished goods.
8. *Measurement System Analysis Studies (MSA).* This usually contains the Gage R&R for the metrology process for inspecting key characteristics, and may necessitate confirmation that gauges used to measure these characteristics are calibrated.
9. *Dimensional Results.* A list of characteristics noted on the ballooned or annotated drawing. This list shows the product design, specification, and measurement results along with the assessment displaying whether characteristics are "Go" or "No Go". Sample size may depend on contract criteria.
10. *Records of Materials / Performance Tests.* This is a summary of every test performed on the part. This summary is communicated as a Design Verification Plan and Report (DVP&R), which can list each individual quality test,

when it was performed, the nominal specification, results and the determination of pass/fail status.

11. *Initial Process Studies*. Statistical process control charts representing the stability and quality control of the manufacturing process of the key characteristics. The intent of this information to demonstrate that critical processes have nominal variation.
12. *Qualified Laboratory Documentation*. All laboratory certifications (e.g. A2LA) of the inspection facilities that conducted any quality measurement or testing.
13. *Appearance Approval Report*. The appearance approval inspection (AAI) report that is applicable to certain types of product
14. *Sample Production Parts*. Samples must be provided from actual manufacturing production runs. The PPAP package will typically show a picture of the samples and determine where they are to be kept.
15. *Master Sample*. A sample that is signed off by customer and supplier. This "golden" artifact may be used to train operators or to calibrate of equipment.
16. *Checking Aids*. When special inspection tools are required, information must be provided including calibration records and other types of dimensional reports that may be specific to the tool.
17. *Customer Specific Requirements*. It is most often the case that the Customer Specific Requirements (CSR) will belong to a PPAP package.
18. *Part Submission Warrant (PSW)*. This form summarizes the entire PPAP package. This form details the reason for submission (new production, engineering design change, process alteration, annual validation, etc.) and describes the level of information that is submitted to the customer.

6.4.2.3 SPC, MSA and Other Statistical Software

There are a host of vendors that provide quality engineering professionals with the tools in trade for statistical and graphical analysis of quality data. From basic SPC and MSA to advanced Design of Experiments Techniques (DOE) the following list provides some examples of the type of statistical and graphical treatments available in the market today, the details of which are beyond the scope of this book.

- Basic statistics
  - Time series plots, scatterplots, grid display, column indicators, point plots, histograms, line charts, etc.
- Regression analysis
  - Linear regression
  - Correlation
  - Confidence and prediction intervals

- Analysis of variance
  - ANOVA
  - Nested designs
  - Mean analysis
  - Correlation studies
  - Degree of Freedom (DoF) analysis
- Design of experiments
  - Factorial design
  - Taguchi methods
- Statistical process control
  - Run chart
  - Pre-Control chart
  - Standard deviation
  - Variables control charts: XBar/R, XBar/S, Individual and Moving Range
  - Time-weighted control charts: CUSUM, EWMA
  - Attributes control charts: P, NP, C, U
  - Pareto chart
  - Multivariate control charts
  - Data tests for assignable causes of variation
  - Non-normal data transformation
  - Distribution curves
  - Process capability (Cpk) and Process performance (Ppk)
  - Acceptance sampling and OC curves
  - Tolerance budgets and risk
- Measurement systems analysis
  - ANOVA
  - Control Charts
  - Part-by-Appraiser plots
  - Gage linearity studies
  - Gage bias studies
  - Gage stability studies
  - Attribute Gage study

6.5 Summary

The quality data analysis and reporting activity is an important element of dimensional metrology. The functionality of this activity is to receive input from measurement process execution and product definition activities, to analyze the part measurement data in terms of production definition requirements, to perform a statistical analysis of the measurement results and present them in the form of a statistical process control report, and to archive whatever measurement values and derived statistics are necessary.

This chapter first introduced computational metrology which is the process of fitting measured data points into substitute geometry and surfaces, and then assessing the results according to design requirements. This process consists of data fitting and filtering. Fitting is the task of associating ideal geometric forms to non-ideal forms. It is used for datum establishment and deviation assessment. Filtering is the task of obtaining scale-dependent information from measured data. Before the computational metrology process starts, correct mathematical models of geometric elements such as point, line, plane, etc. must be used.

A number of research studies have been conducted investigating data fitting algorithms. This chapter has given a thorough review of data fitting criteria including least-sum-of-distances fitting, total-least-squares fitting, two-sided minimax fitting, one-sided minimax fitting, smallest circumscribed fitting, and largest inscribed fitting. Most of the foundational research in data fitting theories is given in the review. These data fitting algorithms are mostly encapsulated in commercial quality data analysis software systems in today's industry. The output of quality data analysis is mostly used for quality control in almost every enterprise department varying from manufacturing process to production planning and resourcing.

In response to this essential need for quality measurement, the typical reaction has long been the creation of local quality measurement structures tailored to specific needs of the users. Many different proprietary vocabularies have evolved to describe essentially universal quality measurement concepts. This has, in turn, effectively prevented the accurate and non-ambiguous flow of quality measurement information from each newly introduced data source. Nowhere does the burden of translation and integration weigh more heavily than on the software and gage providers. Therefore, the need for a standard quality data analysis and reporting data format has emerged. Detailed discussions of available standard data models are given including QMD, DMIS output data, and DML. QMD and DML specifications use XML schema for data modeling. The information defined in these data models is overlapping. They also overlap with STEP AP 219 standards. In the next chapter, a horizontal analysis of all previously introduced data models will be provided.

As quality consciousness improves for manufacturers and data management and analysis technology innovations continue to be developed, quality data is no longer being considered useful for only the quality and production engineering disciplines. With the development of database and data mining technology, quality data is used also for business intelligence. This chapter also offered some discussion of these commercial applications of quality data. With the rapid technology development, industry will enjoy greater benefits in quality production and throughput through improved processing time, accuracy in calculations, and intelligent decision making systems.

References

1. Choi W, Kurfess TR (1999) Dimensional measurement data analysis, Part 1: a zone fitting algorithm. J Manuf Sci Eng Trans ASME 121(2):238–245
2. Diaz C, Hopp Th H (1993) Testing of coordinate measuring system software. In: Proceedings of the American society for quality control measurement quality conference, USA
3. Hopp TH (1993) Computational metrology. Manuf rev 6(4):295–304
4. Taylor BN, Kuyatt CE (1994) Guidlines for evaluating and expressing the uncertainty of NIST measurement results. NIST Technical Note 1297
5. ASME (1985) ANSI/ASME standard PTC 19.1–1985, measurement uncertainty. American Society of Mechanical Engineers, New York
6. Wäldele F et al (1993) Testing of coordinate measuring machine software. Precis Eng 15(2):121–123
7. Busch K, Wäldele F (1991) Testing coordinate measuring machine algorithms–phase II: inquiry on software for coordinate metrology. BCR Report EUR 13418 EN
8. Porta C, Wäldele F (1986) Testing of three coordinate measuring machine evaluation algorithms. BCR Report EUR 10909 EN
9. Cox MG (1992) Improving CMM software quality. NPL Technical Report DITC 194-92
10. Oakland JS (2007) Statistical process control. Butterworth-Heinemann, Oxford
11. Srinivasan V (ed) (2005) Elements of computational metrology. Geometric and algorithmic aspects of computer-aided design and manufacturing: DIMACS Workshop Computer Aided Design and Manufacturing, Oct 7–9, 2003, Piscataway, New Jersey, Janardan R, Smid M, Dutta D. American Math Soc, New Jersey
12. Srinivasan V (1991) A geometer grapples with tolerancing standards. In: Proceedings of the CIRP international working seminar on computer aided tolerancing. Pennsylvania State University, Pennsylvania
13. Srinivasan V (1996) How tall is the pyramid of Cheops?... and other problems in computational metrology. SIAM News 29(3):8–17
14. ISO (1995) ISO/IEC guide 98:1995—guide to the expression of uncertainty in measurement (GUM)
15. Srinivasan V (2007) Computational metrology for the design and manufacture of product geometry: a classification and synthesis. J Comput Inf Sci Eng 7(1):3–9
16. Srinivasan V (2008) Standardizing the specification, verification, and exchange of product geometry: research, status and trends. Comput Aided Des 40(7):738–749
17. Janardan R, Smid M, Dutta D (eds) (2005) Geometric and algorithmic aspects of computer-aided design and manufacturing: DIMACS workshop computer aided design and manufacturing, Oct 7–9, 2003. AMS Bookstore, Piscataway
18. BS (1989) British Standard 7172—guide to assessment of position, size, and departure from nominal form of geometric features, the minimum number of points
19. Feng SC, Hopp TH (1991) A review of current geometric tolerancing theories and inspection data analysis algorithms. National Institute of Standards and Technology (NISTIR 4509), Gaithersburg
20. Forbes AB (1989) Least-square best-fit geometric elements. National Physical Laboratory, Teddington, DITC 140/89 Teddington
21. Etesami F, Qiao H (1990) Analysis of two-dimensional measurement data for automated inspection. J Manuf Syst 9(1):21–34
22. Hillyard RC, Braid IC (1978) Analysis of dimensions and tolerances in computer-aided mechanical design. Comput Aided Des 10(3):161–166
23. Requicha AAG (1983) Toward a theory of geometric tolerancing. Int J Robotics Res 2(4): 45–60
24. Weill R et al (1988) Tolerancing for function. CIRP Ann Manuf Technol 37(2):603–610
25. ANSI (1994) ASME Y14.5M-1994: dimensioning and tolerancing

26. Lin VC, Gossard DC, Light RA (1981) Variational geometry in computer-aided design. Comput Graph (ACM) 15(3):171–177
27. Rossignac JR, Requicha AAG (1986) Offsetting operations in solid modelling. Comput Aided Geom Des 3(2):129–148
28. Yu YC, Liu CR, Kashyap RL (1986) Variational solid model for mechanical parts. In: Proceedings of the Winter Annual Meeting of the American Society of Mechanical Engineers. vol 21. Anaheim, pp 237–245
29. Rogers DF, Fog NR (1989) Constrained B-spline curve and surface fitting. Comput Aided Des 21(10):641–648
30. Etesami F (1988) Tolerance verification through manufactured part modeling. J Manuf Syst 7(3):223–232
31. Hoffmann P (1982) Analysis of tolerances and process inaccuracies in discrete part manufacturing. Comput Aided Des 14(2):83–88
32. Murthy TSR, Abdin SZ (1980) Minimum zone evaluation of surfaces. Int J Mach Tool Des Res 20(2):123–136
33. Traband MT et al (1989) Evaluation of straightness and flatness tolerances using the minimum zone. Manuf Rev 2(3):189–195
34. Cavalier TM, Joshi S (1988) Minimum zone of a set of points. IMSE Working Paper, Pennsylvania State University
35. Preparata FP, Shamos MI (1991) Computational geometry: an introduction. Springer, New York
36. Press WH (2007) Numerical recipes: the art of scientific computing, 3rd edn. Cambridge University Press, Cambridge
37. Lee DT, Drysdale Iii RL (1981) Generalization of Voronoi diagrams in the plane. SIAM J Comput 10(1):269–271
38. Wang ZX, Pang YJ (1996) Generalization of Voronoi diagram by plane seep. J Comput Aided Des Comput Graph 8:114–119
39. Lee DT (1982) Medial axis transformation of a planar shape. IEEE Trans Pattern Anal Mach Intell PAMI-4(4):363–369
40. Ramanathan M, Gurumoorthy B (2005) Constructing medial axis transform of extruded and revolved 3D objects with free-form boundaries. Comput Aided Des 37(13):1370–1387
41. Sherbrooke EC, Patrikalakis NM, Brisson E (1996) An algorithm for the medial axis transform of 3D polyhedral solids. IEEE Trans Vis Comput Graph 2(1):44–61
42. Lai K, Wang J (1988) A computational geometry approach to geometric tolerancing. 16th North American manufacturing research conference. University of Illinois at Urbana-Champaign, Urbana, pp 376–379
43. Chetwynd DG (1985) Applications of linear programming to engineering metrology. Proc Inst Mech Eng Part B Manag Eng Manuf, 199(B2):93–100
44. Kakino Y, Kitazawa J (1978) In situ measurement of cylindricity. Gen Assembly CIRP 28th Manuf Technol 27(1):371–375
45. Murthy TSR (1982) A comparison of different algorithms for cylindricity evaluation. Int J Mach Tool Des Res 22(4):283–292
46. Murthy TSR (1986) A comparison of different algorithms for circularity evaluation. Precis Eng 8(1):19–23
47. Zhang XD et al (2005) Unified functional tolerancing approach for precision cylindrical components. Int J Prod Res 43(1):25–47
48. Carr K, Ferreira P (1995) Verification of form tolerances part II: cylindricity and straightness of a median line. Precis Eng 17(2):144–156
49. Weber T et al (2002) A unified approach to form error evaluation. Precis Eng 26(3):269–278
50. Choi W, Kurfess TR (1999) Dimensional measurement data analysis, Part 2: minimum zone evaluation. J Manuf Sci Eng Trans ASME 121(2):246–250
51. Quality Measurement Data (QMD) (2009) http://www.aiag.org
52. Walker R (1988) GIDEP Alert No. X1-A-88-01. Government-Industry Data Exchange Program, Washington

Chapter 7
Dimensional Metrology Interoperability Issues

Previous chapters have introduced information modeling techniques, dimensional metrology systems and their elements, as well as proprietary and standard data models for each type of dimensional metrology activity. We have addressed the importance of having interoperability within each element of a dimensional metrology system. From an enterprise perspective, the lack of information interoperability becomes a more impeding issue in the competing global market causing the loss of product information integrity in addition to the increase of time and cost in product development. The US automotive industry reported that costs due to translation of measurement data between manufacturing quality systems amounted to over $600 million annually. When considering the cost of data translation between measurement planning and execution systems across the entire manufacturing processes, the number could be easily over billions.

This chapter will provide an in depth discussion on interoperability issues across a dimensional metrology system. The standard data models developed in isolation for each element of dimensional metrology systems (introduced in previous chapters) will be analyzed and compared in detail to produce a roadmap for harmonization of standards efforts in achieving interoperability in dimensional metrology industry.

7.1 Interoperability and Manufacturing Cost

American manufacturers had never had to struggle more so than they do now to remain globally competitive in the cost driven market brought about by the current economic environment. Faced with the disadvantage of having higher labor and energy costs than global competitors, manufacturers are tasked to develop new and creative ways to reduce manufacturing costs while maintaining the quality that defines American manufacturing. At present, no national or international standard exists to provide for the interoperable exchange of data between the various data

Y. Zhao et al., *Information Modeling for Interoperable Dimensional Metrology*,
DOI: 10.1007/978-1-4471-2167-1_7, © Springer-Verlag London Limited 2011

producers and consumers within dimensional metrology systems or manufacturing quality systems (in manufacturing systems dimensional metrology is also commonly known as manufacturing quality systems). Past attempts to address the issue have been driven by piecemeal collaboration efforts comprised of concerned individuals and organizations which has culminated in a plethora of proprietary and discordant specifications. As such, quality system implementers and integrators are tasked with vast and costly data mapping efforts to align their internal data structures to those of each producer and consumer of quality data within their total quality system.

Integrators of manufacturing quality systems are tasked with implementing custom, one-off mapping solutions to port data from each data producer to proprietary storage systems, and again from this proprietary storage system to each quality data consumer. This mapping process has to be replicated for each and every component of the manufacturing quality system. The task of mapping data is non-value added and has high overhead costs that can potentially compromise data integrity.

The following examples outline the need for a common standard by which various component producers and consumers of manufacturing quality systems can communicate data. Each scenario in the following examples depicts a quality system that relies on costly, one-off proprietary custom mapping solutions to achieve interoperability between components of the quality system.

1. A CAD vendor supports GD&T as PMI. A manufacturer's Design Engineering department has specified all information in the CAD model that is sufficient to produce a high-level measurement plan for quality conformance. The manufacturer's quality control department is required to create a CMM part program within a CAIPP system to inspect the manufactured product for conformance to specification. In this scenario the translational mapping is required to port data from PMI to CAIPP systems and other systems within a manufacturer's quality model.
2. A manufacturer has configured all critical variables and attributes to check for each operation in a multistage process within its manufacturing execution software (i.e., CAM software). The plans are issued to the shop-floor workstations through a local area network. The company will use multiple Statistical Quality Control (SQC) systems to acquire the data. Multiple SPC software packages are used from various vendors. In the industry, each SPC package requires a unique translation from each set of SQC data to interface with each SPC package's proprietary data format.
3. A large multi-national corporation has developed its global quality control plans within a centralized Enterprise Resource Planning (ERP) system. Each manufacturing plant has chosen a different Materials Resource Planning (MRP) system and uses it to collect quality control information. To relate data between MRP and ERP systems, the company will have to develop a proprietary system to port the data.

4. A small machine shop uses CAM to produce parts. In addition to programming the cutting paths, they also use the CAM system to perform on machine inspections. As part of their validation process they are required to do Repeatability and Reproducibility (R&R) studies and use off-the-shelf Measurement Systems Analysis (MSA) software system to acquire data and calculate the study results. A custom mapping tool will have to be used to translate data from the CAM system to use in R&R studies and for the MSA software.
5. A job shop manufactures simple parts and receives blueprint drawings from its customer for small lot production runs. They have identified a software vendor that can scan the blueprint and interpret the quality requirements through Optical Character Recognition (OCR) technology. The quality control department is required to perform capability studies on its manufacturing process and uses digital data collection techniques with SPC software. A custom translator is required to port data from the blueprint scanning software to the SPC package.
6. A global manufacturer has multiple production facilities that manufacture the same part. Each plant has developed its own statistical process control programs for quality control. In addition each plant uses a different SPC software vendor. Each facility must create a separate model to translate collected quality data to each SPC package.

The examples above illustrate scenarios where there is a lack of interoperability between common manufacturing quality system components. The different data definition and formats required by each component forces implementers to undertake redundant data harmonization efforts in order to implement the disparate standards within a single manufacturing quality system. In addition to the added costs associated with one-off harmonization efforts, there exists the potential for data integrity to be compromised in the many translations that must be performed to achieve interoperability between components.

7.2 Information Exchange Between Dimensional Metrology Systems

The four pillars of a dimensional metrology system, namely product definition, process planning, process execution, results analysis and reporting, have been introduced in previous chapters as well as their standard data models. Readers who have reached this chapter may notice that these standard data models overlap with each other. For example, design GD&T information is represented in more than three APs in STEP standards alone. Also, measurement features are defined differently in DMIS and STEP AP 219. The main reason that this situation exists is because past quality standards and specifications have been developed in isolation, each targeting a single dimension of a quality system. In previous chapters, we discussed what information is defined in each standard data model. In this section, we provide a systematic comparison of these data models to answer the following questions:

- What information should be defined for each element of the dimensional metrology system,
- What information has been defined, and
- Are the definitions complete?

By answering these questions, we would be able to assess the current status of dimensional metrology interoperability and foresee future research and industrial development activities. Also, we would be able to convey the benefits of using open and non-proprietary data models for the exchange of dimensional metrology data for manufacturing quality and process control.

7.2.1 Product Definition

To support automatic dimensional metrology plan generation using the simplest case, a product consists of a single monolithic part can be selected as an example. Figure 7.1 depicts some key functions that occur during the early stages of the part definition activity. The part must be decomposed into geometric features. Dimensions and tolerances must then be assigned to a geometric feature, or set of features. Datum features must be defined in such a way that they are appropriate both for manufacturing the part and for inspecting it. It is not uncommon that datum features are not the same for manufacturing and for inspecting purposes. Surface texture information must be included in the model, along with relevant information about the orientation or lay of the surface texture to be measured. Such information is typically referred to as Product Manufacturing Information (PMI). Accurately extracting PMI information requires interaction with the manufacturing process plan, which in turn defines the process used to create the surface that is to be measured. Therefore a process definition that defines the manufacturing and measuring process must be interconnected with elements within the product definition. Furthermore, the process requires resources (sensors, fixtures, machines), and therefore a resource definition that supports the process definition must be represented [1]. This ideal situation, however, does not exist in today's industry.

Currently PMI information is available in proprietary software to only a limited extent. There is no CAD product implementation of PMI information using non-proprietary standards. STEP AP 203 edition 2 consists of PMI information but has not yet been fully adopted by CAD vendors. Also, once AP 203 edition 2 is successfully implemented by CAD vendors, the implementation needs to be validated by standards organizations so as to ensure the accuracy.

The current common business model for CAD vendors is to define a closed and proprietary interface, where the process planning vendors (ultimately the users) must pay for access to select portions of geometry + PMI information (most of the time) through an API interface, which may or may not be saved to files. Also, it is

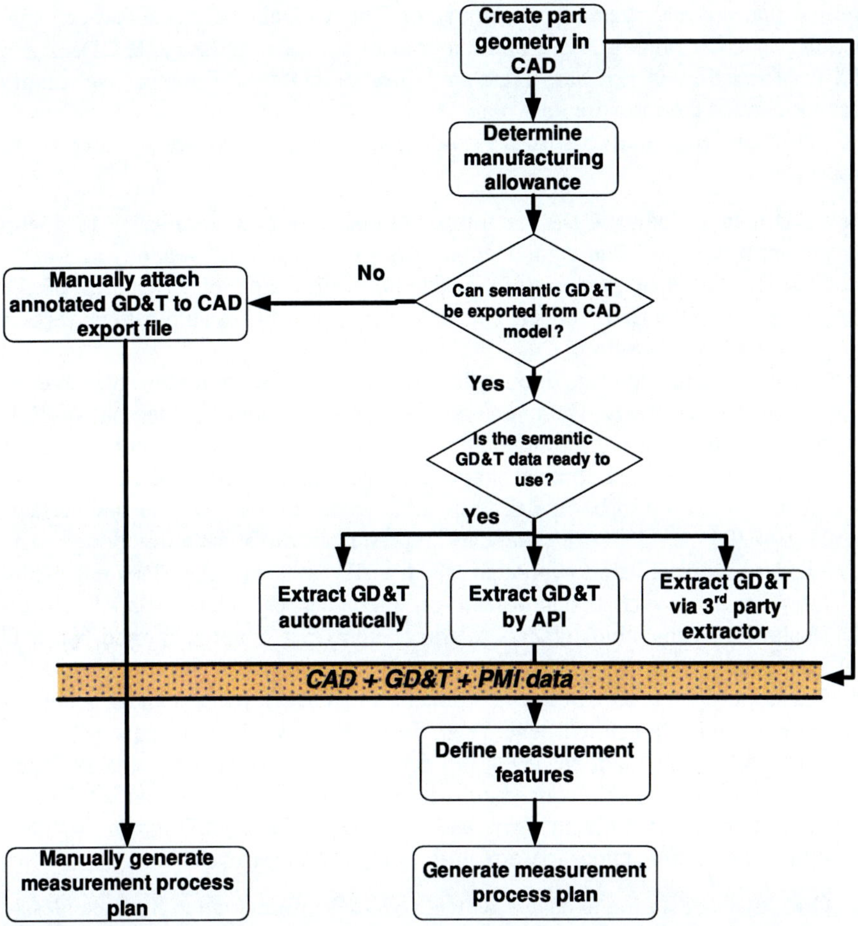

Fig. 7.1 Status quo in generating product definition information for dimensional metrology

very common for end users to require suppliers to read and write design data in native file formats. This type of proprietary file format varies from end user to end user. This may allow each individual end user to create the appearance of interoperability, but interoperability costs are in fact merely passed onto their suppliers, who must support multiple proprietary file formats required by the various end users they support.

From Fig. 7.1, it can be seen that using third party software to extract CAD + GD&T + PMI information is one of the means for process planning vendors to receive product definition information. Many small-to-medium industries depend on this method to abstract GD&T and PMI information to manufacture workpieces designed using systems from different CAD vendors. This method relies on software to translate proprietary design information into the format that process planning vendors want. The accuracy of this software translation is another pressing issue and major error source in industry. If GD&T and PMI information is not associated with design features/geometry, we simply cannot control the measurement plan.

The main issues that exist in the product definition activity are summarised as following:

1. CAD data including GD&T information does not flow seamlessly to downstream processes when components are not from the same vendor.
2. GD&T data is not associated semantically with individual features of the workpiece in the CAD model. This makes it impossible to automate inspection process plan generation.
3. Product Manufacturing Information (PMI) is only limitedly available in proprietary software. There are no CAD product implementations of PMI information using non-proprietary standards. PMI includes elements such as GD&T, surface finish, optical properties, and material properties.
4. GD&T data need to be modelled in CAD data, not just given as annotations. ISO 10303 AP 203 [2] (boundary representation) is the only design data standard representation supported by all CAD systems but it does not model tolerance items such as datum features, tolerances, etc. ISO 10303 AP 224 [3] (feature representation) models tolerance items but is not supported by CAD systems.
5. There are divergences in the interpretation of GD&T paper standards both at national and international level (e.g., ASME Y14.5 and equivalent ISO standard). At the international level, different national GD&T standards exist and they are not completely convergent with each other. At the national level, some major companies differ in their interpretation of the GD&T standards. Interoperability suffers under these realities, but is not destroyed.

One possible solution to enable semantic GD&T information to flow seamlessly to downstream processes is to realize Application Program Interface based (API-based) design-to-process planning. Consider Boeing as an example, who gave away their design kernel software for Advanced Integrated Mathematical Systems (AIMS) to establish the API specification for all Boeing product suppliers in the mathematical representation of design surface and GD&T models. Another major company Honeywell FM&T developed an API system called the Feature-based Tolerancing (FBTol) to assist automated inspection process plan generation for the manufacturing sector in the Department of Energy, USA. This kind of effort can only be accomplished by a handful of major manufacturing industry players. For the rest of the industry, a non-proprietary complete standard data model representing semantic GD&T and design data is indispensable. However, standards efforts are struggling to receive support from key CAD vendors in developing such a data model.

Currently, one of the major standards efforts is to develop a new version of ISO 10303 AP 203 that models tolerance items. The most recent test was carried out by some major CAD vendors to test the annotation GD&T information modelled in

AP 203 edition 2 [4]. The GD&T definition from AP 214 [5] (Core data for automotive mechanical design process) was harmonized into AP 203 edition 2. These GD&T definitions are mainly for annotation purposes; therefore they are not sufficient for automatic generation of dimensional measurement process plans. Further harmonization of GD&T definitions between AP 214 and AP 224 is necessary and the harmonized definitions should be eventually adopted into AP 203. Only in this way will AP 203 be able to provide adequate information for generating measurement process plans.

7.2.2 CAIPP Systems

The generation of measurement process plans is closely related to machining process planning regardless of whether the measurement is carried out in-process of machining or post-process. As shown in Fig. 7.2, process planning for both machining and inspection can be generally divided into macro planning and micro planning. In the macro planning, the choices of machine tools and assigned manufacturing tolerances affect measurement process plan decisions such as when to measure and what to measure, measurement uncertainties, etc. Then, in the micro process planning, detailed machine tool commands, inspection commands, motion commands, reporting and analysis commands are generated and passed onto a vast diversity of measurement equipment.

Chapter 3 introduced high-level measurement process plan generation with a focus on post-process measurement operations. This kind of measurement process planning takes the finished part shapes generated from a CAD system. Then, intermediate workpiece shapes and feature shapes are output from routing planning, inspection planning, and machining planning software modules. The process planning activity is also connected with many types of production planning software such as ERP, QMS, PLM, etc. Many different proprietary formats exist and human intervention is mostly unavoidable in this activity in industry today. There are difficulties not only in information exchange between different measurement process plan software systems, but also in automatic inspection generation. Moreover, the measurement process is required to do more than just inspect the part for conformance to the key dimensions on a drawing in today's manufacturing environment. It also should provide:

- feedback needed for control of the manufacturing process,
- statistical data for the evaluation of conformance to tolerances at the feature level,
- manufacturability feedback to the product definition activity, and
- information or data for machine calibration (such as machine performance, measurement uncertainty, and configuration) from the downstream CNC machine or CMM end to upstream manufacturing process planning.

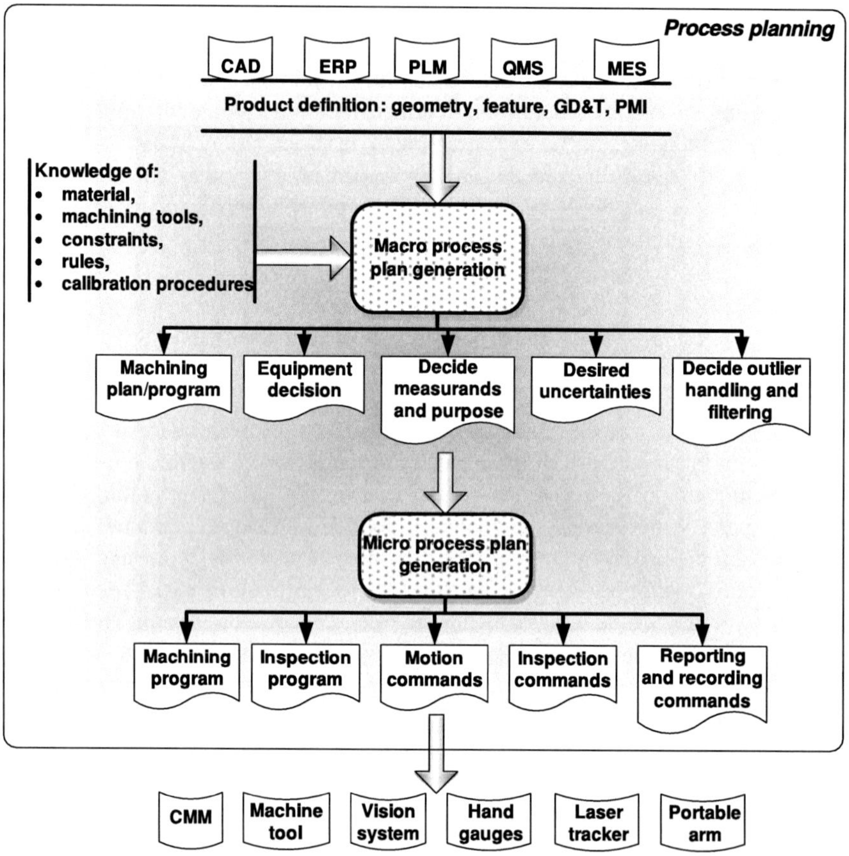

Fig. 7.2 Process planning definition activity

However, the measurement process definition has some major issues hindering the realization of interoperability and automation. Most of the measurement process plan generation is expected to provide device-dependent support for the myriad inspection devices that are available for process execution. It is impossible for medium to large manufacturing companies to employ only one type of inspection device. There is lack of information in digital format to define measuring system capabilities in terms of performance, measurement uncertainty, and configuration. Tolerance definitions are often incomplete, ambiguous, or inaccurate. There is no change management capability or associativity back into the CAD product design model, meaning that there seems to be no way to update/improve a product design when design errors are discovered in measurement process planning. There is also no standard digital format for transmitting knowledge-based manufacturing and inspection rules. It is now done with a lot of "cut and paste" activity in industry. In today's measurement process definition tools, there is currently a lack of DMIS compatibility and a lack of interactive and/or static conformance classes, meaning that there are multiple proprietary formats and a lack of tools allowing user access to the data. To summarize, the top interoperability issues exist in industry are the following:

1. The lack of comprehensive non-shape information available from the product definition activity.
2. No standard GD&T information associated with part design geometry. This issue is a crosscutting issue which exists both in production definition and measurement process definition activities.
3. The lack of a standard mechanism to capture and exchange knowledge including methods, practices, and rules for measurement process planning.
4. Lack of a standard data model for the exchange of information between macro and multiple micro process planning interfaces.
5. No computer-readable and standard resource definitions of measurement equipment capability, capacity, available configuration, performance, measurement uncertainty, sensors, fixtures, rotary tables, etc.
6. Weak end user support for non-proprietary metrology system interface languages.

The reader may notice that the first interoperability issue affects both product definition and process planning activities. This issue is considered a "showstopper" and must be solved if interoperability is to be realized between product definition models, measurement process planning and any downstream activities. As discussed in the above section, a standard data model that is able to represent semantic GD&T and PMI with CAD geometry model is fundamental to any metrology interoperability solution.

The lack of an extensible interface standard that is able to capture and exchange measurement process planning knowledge and the associated rules is another obvious impediment to interoperability. DMIS is the only standard that defines measurement instruction data within the measurement process definition activity. It is a language for controlling dimensional measuring equipment and includes an input and an output language. Part of the DMIS input language defines features, tolerances, sensors, etc. The output language serves both as a log of action commands and settings and a report of results, with actual and nominal point data, features, and tolerances. However, it does not define complete measuring equipment resources. Measuring equipment resource data is necessary to complete the effectiveness of DMIS. An independent testing and certification service is useful in determining a broad set of conformance classes that would function as common knowledge among frequent DMIS users as to which class is required to do which type of job [6]. NIST has developed a DMIS Test Suite 2.2.1 for DMIS version 5.2 [7] to help users and vendors use version 5.2 of the DMIS and to support DMIS conformance testing. DMIS conformance and certification is an on-going effort.

There are various standards that define some measuring equipment capabilities and resource configurations. For example, DMIS includes some definitions of CMM configuration, but it needs to be assessed in relation to the machine configuration definitions. I++DME [8] and Renishaw use XML language to define machine configuration, the completeness and accuracy of these configurations needs to be tested. The ASME B5.59 [9, 10] series should be assessed to explore the applicability of applying these standards to define CMMs configurations. CMM machine type and configurations are defined in the ISO 10360 series [11–16]. However, these definitions are in human readable format. A standard data model in compliance with these standards needs to be developed and validated so that industry could develop implementations in software modules.

7.2.3 Execution Systems

From a high-level perspective, the most important functions of the measurement execution process include to accept input from the measurement process plan and use the input to provide unambiguous instructions to a variety of measurement equipment. These instructions often consist of detailed motion commands and measurement results collection commands. Most of the measurement execution systems also provide preliminary analyzed measurement result data such as fitted geometry or surfaces based on the measurement points. Although these functionalities may sound simple, interactivity issues abound both between the measurement process plan generation and the measurement process execution, and within the execution process itself. Not only there are a huge number of different types of measurement equipment that the execution systems must support, there are an almost limitless number of ways in which a complex part can be inspected. The realistic goal of solving the interoperability issue for measurement execution systems is to achieve a high degree of automation and a minimum amount of manual intervention. Figure 7.3 shows the current status of information exchange between measurement process planning and execution systems.

If the measurement process plan does not result in a complete and unambiguous measurement program, then corrective action must be taken before the measurement process can be executed. If the measurement program is not compatible with the available measurement equipment (i.e. software or controller), there are a multitude of options available for addressing the interoperability problem. Unfortunately, none of them are inexpensive solutions. If the measurement process generated from one process planning software system is not compatible with current measurement equipment software or controller, a company has the following ways to make it work:

1. Translate the measurement program into a format that is compatible with the available equipment.
2. Purchase compatible measurement execution software. In this way, proper training of how to use the software is almost always necessary.
3. Negotiate with the process planning software vendor to make the needed changes for available equipment.

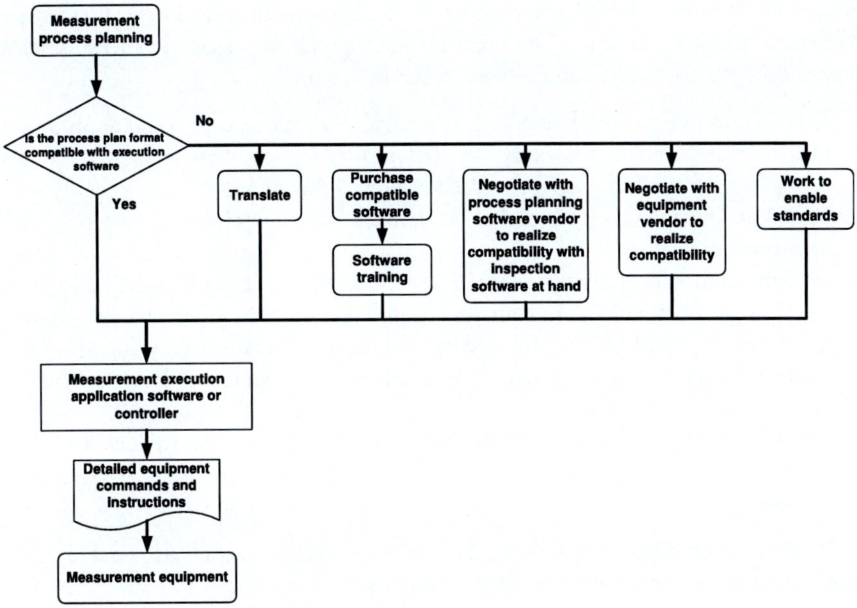

Fig. 7.3 Status quo of information exchange between measurement process planning and execution

4. Replace or augment existing measurement equipment with new equipment that is compatible with the process planning software.
5. Demand standards-compliant dimensional metrology software for both process planning and execution.

The interoperability issue in measurement process execution systems is more important in large, enterprise-level corporations, where a single-vendor solution is impractical if not impossible. The need for interoperable software products that execute the manufacturing and measurement process in a highly automated and equipment-independent fashion becomes critical to an enterprise-level corporation's very survival. Even at the job-shop level, a single-vendor solution can restrict the ability to choose best-in-class equipment for a particular application. It may also require redundant training on new software to enable best-in-class equipment choices.

However, there is no standardization in industry for the detailed equipment commands and instructions shown in Fig. 7.3. There are two publicly available specifications, one of which is formalized as an official ANSI and ISO standard—the equipment module of DMIS Part 2 [17]. The other is the I++DME Interface Specification [8] which is a specification for dimensional measuring equipment information exchange developed by several European automakers and measuring equipment vendors. There are no known product implementations of DMIS Part 2. There are many software implementations of I++DME worldwide, but it is not yet ubiquitous for either CMM software or CMM systems to offer I++DME in their published product offerings. The pressing interoperability issues in measurement execution systems can be summarized as the following:

1. I++DME is not a formal standard. It needs to be released to some appropriate and accredited standards body, so that any company who is interested can provide input toward changes and additions to the standard.
2. I++DME needs to be extended to handle more equipment, sensors, and environments.
3. Implementation barriers of I++DME need to be reduced, such as the entry cost. The I++ group should give sufficient assurances that there will be no requirement that royalties be paid by any individual or company solely for using the I++DME language in their metrology products. This is crucial for a standard to be accepted in industry.
4. There is overlap between I++DME and DMIS Part 2. In order to deal with this, the working committees of I++DME and DMIS need to collaborate rather than competing.

Renishaw and other vendors have I++DME simulators available to enable quick and accurate development of I++DME implementations within measurement plan execution software. The CMM industry and NIST have also developed an I++DME test suite. The I++DME conformance test utility software has not been maintained to the latest version of I++DME, but can still be of value to enable I++DME implementations which can be quickly developed and which are compliant to the specification. There are also some emerging issues in industry when companies trying to embrace I++DME such as no ready-to-use I++DME products; collision avoidance volume definitions in I++DME are too weak; precise sensor shape geometries need to be improved.

7.2.4 Data Analysis and Reporting Systems

Measurement data analysis and reporting systems (as shown in Fig. 7.4) gather measurement results from execution systems and product definition systems to analyze workpiece inspection data in terms of product definition requirements, and to perform a statistical analysis of the measurement results and present them in the form of a statistical process control report or product/process improvement report. These reports are then fed into different software, such as MSA, Quality Information System (QIS), SQC, etc., to be used to improve the product design and process planning and execution activities for future applications. The important functionalities of measurement data analysis and reporting systems include:

- to gather sensor or measurement result data,
- to gather traceability information from manufacturing processes,

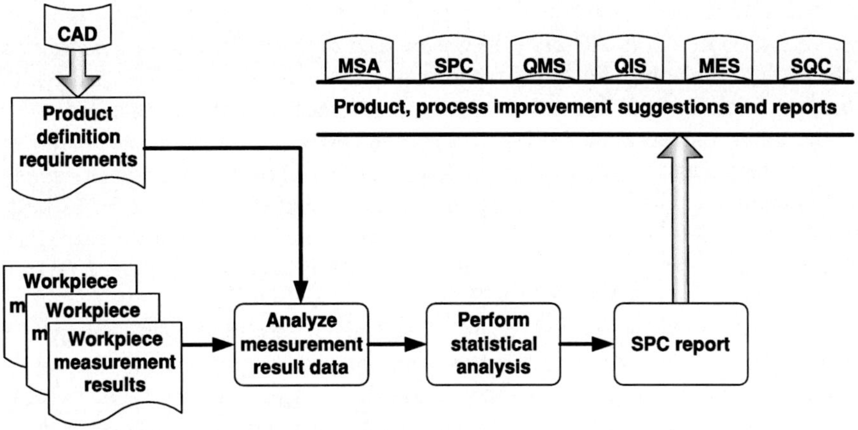

Fig. 7.4 Measurement data analysis and reporting process

- to perform statistical analysis, and
- to produce statistical reports to be used for possible design/process improvement suggestions.

Apart from these functionalities, measurement data analysis and reporting systems also need to fulfil requirements from different types of software systems. Some of these requirements include different output formats for different software systems, proper data reduction methods without losing critical information, and suitable measurement result data storage techniques.

From an information exchange standards perspective, the main existing issues in the analysis and reporting activity are:

1. lack of understanding and definition of how measurement results and summary statistics can be used to improve the manufacturing process, e.g., current measurement activities are still largely used to accept or reject parts, instead of as a feedback to manufacturing process and part design improvements,
2. lack of a uniform data model for traceability,
3. lack of consistency of statistical calculation methods and definitions,
4. present lack of a standard data format for measurement data and single part report, though DML is expected to be promoted to an international standard soon, and
5. lack of methods to report measurement data in the semantics of business systems.

The MEPT team of the AIAG group has created the DML data model which defines measurement feature actuals and nominals for a CMM, sufficient for complete reanalysis of derived values, such as feature dimensions. In partnership with the AIAG, the Dimensional Metrology Standards Consortium (DMSC) is progressing DML to ANSI and ISO standardization. DML is having moderate usage largely in North America. A format for CMM measurement results is defined within DMIS, and has enjoyed some usage, wherever DMIS is used. Within the STEP effort, AP 219 [18] was defined to cover all important metrology information, including, but not limited to, measurement results. The latest ISO standard version of AP 219 only defines measurement results information.

Harmonization between DMIS, AP 219, and DML for providing a standardized measurement data format is essential. There are multiple standards/specifications that define traceability data such as DMIS, DML, and ISO 10303 AP 238 [19]. However, the link between traceability and measurement data is insufficient. Part of the current effort on DML is to ensure that DML is consistent with both GD&T paper standards like ASME Y14.5, STEP AP 219 and DMIS.

There is no comprehensive standard science or standard methodology for adjusting a manufacturing process based on analysis of quality data. To realize this, an unambiguous statement of the causal link between events/trends in measurement results and elements of the manufacturing processes is necessary. As a result, the causal link between quality control results and the process is only known by human experts, so human intervention is needed to carry out appropriate process adjustments manually. There are also multiple standards/specifications to perform statistical analysis of quality data, such as ASQ [20] and ISO 16949 [21]. The standardization and harmonization of these standards/specifications is necessary.

7.2.5 Crosscutting Interoperability Issues

Among all the interoperability issues discussed in the above sections, one of them is a crosscutting issue that currently has an adverse effect on every aspect of the dimensional metrology process—CAD data is not associated adequately with GD&T and PMI information. GD&T and PMI information cannot flow seamlessly to downstream processes when system components are needed from different vendors.

There are two major factors that caused this crosscutting issue to exist. One is the technology difficulty of providing semantic GD&T and PMI with geometric models in CAD systems, which is discussed in the following paragraphs. Among all the standards, ISO 10303 standards are the major effort towards resolving the interoperability issue in both manufacturing and measurement processes in industry. It is also the only international standards that is working towards defining a complete data model to represent semantic GD&T and PMI information with design geometry. Three parts of this standard provide standardization of information flow between CAD and machining planning (AP 203 and AP 224), CAD and measurement process planning (AP 203 edition 2 and AP 219), machining planning and measurement planning (AP 219 and AP 238). To solve the crosscutting issue, the AP 203 edition 2 standard is the key. Currently, the second edition of AP 203 is approaching release. Major CAD vendors—CATIA, Pro/E (CREO), and NX—have participated in implementing and validating this standard.

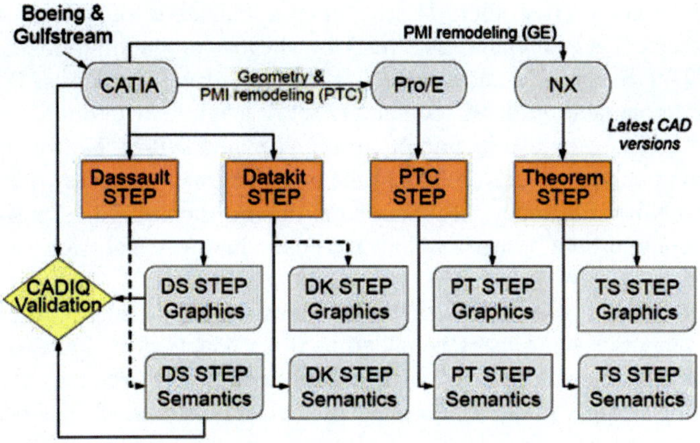

Fig. 7.5 PMI validation test data ancestry (Courtesy of ITI TranscenData)

Also, projects determining how accurate each CAD vendors' implementation of AP 203 is have been carried out. The most significant validation effort (shown in Fig. 7.5) is carried out by NIST, ITI TranscenData and Advanced Dimensional Management LLC. Throughout the validation process, a series of data modelling issues were discovered including:

- ambiguous linkages between annotations which share geometry,
- missing entities, attributes and relationships.

Another problem stems from various political issues between software vendors and users. There is no shared vision between vendors and users for what is considered best practice due to program diversity in the market. There is a lack of consensus on whether exclusive open-source, non-proprietary, standards-based hardware and software is a more effective option than a single-supplier network, proprietary hardware and software. There are many cultural issues that prevent a shared vision from being adopted. Equipment and software vendors are resistant towards the adoption of standards. Part of the reason is the multitude of competing and conflicting standards and practices. For each aspect of dimensional metrology systems, there is either no standard or no conformance tests exist to verify compliance to the standard. Therefore, it is very hard for industry to initiate implementation in their product to adopt standards. Even though some compelling research has shown that interoperability issues are responsible for up to $1.05 billion/year in the United States automobile industry alone [22]; there is a lack of consensus in industry on whether the exclusive use of open-source, non-proprietary, standards-based hardware and software is a more effective option than single-supplier network proprietary hardware and software. Vendors feel compelled by economic necessity to protect their proprietary information in order to offer improved products that are differentiated from those of their competitors.

From their perspective, there is no economic incentive to offering standards-based products. The perception that vendors will lose product differentiation is at least partly false, as can be shown easily through an example. Clearly, PC printers are now interoperable with PC computers: only a minimal effort is required to install and begin using a new printer from any manufacturer. However, printing quality and price vary widely, allowing the customer many choices with regard to quality durability, efficiency, etc. Standards organizations need to push forward this philosophy to both dimensional measurement hardware/software vendors and consumers.

Standards are not typically in the best interests of the vendor, particularly for the large vendor. Having users beholden to the products of a single vendor virtually eliminates competition and invites a more profitable product pricing structure. Smaller vendors may be interested in standards, but small vendors want to eventually become large vendors, so the interest may be short-lived. OEM pressure and support is the secret to the success of most if not all standards and interoperability solutions. If enough consumers demand an open, non-proprietary standard or any other kind of solution, the vendors must get on board or be left behind. The more progressive vendors try to get in on the ground floor of new developments in these areas so that they are ahead of their competitor. In order to push forward the standard efforts, the users, manufacturing producers, such as airplane manufacturers, automobile manufacturers, army research organizations, must play the leading role and demand standards-based hardware and software. As for the standards organizations, they need to gather sufficient information from major dimensional metrology vendors to determine their business and organizational objectives.

Although due to the complexity of ISO 10303 standards, the lack of implementation and some negative views from the dimensional metrology community, STEP is still the most comprehensive standard that deals with this crosscutting interoperability issue.

7.3 Road Map of Standards Harmonization for Achieving Interoperability

In large measure, the afore-discussed interoperability issues can be depicted in the following figure—Fig. 7.6. The horizontal dashed lines represent the boundaries that prevent information from flowing seamlessly within and between elements of dimensional metrology systems. Dimensional metrology systems encompass a large number of software and hardware systems. The interoperability issues that exist in dimensional metrology systems as introduced in the above sections are numerous and cannot be resolved in a short time.

However, due to the potential and substantial payoff, it is worthwhile to exert efforts to achieve interoperability within dimensional metrology systems. As it is shown in Fig. 7.7, without all the boundaries between different software/hardware

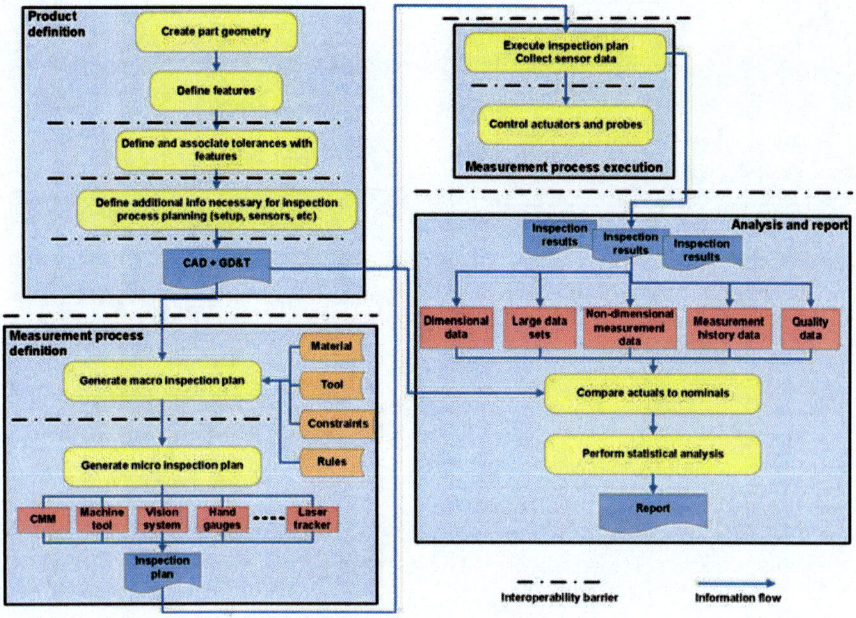

Fig. 7.6 Current state of dimensional metrology interoperability issues

systems the dimensional metrology system can be much more efficient and adequate real-time closed-loop control of the machining process with measurement data feedback is achievable.

Much has been said about the need for standardization and the open exchange of information. Perhaps nowhere is the need more compelling than in the realm of quality measurement. Quality is the lynchpin for success in every enterprise.

In response to this essential need for quality measurement, the typical reaction has long been the creation of a local quality measurement enclave tailored to parochial needs of the users. With that, many different colloquial vocabularies have evolved to describe what are essentially universal quality measurement concepts, effectively precluding the free flow of quality measurement information from each newly introduced data source. The concomitant tax for data translation and reintegration can quickly consume all available resources leaving none for more value-added activities.

As is the case with any business, it is the benefits that really speak to the business case for embracing such a standard.

In a classic case of 20% of the effort yielding 80% of the benefit, it is the standardized export format for quality related information that presents the greatest business opportunity for industry. The actual cost resulting from the lack of standardization is enormous and, in terms of the impediment it places on innovation, this lack of standardization is not even quantifiable. At the very least, it consumes a significant portion of IT resources in most enterprises, and it consumes all of them in others.

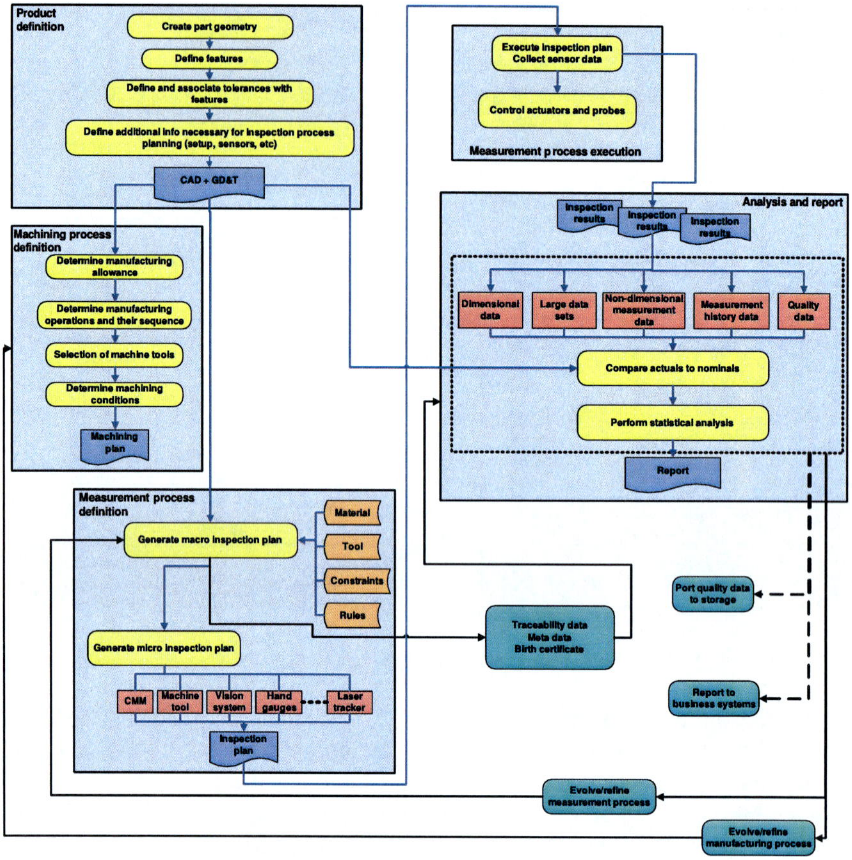

Fig. 7.7 Future vision of dimensional metrology system

Here is a summary of the benefits of standardization:

- Eliminates wasted resources, money, and time in data integration tasks.
- Redirects these savings to value-added activities, enhancements, etc.
- Allows Solutions Providers and Metrology manufacturers to redirect more energy to new development.
- Enables metrology solutions to communicate with other solutions, making both solutions more useful.
- Permits manufacturers to focus more on core business.
- Moves away from specific dependencies and proprietary schemas that require separate technical support.

The value is clear, but the political challenges are great. Active participation in the standards community is critical to the success of interoperability; however in today's competitive environment with all companies doing more with less, the resources for this work are extremely limited. Many times we have heard that anything can be done with enough time and money (both of which are scarce in most industries), but firms also need to have the political will to work with others and consider the larger picture. Many of today's consortiums are not adequately funded or disciplined to produce the type of collaborative framework necessary for significant progress. Many vendors sit on the sidelines to watch the effort without direct participation, waiting to see if a given standard will reach critical mass and ultimately establish some level of adoption. Many standards exist. Some are used some are not. One thing we know for certain is that an interoperability standard only provides value when it is needed and subsequently used.

7.4 Summary

Interoperability issues that exist within each of the four pillars of a dimensional metrology system, namely product definition, process planning, process execution, results analysis and reporting, and have been discussed in detail in this chapter. To support automatic dimensional metrology plan generation, the part must be decomposed into geometric features. Dimensions and tolerances must then be assigned to a geometric feature, or set of features. Datum features must be defined in such a way that they are appropriate both for manufacturing the part and for inspecting it. PMI information such as surface texture must be included in the model, along with relevant information about the orientation or lay of the surface texture to be measured. This information must be defined completely and accurately in a CAD data model. Currently PMI information is available to only a limited extent in a limited number of proprietary systems. There is no CAD product implementation of PMI information using non-proprietary standards. STEP AP 203 edition 2 consists of PMI information but has not yet been fully adopted by CAD vendors. Also, once AP 203 edition 2 is successfully implemented by CAD vendors, the implementations need to be validated by standards organizations so as to ensure their accuracy.

The measurement process planning activity is connected with many types of production planning software. Many different proprietary formats exist and human intervention is mostly unavoidable in the process planning stage in industry today. There are difficulties not only in information exchange between different measurement process plan software systems, but also in automatic inspection generation. The lack of an extensible interface standard that is able to capture and exchange measurement process planning knowledge and rules is an obvious impediment to interoperability. DMIS is the only standard that defines measurement instruction data within the measurement process definition activity. However, it does not define complete measuring equipment resources. The I++DME specification defines a machine configuration model; however, the completeness and accuracy of this model needs to be tested. I++DME does not have complete information for defining the measurement processes. Other measurement equipment configuration standards such as the ASME B5.59 series are only in human readable format. A standard computer readable data model in compliance with these standards needs to be developed and validated so that industry could develop implementations in software modules.

Once a measurement plan is generated, it must be properly executed through the measurement execution process. The most important functions of this process include accept input from the measurement process plan and use the input to provide unambiguous instructions to a huge number/types of measurement equipment. However, there is no proper standardization for this process. There are two publicly available specifications—DMIS Part 2 and I++DME. There are no known product implementations of DMIS Part 2 in industry. There are many software implementations of I++DME worldwide, but it is not yet ubiquitous for either CMM software or CMM systems to offer I++DME in their published product offerings. Measurement data analysis and reporting systems gather measurement results from execution systems and product definition systems to analyze workpiece inspection data in terms of product definition requirements, and to perform a statistical analysis of the measurement results and present them in the form of a statistical process control report or product/process improvement report. DML was developed to store measurement result data and had moderate usage largely in North America. A format for CMM measurement results is defined within DMIS, and has enjoyed some usage, wherever DMIS is used. STEP AP 219 has very limited definitions for storing measurement results. Harmonization between DMIS, AP 219, and DML for providing a standardized measurement data format is essential.

Among all the interoperability issues discussed in the above sections, one of them is a crosscutting issue that currently has an adverse effect on every aspect of the dimensional metrology process—CAD data is not properly associated adequately with GD&T and PMI information. GD&T and PMI information cannot flow seamlessly to downstream processes when components are from different vendors. In order to solve this issue, vendors, end users and standards organizations must work together to solve the political and cultural issue first. The end users must play the leading role and demand standards-based hardware and software. As for the standards organizations, they need to gather sufficient information from major dimensional metrology vendors to determine their business and organizational objectives. The vendors need to realize that there is economic incentive to offer standards-based products. The perception that vendors will lose product differentiation is incorrect.

References

1. IMTI (2006) A roadmap for metrology interoperability integrated manufacturing technology initiative. IMTI, Gaithersburg
2. ISO (2007) ISO 10303-203: Industrial automation systems and integration—Product data representation and exchange—Part 203: application protocols: configuration controlled 3D design
3. ISO (2006) ISO 10303-224: Industrial automation systems and integration—product data representation and exchange—Part 224: application protocol: mechanical product definition for process planning using machining features

4. STEP Tools Inc. (2009) Available from: http://www.steptools.com/. July 2009
5. ISO (2001) ISO 10303-214: Industrial automation systems and integration—product data representation and exchange—Part 214: application protocol: core data for automotive mechanical design processes
6. Chalmers RE (2002) Metrology for manufacturing means business. Manuf Eng 128(4):58–64
7. NIST. DMIS Test Suite 2.1.4. (2009) Available from: http://www.isd.mel.nist.gov/projects/metrology_interoperability/dmis_test_suite.htm
8. I++DME (2005) Dimensional measurement equipment interface, the international association of coordinate measuring machine manufacturers
9. ANSI (2009) ANSI/ASME B5.59-1: Data specification for machine tool performance tests (draft)
10. ANSI (2009) ANSI/ASME B5.59-2: Data specification for properties of machining and turning centers (draft)
11. ISO (2000) ISO 10360-1:2000: Geometrical product specifications (GPS)—acceptance and reverification tests for coordinate measuring machines (CMM)—Part 1: Vocabulary
12. ISO (2009) ISO 10360-2:2009: Geometrical product specifications (GPS)—acceptance and reverification tests for coordinate measuring machines (CMM)—Part 2: CMMs used for measuring linear dimensions
13. ISO (2000) ISO 10360-3:2000: Geometrical product specifications (GPS)—acceptance and reverification tests for coordinate measuring machines (CMM)—Part 3: CMMs with the axis of a rotary table as the fourth axis
14. ISO (2000) ISO 10360-4:2000: Geometrical product specifications (GPS)—acceptance and reverification tests for coordinate measuring machines (CMM)—Part 4: CMMs used in scanning measuring mode
15. ISO (2010) ISO 10360-5:2010: Geometrical product specifications (GPS)—acceptance and reverification tests for coordinate measuring machines (CMM)—Part 5: CMMs using single and multiple stylus contacting probing systems
16. ISO (2001) ISO 10360-6:2001: Geometrical product specifications (GPS)—acceptance and reverification tests for coordinate measuring machines (CMM)—Part 6: estimation of errors in computing Gaussian associated features
17. ANSI (2003) Dimensional measuring interface standard—Part 2 : object interface specification, consortium for advanced manufacturing—international
18. ISO (2007) ISO 10303-219: Industrial automation systems and integration—product data representation and exchange—Part 219: application protocol: dimensional inspection information exchange
19. ISO (2004) ISO 10303-238: Industrial automation systems and integration—product data representation and exchange—Part 238: application protocols: application interpreted model for computerized numerical controllers
20. American Society for Quality (ASQ) (2009) Available from: http://www.asq.org/.July 2009
21. ISO (2009) ISO/TS 16949:2009: Quality management systems—particular requirements for the application of ISO 9001:2008 for automotive production and relevant service part organizations
22. Tassey G, Brunnermeier SB, Martin SA (1999) Interoperability cost analysis of the U.S. automotive supply chain. Research Triangle Institute. RTI Project Number 7007-03

Chapter 8
Dimensional Metrology for Manufacturing Quality Control

This chapter presents the current quality control techniques used for dimensional measurement data used in small to medium sized and global manufacturing industries. Quality is defined as strict and consistent adherence to measureable and verifiable standards to achieve uniformity of output that satisfies specific customer or user requirements. It is exactly this striving for quality within manufacturing today that separates one business from the other with respect to production costs, product reliability, and brand reputation. There are many approaches to establishing a quality program and many tools to choose from. This chapter describes these approaches and tools in order to provide a backdrop for the types of information that are central to quality analysis and data reporting. Information modeling for different manufacturing industries is discussed for different types of quality control at the end of this chapter.

8.1 Six Sigma and Dimensional Metrology

The goal and process of achieving less than 3.4 defects/million opportunities in production is commonly known as a six sigma program. Since its advent within Motorola in 1981, the six sigma approach to defect reduction is used by more than two thirds of the Fortune 500 companies [1]. This discipline is viewed by many as a business process management strategy that draws extensively on quality improvement techniques such as statistical quality control (SQC) and total quality management (TQM).

The Six Sigma philosophy holds that the first priority of a manufacturing quality improvement program is to achieve process stability through continuous improvements that are intended to reduce variation, thereby resulting in predictable behavior. Through identification of characteristics that can be measured, analyzed and controlled, quality improvements can be sustained throughout the manufacturing enterprise.

Y. Zhao et al., *Information Modeling for Interoperable Dimensional Metrology*, DOI: 10.1007/978-1-4471-2167-1_8,

The primary project methodology in a six sigma program employs a phased approach derived from the Deming plan-do-check-act (PDCA) cycle for continuous process improvement. It is a 5 phase cycle whose acronym is DMAIC. The DMAIC cycle is described as:

- Define the problem with emphasis on stakeholder consideration, and the project goals.
- Measure key aspects of the current process and collect relevant data.
- Analyze the data to investigate and verify cause-and-effect relationships. Determine what the relationships are, and attempt to ensure that all factors have been considered. Seek out root causes of defects.
- Improve or optimize the current process based upon data analysis using techniques such as design of experiments, poka-yoke or mistake proofing, and standard work to create a new, future state process. Set up pilot runs to establish process capability.
- Control the future state process to ensure that any deviations from target are corrected before they result in defects. Implement control systems such as statistical process control, production boards, and visual workplaces, and continuously monitor the process.

Within the DMAIC cycle a Six Sigma initiative will rely on many quality management techniques such as:

- Process Capability Studies
- ANOVA (Analysis of Variance)
- Gage R&R (Repeatability and Reproducibility)
- Control Charts (XbarR, XbarS, EWMA, etc.)
- Correlation Studies
- Histograms
- Pareto Analysis
- Root Cause Analysis
- Design of Experiments

From a dimensional metrology perspective it is clear that the Six Sigma method can be applied to quality problems associated with deviations of manufactured items from design intent. The purpose of inspection and measurement in manufacturing is to provide data that can be used both for validation of conformance to specification and for providing data for process analysis when deviation from design nominal dimensions is found.

Six Sigma identifies several key roles for its successful implementation [2].

- Executive Leadership includes the CEO and other members of top management. They are responsible for setting up a vision for Six Sigma implementation. They also empower the other role holders with the freedom and resources to explore new ideas for breakthrough improvements.

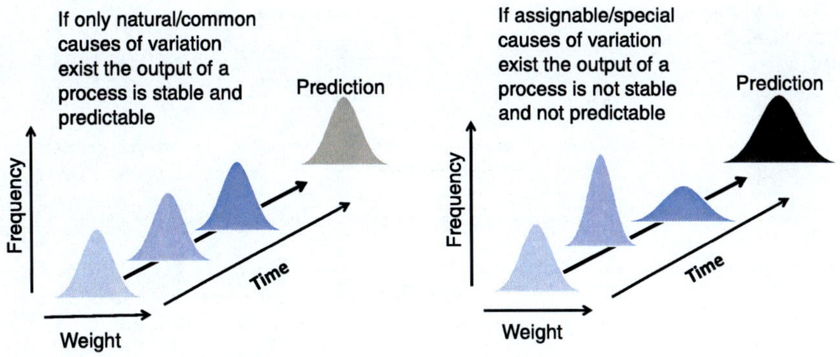

Fig. 8.1 Natural versus Assignable causes of variation through time

- Champions take responsibility for Six Sigma implementation across the organization in an integrated manner. The Executive Leadership draws them from upper management. Champions also act as mentors to Black Belts.
- Master Black Belts, identified by champions, act as in-house coaches on Six Sigma. They devote 100% of their time to Six Sigma. They assist champions and guide Black Belts and Green Belts. Apart from statistical tasks, they spend their time on ensuring consistent application of Six Sigma across various functions and departments.
- Black Belts operate under Master Black Belts to apply Six Sigma methodology to specific projects. They devote 100% of their time to Six Sigma. They primarily focus on Six Sigma project execution, whereas Champions and Master Black Belts focus on identifying projects/functions for Six Sigma.
- Green Belts are the employees who take up Six Sigma implementation along with their other job responsibilities, operating under the guidance of Black Belts.

8.2 Quality Control for Manufacturing Industry

In the complexity of applying production operations against raw materials to make finished goods it is impossible to avoid variation of actual dimensions from design dimensions. There are many sources of variation that contribute to the need for quality control.

8.2.1 Process Variation

Variation can be both natural and special. Natural variation can be considered as stemming from background noise in the manufacturing system. These causes of variation are inherent to the system and lack significance when encountering low or high values (Fig. 8.1). The distribution of dimensional values from a system

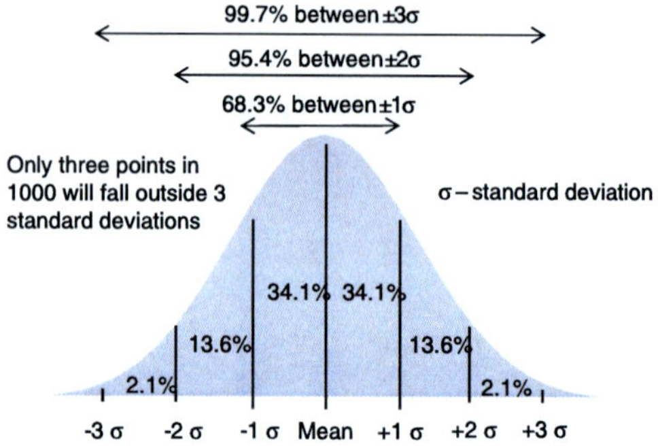

Fig. 8.2 *Bell shaped curve* with standard deviation spread

that is subject to natural variation only will be represented by a Gaussian curve (as shown in Fig. 8.2), thus reflecting the random dispersion from a process mean or central tendency. Examples of these Natural or ''common'' fluctuations include:

- Lack of consistency of raw materials
- Vibration in industrial processes
- Ambient temperature and humidity
- Normal wear and tear
- Variability in settings
- Computer response time

Special or ''Assignable'' causes of variation are described as those that exist outside of natural variation. In many cases they are seen as identifiable events or signals from the system that may be reflective of new, emergent or previously neglected phenomena. Examples of these special causes of variation include:

- Poor adjustment of equipment
- Operator error
- Broken tools
- Faulty controllers
- Machine malfunction
- Computer crashes
- Poor batch of raw material
- Power surges

In any quality control program the first task is to identify the functional characteristics along with their nominal values and tolerances that are necessary from a design perspective for form and function of the manufactured part. Once this is accomplished, a quality control plan can be formulated in which an inspection program can be identified. It is in this process that a dimensional measurement plan and the associated dimensional metrology equipment are specified.

A quality control plan will typically incorporate the tools of the statistical process control method in which variable and attribute data for product conformance characteristics are measured.

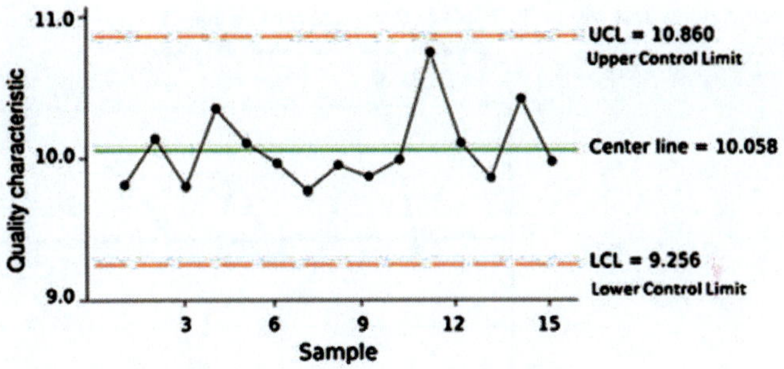

Fig. 8.3 Natural variation about nominal through time on SPC control chart

8.2.2 Control Chart Theory

Since quality inspection is not considered a value-added process by most businesses the cost of quality has been driven down by the use of statistical methods.

This began when Dr. Walter A. Shewhart wrote Economic Control of Quality of Manufactured Product in 1931 [3]. As a physicist, engineer and mathematician working for the Western Electric Company's Hawthorne Works plant, he had recognized the only source of quality control was limited to inspecting finished products and removing defective items. This is sometimes known as acceptance sampling. Others refer to this quality control technique as ''inspecting quality into the product". Shewhart recognized the natural variation in the manufacturing process and derived a basis for developing sampling plans for quality control based on the well-established mathematical principles of probability and statistics. By understanding the extrapolated relationship between the characteristics of a specific sample group within a larger population (e.g. production lot), he conceived the control chart as a time ordered line chart displaying the average and range of production samples. In addition to plotting these points against the process mean or ''central tendency", Shewhart also incorporated the concept of control limits based on standard deviation calculations of the entire sample population. Figure. 8.3 is an example. Example SPC chart showing random variation about a nominal value.

8.2.3 Data Tests

As part of the control chart theory Western Electric also worked to identify pattern recognition on the Shewhart control charts [4]. These behaviors included:

- Cycles
- Trends
- Freaks
- Mixtures

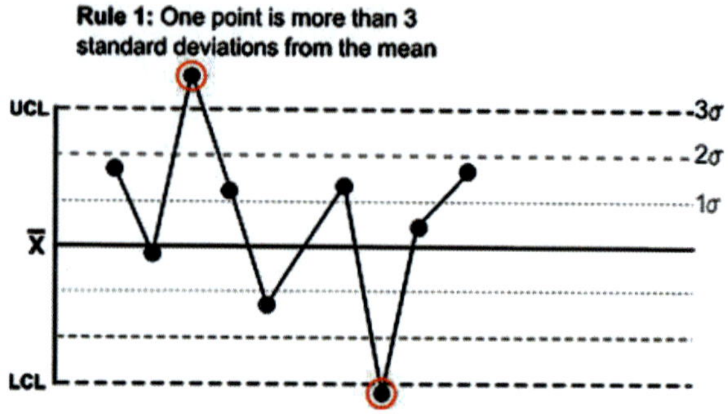

Fig. 8.4 Extreme point test on control chart

- Grouping or "bunching" of measurements
- Gradual change in level
- Sudden shift in level
- Instability (abnormally large fluctuations)
- Stratification (abnormally small fluctuations)
- Interactions (two or more variables acting together)
- Systematic variation
- Tendency of one chart to follow another

In 1981 Dr. Lloyd S. Nelson, expanded the Western Electric Rule definition to form the Nelson Rules. Any failure of these rules would indicate the presence of assignable causes of variation [5].

- Extreme Points Test (Rule 1)
  This test watches for extreme subgroups beyond the control limits. This test applies to both Xbar and R control charts. The rule is as follows: The existence of a single point beyond a control limit signals the presence of an out-of -control condition. An example of data flagged by this Rule 1 is shown in Fig. 8.4.
- Run Above or Below the Centerline Test (Rule 2)
  This test watches for 7, 8 or 9 consecutive subgroups above or below the centerline and applies to both the control charts. This test is defined by a number of successive points that fall above or below the centerline. The presence of such a run is strong evidence that the process mean or variability has shifted from the centerline. An example of data that Rule 2 will flag is shown in Fig. 8.5.
- Linear Trend Test (Rule 3)
  This test watches for six subgroups in a row steadily increasing or decreasing. This test applies to control charts and fails when there is a systematic increase or decrease trend in the process. Neither the zones nor the centerline come into play in this test. An example of data that Rule 3 will flag is shown in Fig. 8.6.
- Oscillatory Trend Test (Rule 4)
  This test watches for fourteen subgroups in a row alternating up or down and applies to control charts. When 14 successive points oscillate up and down a systematic trend in the process is signaled. Again, neither the chart centerline

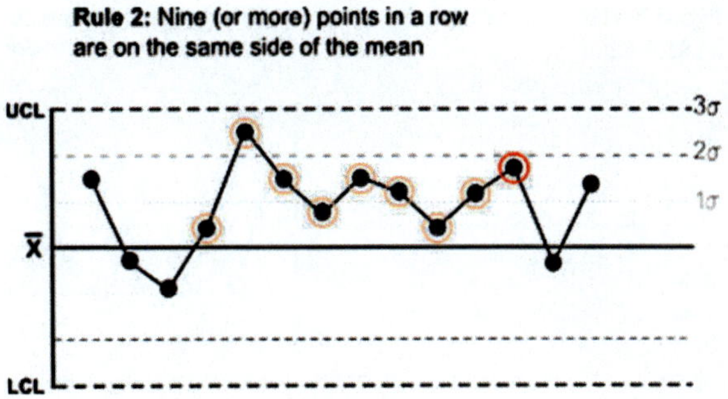

Fig. 8.5 Run test on control chart

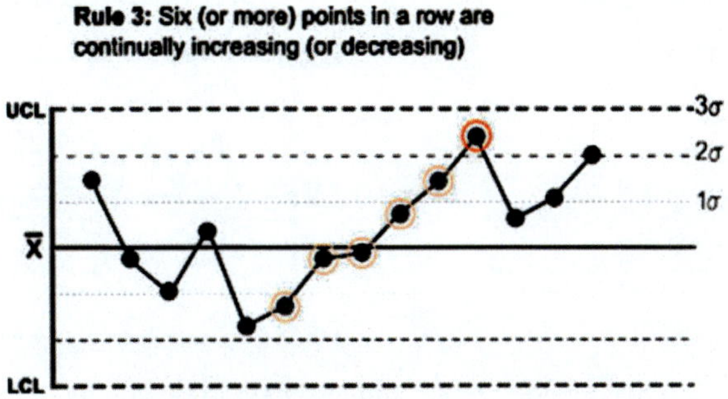

Fig. 8.6 Linear trend test on control chart

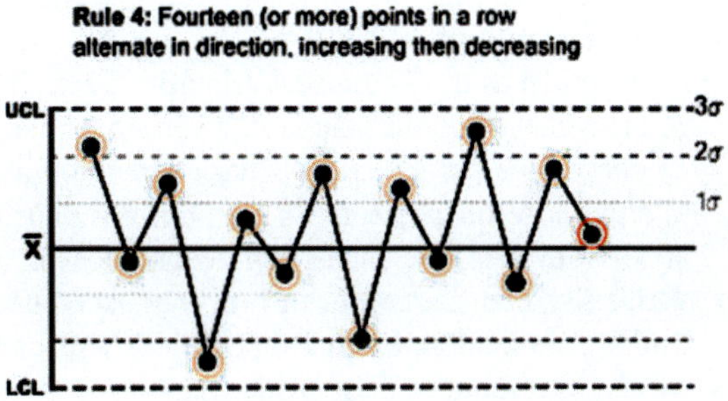

Fig. 8.7 Oscillatory test on control chart

nor the zones come into play for this test. An example of data that Rule 4 will flag is shown in Fig. 8.7.

- Two Sigma Test (Rule 5)
  This test watches for two out of three subgroups in a row in outside of two standard deviations. It is based on the specific control chart zones and therefore only applies to the Xbar chart. The rule is this: The existence of two of any three successive points outside of two standard deviations signals the presence of an out-of-control condition. An example of data that Rule 5 will flag is shown in Fig. 8.8.
- One Sigma Test (Rule 6)
  This test watches for four out of five subgroups in a row outside of one standard deviation. It is based on the specific control chart zones and therefore only applies to the Xbar chart. The rule is this: The existence of four of any five successive

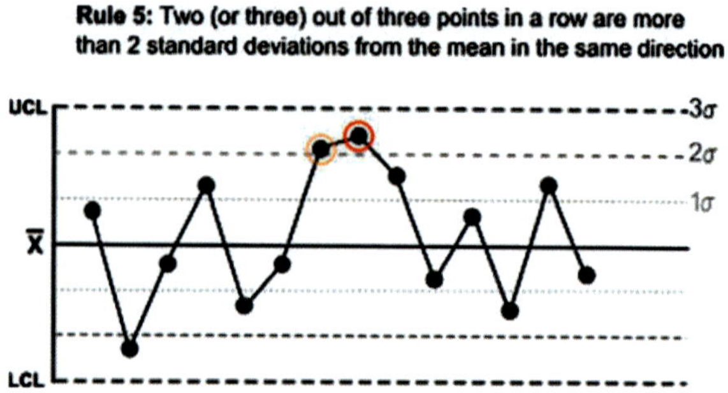

Fig. 8.8 Two sigma test on control chart

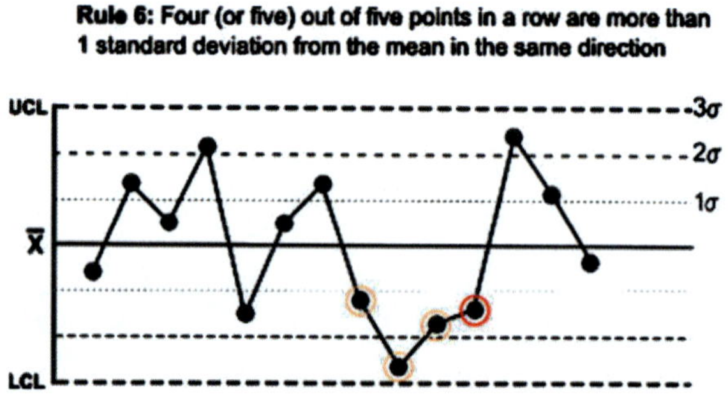

Fig. 8.9 One sigma test on control chart

points outside of one standard deviation signals the presence of an out-of-control condition. An example of data that Rule 6 will flag is shown in Fig. 8.9.

- Stratification Test (Rule 7)
  Also known as the Reduced Variability Test. This test watches for 15 subgroups in a row in within one standard deviation, above and below the centerline. When 15 successive points on the Xbar chart fall within one standard deviation only, to either side of the centerline, an out-of-control condition is signaled. This can arise from improper sampling techniques or a change (decrease) in process variability that has not been properly accounted for in the X-bar chart control limits. An example of data that Rule 7 will flag is shown in Fig. 8.10.
- Mixing/Overcontrol Test (Rule 8)
  This test watches for eight subgroups in a row on both sides of the centerline outside of one standard deviation. The rule is: Eight successive points on either side of the centerline outside of one standard deviation signals an out-of-control condition. This test failure could mean more than one process being plotted on a single chart (mixing) or perhaps overcontrol (hyper-adjustment) of the process. An example of data that Rule 8 will flag is shown in Fig. 8.11.

8.2.4 Taguchi Method

In the early 1990s Genichi Taguchi developed an approach to quality control that began with a focus on the costs of poor quality. He understood that quality losses result mainly from product failure after sale and that product "robustness" was

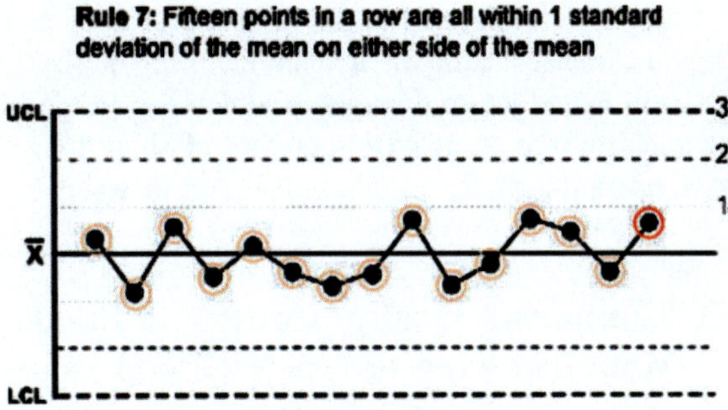

Fig. 8.10 Stratification test on control chart

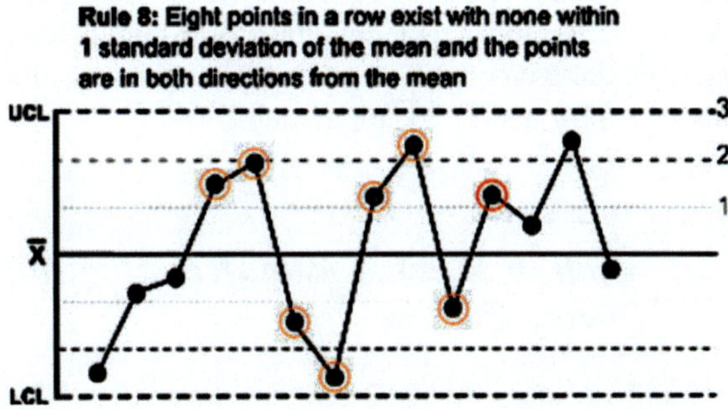

Fig. 8.11 Overcontrol test on control chart

more of a function of product design than online control, however stringent a manufacturing process [6, 7].

The approach to robust design recognizes that a work-in-progress may be subject to wide variations in manufacturing processes and that products when used are subject to wide conditional variations.

By recognizing the tolerance stack-up of components in any assembled product and the reduction in perceived quality of parts that behave in less than a perfect manner, Taguchi rejected the commonly held quality philosophy of zero defects, especially the pass/fail, in-spec/out-of-spec style of thinking. Robustness derives from consistency.

In the Taguchi approach to quality, variations must be driven from the system and a design target must be achieved. This begged the question of which design target was appropriate and led to the development of Designed Experiments in which an orthogonal array of system input parameters are varied and several trials are conducted in order to maximize a given output.

For example, when the Ford Motor company set about designing the anti-lock braking system, they employed the Taguchi method and set up an experiment designed to minimize braking distance. Several component characteristics were varied in combinatorial permutations. Various spring tensions, fluid viscosities and pad materials were used in the experiment. What resulted was the identification of the appropriate combination that minimized braking distance. A happy coincidence also resulted in a system that weighed less and reduced the overall cost.

Take these ideas back to the factory and it is easy to see how maximizing the ''signal-to-noise" ratio in a manufacturing process can be accomplished by a design to manufacture strategy in which the designer seeks to eliminate variation generated by the complex interaction of shop-floor quality factors such as operators, operating methods, equipment and material

8.3 Comparing Quality Control in Small and Medium Manufacturing to Large Global Industry

Significant differences exist in the necessity, approach, toolsets and resources between large and small manufacturing quality control programs. This section describes these environments, the differences and similarities and the tools used as provided from one sector to another.

8.3.1 Small to Medium Manufacturing Industry Quality Control

Every company must satisfy customers, stakeholders and employees to survive. Day-to-day details often divert attention from what is good for the company. Conflicts in priorities and contention for resources combine to form a huge barrier to organizational excellence. Smaller businesses also have a narrow buffer to shelter customers from error and waste. In a small business, quality planning and business planning are synonymous. The best time to start a quality program is during the initial planning for the business when designing quality into product and service delivery is essential.

A review of the relative strengths and weaknesses of small firms reveals that the TQM principles such as employee participation and flexibility could be more successfully applied in small firms than in large ones [8]. For example, small businesses tend to encourage innovation and can supply products at lower cost due to low overhead. On the other hand, researchers argue that small firms lack clout with suppliers and lack sufficient capital. They also lack professional managerial expertise, which accounts for about 90 percent of small business failures [9]. These are critical aspects for implementing TQM. For example, lack of clout with suppliers could impact a firm's ability to dictate the quality of incoming material. Also, lack of capital may limit a small firm from investing in high quality processes. Furthermore, knowledgeable and committed management is essential for successfully implementing TQM. Human resource management priorities and the practices of small firms also differ from those of large firms [10]. Small business owners and managers tend to view human resource management strategies as being less important than finance, marketing, and planning [11]. Small business managers do not perceive incentives to be critical to improving productivity. These findings may lead one to believe that the experiences of small firms with TQM implementation may differ widely from large firms. Thus, the size of an organization does, indeed, influence the effectiveness of various quality management practices.

Large original equipment manufacturers (OEMs) are increasingly pushing production up the supply chain. Along with these production mandates, the tier 1 and tier 2 suppliers to the large manufacturers are also required to carry the quality directives as provided by their customers. Quality information flows through the supply chain. From the smallest component in an assembly to the finished good, quality is a necessary component of a manufacturing operation in order to assure fit and function.

In the mid-market business environment, companies are faced with the need to be responsive to customer demands. Family owned job-shops provide goods to downstream consumers that have a wide array of needs. This results in multiple quality control measures. Some may employ traditional quality lab environments in which samples are brought to the inspector for measurement. The inspector is responsible for maintaining the lab's equipment procedures and ultimately evidence of conformance to specification. Quite often the smallest of shops have only a single individual responsible for quality inspection. Although quality control theory may well be understood, these companies typically have few resources to create robust quality control initiatives; therefore, they tend to find efficiencies in data acquisition and reporting. The movement from mechanical gauging to digital data collection has continued for the last several decades as metrology equipment providers design better tools for taking measurements. Micrometers, Calipers, indicators are now produced with output and even wireless communication to desktop computers that host inexpensive quality management software.

The small manufacturers tend to focus on production and establish sufficient quality programs to gain contracts and ensure adequate quality for customer acceptance and contract renewal. They tend not to have continuous improvement programs but rather rely on standard technique. Some, however, do have visionary programs that recognize the competitive advantage of improving quality, especially as it relates to fostering a healthy reputation among clients.

Many small manufacturing companies, as component suppliers, are pressured by their customers to improve their quality assurance. Component suppliers may also be subcontractors themselves and in turn need the same assurance from their suppliers. This pressure through the supply chain has led many suppliers to rethink their processes [12].

Successful implementation of TQM has been seen to hinge upon a human resource policy that is based on effective communications, teamwork, empowering of employees and the reinforcement of commitment [13]. This means creating trust between employees and management and also in internal and external customer–supplier relations. Meeting customer needs depends upon understanding what these are, and many TQM programs express internal relationships in terms of everyone having a customer who relies upon your part of the product or service delivery. Thus, successful external customer relationships hinge upon the effectiveness of internal relationships, which are facilitated by teamwork and communications. It is easy to recognize the need for horizontal integration in large functionally organized companies, where communications across functional boundaries are often physically and culturally inhibited. People feel strong loyalties to their particular area and 'turf wars' are not uncommon between functional heads. Vertical integration can be problematic too, as is often the case. In particular, management-employee relations in manufacturing have traditionally been adversarial [14].

Part of the strategy involves the development of selective partnerships to drive down costs and open the door to increased quality and value. Change also arises from the development of supply chain relationships between original equipment manufacturers and their suppliers, involving closer, ongoing relationships with a smaller number of suppliers. This has, for instance, led to greater involvement by first-tier suppliers in particular in the design of the component that they will be providing. In this way, design features can be identified which, if modified, would avoid quality problems later in manufacture [15].

The need to adopt a longer-term perspective is particularly problematic in smaller companies, where human and financial resources are often stretched seemingly to the limit, particularly in the tight profit margin environment of most component suppliers. Integration across functions is harder in large firms than smaller ones, as in small firms the authority and influence of top management are more immediate. In manufacturing, owner managing directors who have developed the company from inception usually have in-depth process knowledge or technical skills that are keys to the company's operations [13].

Over the years, several programs have forced the hand of small and medium sized manufacturers to standardize and institute better quality programs. For example, AIAG originally established as a consortium of the Big Three American car companies (GM, Ford, Chrysler), created a set of quality guidelines for use by suppliers. A standard guide for SPC sampling and charting allowed the market to standardize on calculation and data visualization methods. A specification for Measurement Systems analysis (MSA) provides a consistent and systematic approach to determining the variation inherent in measurement systems such as gage repeatability, appraiser reproducibility, gage linearity, bias and stability. The Part Production Approval Process established a set of guidelines on the quality analysis requirements for validating new or changing existing manufacturing processes.

Indeed, the AIAG raised the bar for industry with its release of the QS-9000 program in which manufacturers were required to establish quality procedures and prove them out through a detailed audit process whose ultimate goal was certification.

8.3.2 AS9100

In the Aerospace industry, Boeing has similarly established a specification for quality control and first article inspection (FAI). The AS9100, Aerospace Basic Quality System Standard, was developed by a group of US aerospace prime contractors, including Allied-Signal, Allison Engine Company, Boeing, General Electric Aircraft Engines, Lockheed Martin, McDonnell Douglas, Northrop–Grumman, Pratt and Whitney, Rockwell Collins, Sikorsky Aircraft, and Hamilton Sundstrand. AS9100 was developed and issued under the auspices of the Society of Automotive Engineers.

The intent and concept behind AS9100 are similar to Boeing's D1-9000. The standard is based in ISO 9000, with 27 additional requirements unique to the aerospace industry. The intent is to standardize and streamline many of the other aerospace quality management standards.

Representing the first international effort to formulate a quality management system standard for the aerospace industry, the two-year-old AS9100 is beginning to show its long-term value as an updated specification for quality control practices. The standard supplements ISO 9001 by addressing the additional expectations of the aerospace industry. Already, reports along this complicated manufacturing chain attest to among other benefits AS9100s contribution to more consistent verification methods and fewer verification audits. Initially released in October 1999 by the Society of Automotive Engineers in the Americas and the European Association of Aerospace Industries in Europe, and shortly thereafter by standards organizations in Japan and Asia, AS9100 was a cooperative effort of the International Aerospace Quality Group. As such, it combines and harmonizes requirements outlined in the SAE's AS9100 and Europe's prEN9000-1 standards. Recently, AS9100 was revised to align with ISO 9001:2000 [16].

AS9100 defines additional areas within an aerospace quality management system that must be addressed when implementing an ISO 9001:2000-based quality system. Typically, these requirements are included within robust aerospace quality systems. The industry experts who wrote the standard and the representatives who approved it all agree that these additions are essential to ensure product, process and service safety and quality. The AS9100 standard provides guidance for managing variation when a "key characteristic" is identified. Keys are features of a material, process or part in which the variation has a significant influence on product fit, performance, service life or manufacturability. AS9100 requires that an organization establish and document a configuration management process.

Planning product realization is essential for effective and efficient processes. The standard emphasizes planning for in-process verification when a product cannot be verified at a later point. Tooling design must also be considered when process control methodology is used to ensure that process data will be captured. The AS9100 standard includes extensive supplementation in design-and-development functions. This is not surprising given the complexity of aerospace products and customers' expectations for reliable performance during a protracted period of time. The European prEN9000-1 standard provided many of these additions. Both standards cover planning for design-and-development activities and ensuring interim control points during the design process. Design outputs are supplemented to provide identification of key characteristics, and the data essential for the product that will be identified, manufactured, inspected, used and maintained is detailed. Notes are included for both design-and-development verification and validation highlighting traditional areas of emphasis. Additionally, AS9100 provides information on areas of verification documentation and validating testing and results [17].

Managing suppliers throughout the aerospace supply chain remains a major challenge for the industry. The chain is very long, and within the supply base, there are sources that serve multiple industries. Because the industry is so dependent upon this supply chain, it isn't surprising that AS9100 includes a number of additional expectations for identifying and maintaining suppliers. Supplier approval is just one step in the process of managing suppliers. The industry typically relies upon one of three methods for product acceptance. An organization might conduct a receiving inspection, perform the inspection at the supplier's facility or formally delegate product acceptance to the supplier. Procedures for determining the method of supplier control are required, as are the processes used when employing these methods. However, no element of supplier control is more important than understanding that a supplier is responsible for managing its suppliers and sub tier suppliers. This includes performing special processes that are frequently subcontracted to processing houses. The supplier must use customer-approved sources; however, ensuring that the processing is properly performed is the supplier's responsibility.

Manufacturing a product as sophisticated as an airplane or space vehicle requires special attention during the production processes. It's important, for example, to ensure that the correct revision of the engineering documentation is being used and documented within the work instructions, and that work performance is recorded. This frequently requires a specific reference to the person performing the work. Controlling production processes is essential to demonstrate that operations have been correctly performed. This is especially important when conducting special processes that don't lend themselves to after-the-fact inspection techniques [17].

The industry frequently relies upon tooling and other production equipment, including computer-controlled machines, to fabricate and assemble products. This equipment often forms the basis for product acceptance. In these cases, it's essential to demonstrate the integrity of these tools and machines and to develop a process that will ensure adequate oversight of the entire process. Aircraft are designed to perform for 50 years or more, and properly maintaining the aircraft is essential for continued safe operation. Thus, servicing requirements are an important part of the total quality system. These include maintenance and repair manuals as well as the actual servicing work. Again, record-keeping is important in documenting the work performed, the equipment used and the people doing the work. Some products require traceability of part or all of their components. This requirement may be imposed by contract, regulatory agency or internal need. In any case, AS9100 provides the essentials of an effective traceability program. Using measuring devices of known accuracy (and this may include computer-assisted measuring and test equipment) is essential in the verification process. Maintaining a calibration history of this equipment and documented proof that it's reviewed and verified periodically underlies the entire metrology system.

Diagnosing the quality management system's health and using this information to guide improvement activity is important for efficiency and effectiveness. Internal audits performed by competent personnel are a vital input into this health measurement system. AS9100 provides some additional expectations regarding internal quality audits. Detailed first-article inspections are frequently performed to demonstrate product conformance to engineering requirements. Documenting the actual inspection and test results is an established method of demonstrating initial item acceptance [16, 17].

When things don't go as planned, AS9100 gives directions for controlling and disposing nonconforming material. This includes specific requirements for contacting the customer for authorization when using or repairing a product that doesn't conform to engineering requirements. Organizations within the industry differ in their compliance to AS9100 verification requirements. Some use their own external auditors to verify suppliers' quality management systems. Others share the results of their quality system audits with suppliers in the industry. Most provide suppliers with copies of external audits. Most permit suppliers to share the audit results with other customers, too.

Increasingly, the industry is using the results of third-party registrars as a means of demonstrating a quality management system's compliance to AS9100. The Americas Aerospace Quality Group (AAQG), working with the Registrar Accreditation Board, has established a process and requirements for auditors performing audits to AS9100 and registrars granting supplemental registrations. The process includes additional training and practical experience and ensures that auditors are competent and that registrars are experienced in the industry. The AAQG has created a Registrar Management Committee to oversee this important function. Its methodology is defined in SAE AIR5359. Europe and Asia are developing equivalent methods.

The United States Federal Aviation Administration has determined that AS9100 is "a comprehensive quality standard containing the basic quality control/assurance elements required by the current Code of Federal Regulations (CFR), Title 14, Part 21." Both the US Department of Defense and NASA have reviewed the standard and have published guidance material on using the standard for contractual requirements. As AS9100 becomes established within the industry, the standard's benefits become apparent. Two obvious ones are a reduction in multiple expectations and a consistency in verification methodology. Both prime manufacturers and their suppliers are pleased with the results. Suppliers report a reduction in verification audits and an increased consistency in expectations. As a direct result, suppliers' customers are seeing a reduction in oversight costs and an improvement in supplier performance [18].

As indicated, larger OEMs are extremely concerned with the performance of their suppliers in that their performance is crucial to the overall quality of their output.

In all cases the following is recognized as the distinctive approach expected in common:

- It is important to control the process and not the product.
- Controlling the human process is as vital, if not more so, as controlling the technical process.
- Quality is the responsibility of top management.
- Management must foster the participation of the workforce to develop a quality culture.
- Education and training are needed for changing attitudes and enhancing competence.
- Emphasize prevention of defects, not inspection after the event.
- Quality improvement is a process built up over time and not an instant cure.
- Functional integration is an important ingredient of TQM.
- Quality is a company-wide activity.

8.3.3 Global Manufacturing Industry Quality Control

In today's postmodern factories the systems approach embeds the physical process of making things to create business value. The Deming Cycle and kaizen are institutionalized practices. SQC techniques build quality and productivity into the manufacturing process. Numerically controlled machines and robots rapidly change tools, fixtures, molds as automated equipment reduces non-producing time by highly repeatable processes that are designed to ''get it right the first time''. Changeover times have been dramatically decreased in today's agile manufacturing environments.

The cooperative efforts have further been enhanced through organizations such as CAM-I group where automation producers, multi-national manufacturers and accountants have developed new cost accounting procedures that focus on resource consumption, capacity and throughput [19].

When Henry Ford famously stated that ''The customer can have any color as long as it's black'', he understood that flexibility costs time and money. Standardization enables the low cost model. In the early 1920s, when the model T was in its full glory, Ford decided to control the entire process of making and moving all supplies and parts needed by his new plant, the gigantic River Rouge. He built his own steel mill and glass plant. He founded rubber plantations in Brazil. He brought the railroad in to carry the finished cars across the country after rolling off the assembly line. This created a monstrous conglomerate that was expensive, unmanageable and horrendously unprofitable. When the Japanese introduced just-in-time concepts for supply chain inventory control the factory logistics model required an end-to-end overhaul starting from the end backwards and managed as an integrated flow.

In either case, manufacturing needed to concern itself with the responsibility of integrating people, materials, machines and time. Whether parts are outsourced or processed internally, quality procedures and conformance practices were keys to survival. Today's larger manufacturers employ several tools beyond the simple SQC control charts of the Shewhart era.

TQM is a principle as established by Deming, The basis of TQM is to reduce the errors produced during the manufacturing or service process, increase customer satisfaction, streamline supply chain management, aim for modernization of equipment and ensure workers have the highest level of training.

The Plan Do Check Act cycle introduced the concept of quality circles. A quality circle is a volunteer group composed of workers usually under the leadership of their supervisor. These individuals are trained to identify, analyze and solve work-related problems and present their solutions to management in order to improve the performance of the organization, and motivate and enrich the work of employees. They bring back the concept of craftsmanship, which when operated on an individual basis is uneconomic, but when used in group form (as is the case with quality circles) it can be devastatingly powerful and enables the enrichment of the lives of the workers and creates harmony and high performance in the workplace. Typical topics are improving occupational safety and health, improving product design, and improvement in the workplace and manufacturing processes.

Quality circles are formal groups that spring up from large quality conscious manufacturers. Members are typically cross-functional by design. They meet at least once a week on company time and are trained by competent persons (usually designated as facilitators) who may be personnel and industrial relations specialists trained in human factors and the basic skills of problem identification, information gathering and analysis, basic statistics, and solution generation [20].

The toolset is often referred to as the seven basic tools of quality. These are a fixed set of graphical analyses that are identified as being the most useful for quality control. They are called basic because they are suitable for people with little formal training in statistics and because they can be used to solve the vast majority of quality-related issues [21, 22].

The tools are:

- The cause-and-effect or Ishikawa diagram
- The check sheet
- The control chart
- The histogram
- The Pareto chart
- The scatter diagram
- Stratification (alternately flow chart or run chart)

We have discussed the control chart in great depth earlier in this chapter.

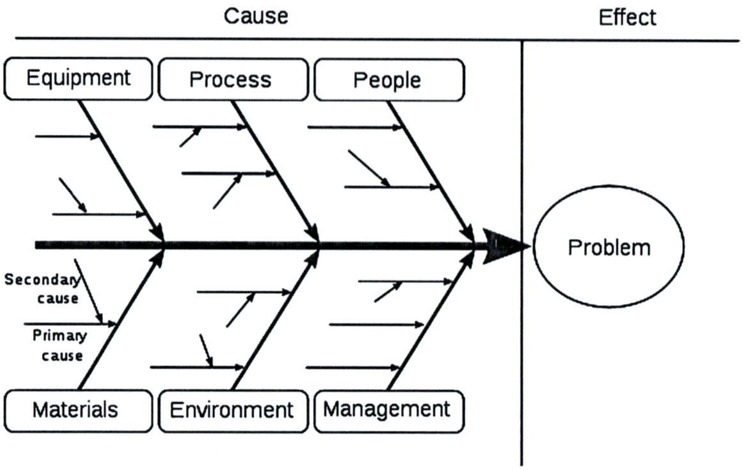

Fig. 8.12 Ishikawa fishbone cause/effect diagram

The check sheet is a simple document that is used for collecting data in real-time and at the location where the data is generated. The document is typically a blank form that is designed for the quick, easy, and efficient recording of the desired information, which can be either quantitative or qualitative. When the information is quantitative, the check sheet is sometimes called a tally sheet.

A defining characteristic of a check sheet is that data is recorded by making marks (“checks”) on it. A typical check sheet is divided into regions, and marks made in different regions have different significance. Data is read by observing the location and number of marks on the sheet. There are five basic types of Check Sheets:

- Classification: a trait such as a defect or failure mode must be classified into a category.
- Location: the physical location of a trait is indicated on a picture of a part or item being evaluated.
- Frequency: the presence or absence of a trait or combination of traits is indicated. Also number of occurrences of a trait on a part can be indicated.
- Measurement Scale: a measurement scale is divided into intervals, and measurements are indicated by checking an appropriate interval.
- Check List: the items to be performed for a task are listed so that, as each is accomplished, it can be indicated as having been completed.

Ishikawa diagrams (also called fishbone diagrams or cause-and-effect diagrams) are diagrams that show the causes of a certain event. An example is shown in Fig. 8.12. Common uses of the Ishikawa diagram are product design and quality defect prevention, to identify potential factors causing an overall effect. Each cause or reason for imperfection is a source of variation. Causes are usually grouped into major categories to identify these sources of variation. The categories typically include:

- People: anyone involved with the process.
- Methods: how the process is performed and the specific requirements for doing it, such as policies, procedures, rules, regulations and laws.
- Machines: any equipment, computers, tools etc. required to accomplish the job.
- Materials: raw materials, parts, pens, paper, etc. used to produce the final product.
- Measurements: data generated from the process that are used to evaluate its quality.
- Environment: the conditions, such as location, time, temperature, and culture in which the process operates.

Causes in the diagram are often categorized, such as to the 6 M's, described below. Cause-and-effect diagrams can reveal key relationships among various variables, and the possible causes provide additional insight into process behavior.

Causes can be derived from brainstorming sessions. These groups can then be labeled as categories of the fishbone. They will typically be one of the traditional categories mentioned above but may be something unique to the application in a specific case. Causes can be traced back to root causes with the 5 Whys technique.

Typical categories are:

- The 8 Ms (used in manufacturing)
  - Machine (technology)
  - Method (process)
  - Material (includes raw material, consumables and information)
  - Man Power (physical work)/Mind Power (brain work): Kaizens, Suggestions
  - Measurement (inspection)
  - Milieu/Mother Nature (environment)
  - Management/Money Power
  - Maintenance
- The 8 Ps (used in service industry)
  - Product = Service
  - Price
  - Place
  - Promotion
  - People
  - Process
  - Physical Evidence
  - Productivity and Quality

A Pareto chart is a type of chart that contains both bars and a line graph, where individual values are represented in descending order by bars, and the cumulative total is represented by the line.

The left vertical axis is the frequency of occurrence, but it can alternatively represent cost or another important unit of measure. The right vertical axis is the cumulative percentage of the total number of occurrences, total cost, or total of the particular unit of measure. Because the reasons are in decreasing order, the cumulative function is a concave function.

The purpose of the Pareto chart is to highlight the most important among a (typically large) set of factors. In quality control, it often represents the most common sources of defects, the highest occurring type of defect, or the most frequent reasons for customer complaints, and so on.

The Pareto principle (also known as the 80–20 rule, the law of the vital few, and the principle of factor scarcity) states that, for many events, roughly 80% of the effects come from 20% of the causes. Business management thinker Joseph M. Juran suggested the principle and named it after Italian economist Vilfredo Pareto, who observed in 1906 that 80% of the land in Italy was owned by 20% of the population; he developed the principle by observing that 20% of the pea pods in his garden contained 80% of the peas [3]. It is a common rule of thumb in business; e.g., ''80% of your sales come from 20% of your clients.'' Mathematically, where something is shared among a sufficiently large set of participants, there must be a number k between 50 and 100 such that ''k% is taken by (100−k)% of the participants''. The number k may vary from 50 (in the case of equal distribution, i.e., 100% of the population have equal shares) to nearly 100 (when a tiny number of participants account for almost all of the resource). There is nothing special about the number 80% mathematically, but many real systems have k somewhere around this region of intermediate imbalance in distribution.

8.3.4 ISO/TS 16949

The ISO/TS16949 is an international standard aiming to the development of a quality management system that provides for continual improvement, emphasizing defect prevention and the reduction of variation and waste in the supply chain. TS16949 applies to the design/development, production and, when relevant, installation and servicing of automotive-related products. It is based on ISO9001 and supersedes the QS9000 certifications. The requirements are intended to be applied throughout the supply chain. Most automotive manufacturing vehicle assembly plants are encouraged to seek ISO/TS16949 certification.

8.4 Information Modeling for Manufacturing Quality Control

Software is increasingly determining the nature of the experiences customers, employees, partners and investors have with a company, its products and services and its operations. Positive software mediated transactions are critical for retaining customers, motivating employees, and collaborating effectively with partners [6].

Many companies have accumulated an unwieldy number of incompatible, customized software systems to handle the same applications. The CIO at General Motors has estimated the organization has installed more than 7800 distinct software systems worldwide. When these systems are not compatible, transferring data, information and knowledge become nearly impossible. It is quite often the case that many of these installed applications and databases were built for individual business lines and simply do not talk to each other. In today's market, competitive advantage depends on the nature and sophistication of a company's information infrastructure. Businesses run not only through effective use of property, machines and people, they rely heavily on its data sources, databases and operating systems.

Managerial decisions depend on the availability of high quality information supplied by application software. Consider a company's primary supplier relationships and the software that mediates these interfaces. More and more information is being exchanged and each party increasingly relies on the other's information systems. It is precisely these interfaces that need sufficient structure in order to ensure low information loss.

Incoming raw material may be validated for conformance to specifications upstream in the supply chain. In the past the quality control information was typically provided in a paper report. Today, it is more often the case that the information is supplied electronically. Many formats are available including spreadsheets, text files, and other documents that serve to haul the data downstream from creator to consumer. Lack of standardization makes persistent analysis difficult at best. Unless a common format is established and agreed upon, a manufacturer can end up with many different forms of the same type of information. Integration with internal systems becomes a challenge.

Inside the factory things can be just as bad. Historically IT organizations were set up to manage an information infrastructure designed around a centralized mainframe. However, over the last 20 years these same organizations have seen a mass proliferation of decentralized and distributed computing systems, including client server architectures that have interfaces with intranets and the Internet. Although some packaged applications such as enterprise resource planning (ERP) have alleviated some standardization pressures that require information modeling exercises to design cross platform compatibility, these systems may not be robust in certain lines of business applications including engineering specialties such as computer aided design, manufacture and quality control.

8.4.1 Statistical Process Control Data Model

Statistical process control (SPC) relies on graphical presentation of inspection data through the use of control charts. SPC is the application of statistical methods to the monitoring and control of a process to ensure that it operates at its full potential to produce conforming product. Under SPC, a process behaves predictably to produce as much conforming product as possible with the least possible waste. While SPC has been applied most frequently to controlling manufacturing lines, it applies equally well to any process with a measurable output. Key tools in SPC are control charts, a focus on continuous improvement and designed experiments.

Much of the power of SPC lies in the ability to examine a process and the sources of variation in that process using tools that give weight to objective analysis over subjective opinions and that allow the strength of each source to be determined numerically. Variations in the process that may affect the quality of the end product or service can be detected and corrected, thus reducing waste as well as the likelihood that problems will be passed on to the customer. With its emphasis on early detection and prevention of problems, SPC has a distinct advantage over other quality methods, such as inspection, that apply resources to detecting and correcting problems after they have occurred.

In addition to reducing waste, SPC can lead to a reduction in the time required to produce the product or service from end to end. This is partially due to a diminished likelihood that the final product will have to be reworked, but it may also result from using SPC data to identify bottlenecks, wait times, and other sources of delays within the process. Process cycle time reductions coupled with improvements in yield have made SPC a valuable tool from both a cost reduction and a customer satisfaction standpoint.

8.4.1.1 SPC Quality Indices

Quality statistics give the machine operator or quality engineer a current reading of relevant numerical information.

The following items are examples of traditional quality statistics:

- Subgroup number (k).
- Subgroup size (n).
- ObsCnt = Observation Count = number of individual observations made.
- XDBar = $\overline{\overline{X}}$= X-double bar = the mean of the subgroup averages.
- Rbar = average of the subgroup ranges.
- s (all) = $\sqrt{\frac{\sum (X-\overline{\overline{X}})^2}{n-1}}$ = standard deviation of measurement population
- RBar/d2 = estimated standard deviation
- Min = minimum observation value
- Max = maximum observation value
- Mean ± 3,4,6s = control limits for XBar chart
- Defect Ratio = percentage of measurements outside of tolerance
- PPM = parts per million outside of tolerance
- Cp = potential process capability index = (UTL-LTL)/6 s, where s = standard deviation = RBar/d2, (UTL = Upper Tolerance Limit, LTL = Lower Tolerance Limit)

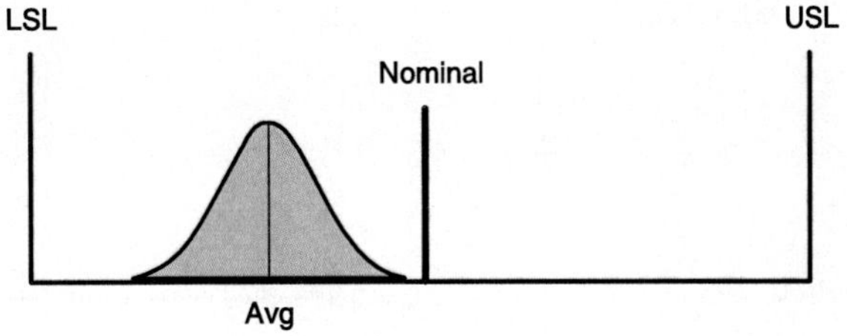

Fig. 8.13 Case 1

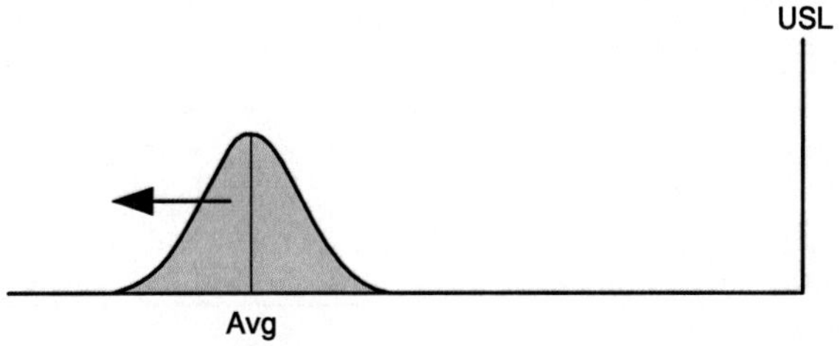

Fig. 8.14 Case 2

- Cpk = potential process performance index = Zmin/3, where Zmin = min $\{(\text{UTL} - \bar{\bar{X}})/\text{s}, (\bar{\bar{X}} - \text{LTL})/\text{s}\}$, where s = RBar/d2,
- CR = capability ratio = 1/Cp
- CPL = $(\bar{\bar{X}} - \text{LTL})/3\text{s}$, where s = RBar/d2
- CPU = $(\text{UTL} - \bar{\bar{X}})/3\text{s}$
- Pp = actual process capability index = (UTL−LTL)/σs, where $\sigma = \sqrt{\frac{\sum (X-\bar{\bar{X}})^2}{n-1}}$
- Ppk = actual process performance index = Zmin/3, where Zmin = min $\{(\text{UTL} - \bar{\bar{X}})/\text{s}, (\bar{\bar{X}} - \text{LTL})/\text{s}\}$.

8.4.1.2 Special Cases for Process Capability Calculations

Given: Engineering Specifications = 60 ± 5; USL = UTL = 65, LSL = LTL = 55, $\sigma \approx s = 2.3232$ (USL = Upper Specification Limit, LSL = Lower Specification Limit). Since Cp depends on the unknown value of σ, we will use an estimate of σ (which is s), to estimate Cp.

- Step 1: Calculate the engineering Tolerance.

  Engineering tolerance is 65−55 = 10.

- Step 2: Estimate capability.

  Process capability = $6\,\sigma \approx 6\,\text{s} = 6 \times 2.3232 = 13.9392$.

- Step 3: Estimate Cp.

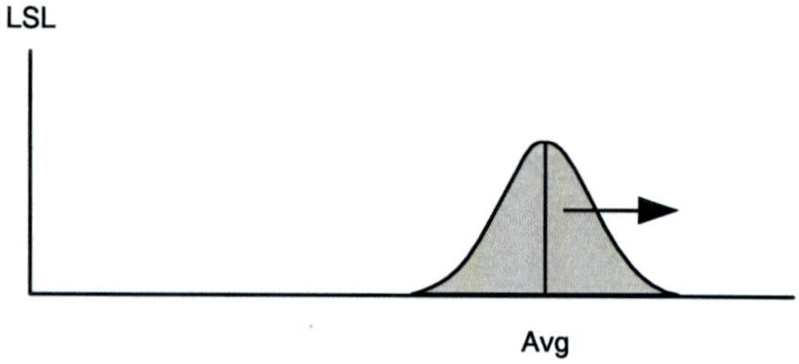

Fig. 8.15 Case 3

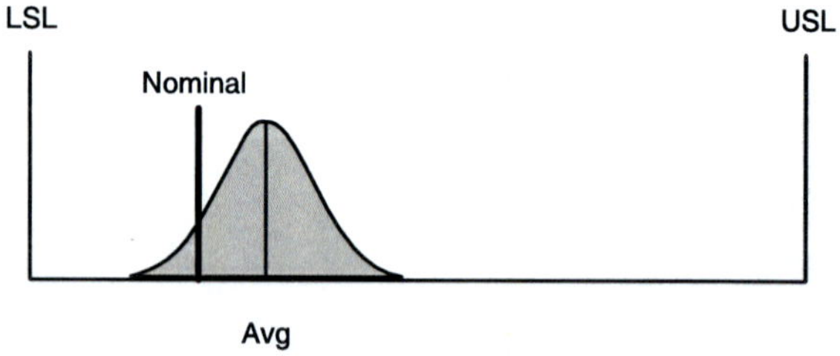

Fig. 8.16 Case 4

$$\text{Cp} \approx \frac{10}{13.9392} = 0.72$$

- Step 4: Estimate Cpk.

$$\text{Cpk} = \text{Smaller of } \left[\frac{USL - Avg}{3\sigma}, \frac{Avg - LSL}{3\sigma}\right]$$

Given engineering specifications = 60 ± 5, USL = 65, LSL = 55. Avg = 60.15, $\sigma \approx s = 2.3232$.

$$\text{Cpu} = \frac{USL - Avg}{3\sigma} \approx \frac{65 - 60.15}{3 \times 203232} = \frac{4.85}{6.9696} = 0.70$$ ← Smaller of the two

$$\text{Cpl} = \frac{Avg - LSL}{3\sigma} \approx \frac{60.15 - 55}{3 \times 2.3232} = \frac{5.15}{6.9696} = 0.74$$

Calculating Cpk for Specific Cases:

- *Case 1* (Fig. 8.13): Upper and lower specifications are provided and engineering nominal (or target) is centered between the specification limits.

$$\text{Cpk} = \text{Smaller of} \left[\frac{USL - Avg}{3\sigma}, \frac{Avg - LSL}{3\sigma}\right]$$

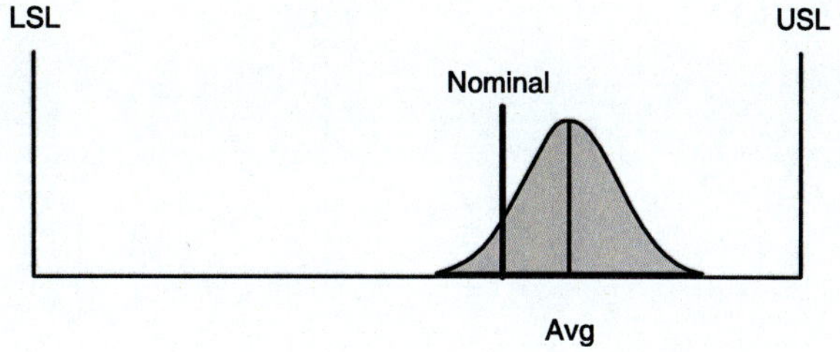

Fig. 8.17 Case 5

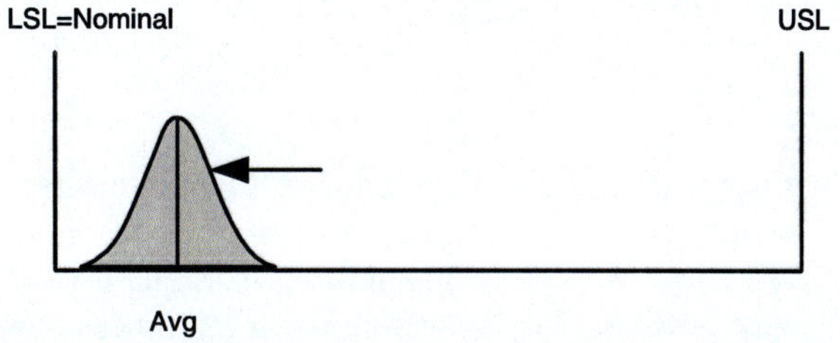

Fig. 8.18 Case 6

- *Case 2* (Fig. 8.14): A lower physical bound is used as the lower specification limit, or no lower specification exists. It is assumed that smaller feature measurements are always superior to larger values.

$$\text{Cpk} = \text{Cpu} = \frac{USL - Avg}{3\sigma}$$

- *Case 3* (Fig. 8.15): An upper physical bound is used as the upper specification limit, or no upper specification exists. It is assumed that larger feature measurements are always superior to smaller values.

$$\text{Cpk} = \text{Cpl} = \frac{Avg - LSL}{3\sigma}$$

- *Case 4* (Fig. 8.16): Upper, lower, and engineering nominal (or target) specifications are given, but nominal is closer to the lower specification than the upper specification. Cpk is maximized when the process average equals the nominal specification. Cpk is positive when the process average lies between the upper and lower specification limits, and is 0.0 when the process average equals either LSL or USL. When nominal is not centered between the upper and lower specification limit, a higher Cp is required to meet a Cpk of 1.33 than if the nominal had been centered.

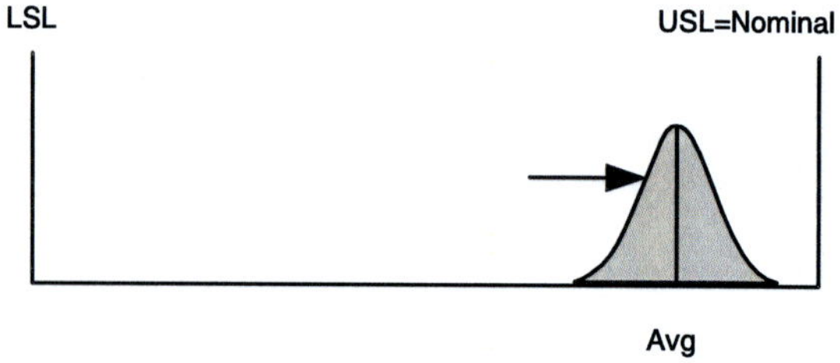

Fig. 8.19 Case 7

$$\text{Cpk} = \text{Smaller of}\left[\frac{Avg - LSL}{3\sigma}, \left(\frac{USL - Avg}{3\sigma}\right)\left(\frac{Nom - LSL}{USL - Nom}\right)\right]$$

- *Case 5* (Fig. 8.17): Upper, lower, and engineering nominal (or target) specifications are given, but nominal is closer to the upper specification than the lower specification. Cpk is maximized when the process average equals the nominal specification. Cpk is positive when the process average lies between the upper and lower specification limits, and is 0.0 when the process average equals either LSL or USL. When nominal is not centered between the upper and lower specification limit, a higher Cp is required to meet a Cpk of 1.33 than if the nominal had been centered.

$$\text{Cpk} = \text{Smaller of}\left[\left(\frac{Avg - LSL}{3\sigma}\right)\left(\frac{USL - Nom}{Nom - LSL}\right), \frac{USL - Avg}{3\sigma}\right]$$

- *Case 6* (Fig. 8.18): Upper, lower, and engineering nominal (or target) specifications are given, but the nominal is equal to the lower specification limit and there are no physical bounds limiting measurements from going below nominal.

$$\text{Cpk} = \frac{Avg - LSL}{3\sigma}$$

For this case and the following case only, a large Cpk is not desirable. The optimal Cpk is 1.33, and Cp should be maximized instead of Cpk.

- *Case 7* (Fig. 8.19): Upper, lower, and engineering nominal (or target) specifications are given, but the nominal is equal to the upper specification limit and there are no physical bounds limiting measurements from going above nominal.

$$\text{Cpk} = \frac{USL - Avg}{3\sigma}$$

For this case and the preceding case only, a large Cpk is not desirable. For most operations optimal Cpk is 1.33, and Cp should be maximized instead of Cpk.

When no nominal is given, a manufacturing target should be established—generally halfway between the upper and lower specifications. In such instances, use case 1, 4, 5, 6, or 7, as appropriate.

Cases 4, 5, 6, and 7 are encountered in manufacturing on a daily basis. Engineers give design guidance to manufacturers when nominal is intended to be off-centered and is so desired to achieve optimum product performance in the market-place. Likewise, operators machining features to maximum material condition (MMC) may help to minimize scrap and add serviceable life to many high-cost parts. Therefore, it becomes advantageous for manufacturing to know and understand where to center a process on what optimum target value, and when to aggressively strive for improving Cp while holding Cpk to a relatively lower, constant index. The value of a capable measurement system cannot be overstated, especially for these cases. Gage variation studies should be performed to add confidence in accepting and rejecting process output targets close to specification limits. There are, of course, cost considerations and tradeoffs, but setting the proper capability goals can help the producer (as well as the customer) achieve superior quality and performance. Data and information feedback to Engineering and manufacturing will enhance current and future products. Along with other information, the use of statistical control charts and capability data are vital pieces of the communication process [18].

8.4.2 Advanced Product Quality Plan Data Model

Advanced product quality planning (APQP) [23] is a framework of procedures and techniques used to develop products in industry, particularly the automotive industry. It is quite similar to the concept of design for six sigma (DFSS). It is a defined process for a product development system for General Motors, Ford, Chrysler and their suppliers. According to AIAG, the purpose of APQP is "to produce a product quality plan which will support development of a product or service that will satisfy the customer."

APQP is a process developed in the late 1980 s by a commission of experts gathered from the 'Big Three' US automobile manufacturers: Ford, GM and Chrysler. This commission invested five years to analyze the then-current automotive development and production status in the US, Europe and especially in Japan. At the time, the success of the Japanese automotive companies was starting to be remarkable in the US market. APQP is utilized today by these three companies and some affiliates. Tier I suppliers are typically required to

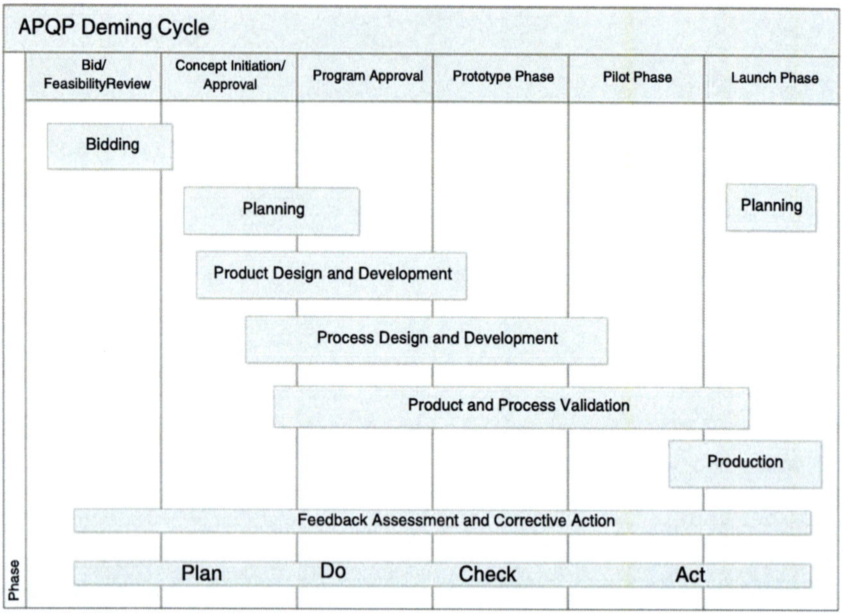

Fig. 8.20 APQP Deming cycle

follow APQP procedures and techniques and are also typically required to be audited and registered to ISO/TS 16949.

The APQP process is defined in the AIAG‘s APQP Manual, which is part of a series of interrelated documents that the AIAG controls and publishes. The basis for the make-up of a process control plan is included in the APQP Manual. The APQP provides a five stage process (as shown in Fig. 8.20) for establishing a product quality program.

- Phase 1—Plan and Define Program
  Determine customer needs, requirements and expectations using tools, such as quality function deployment (QFD), to review the entire quality planning process in order to enable the implementation of how to define and set the inputs and the outputs of a quality program.
- Phase 2—Product Design and Development
  Review the inputs and execute the outputs, which include failure mode effect analysis (FMEA), design for manufacture and assembly (DFMA), design verification, design reviews, material and engineering specifications.
- Phase 3—Process Design and Development
  Address features for developing manufacturing systems and related control plans. These tasks depend on the successful completion of phases 1 and 2 with executed outputs.

- Phase 4—Product and Process Validation
  Validate the manufacturing process and its control mechanisms through production setup while evaluating process conditions and production requirements through the analysis of pilot phase outputs.
- Phase 5—Launch, Feedback, Assessment and Corrective Action
  Focus on reducing variation and continuously improving outputs and links to customer expectations.

Control plan methodology is a key concept in the APQP specification. It describes the use of control plans and relevant data required to construct and determine control plan parameters and stresses the importance of the control plan in the continuous improvement cycle.

APQP defines the following control plan element descriptions:

- Control Plan Type—Prototype, Pre-Launch, Production
- Control Plan Number—for tracking purposes
- Part Number Latest Change Level
- Part Name and Description
- Supplier and Plant
- Supplier Code
- Key Contact and Phone Number
- Core Team
- Supplier Plant Approval Date
- Date of Original Plan
- Date of Current Revision
- Customer Engineering Approval and Date
- Process Name, Operation and Description
- Machine, Device, Jig, Tools for Manufacturing
- Product Characteristics
- Process Characteristics
- Special Characteristic Classification—e.g. Critical, Key, Safety, etc.
- Product and Process Specification and Tolerances
- Evaluation and Measurement Techniques
- Sample Size and Frequency
- Control Method
- Reaction Plan

8.4.3 OAGi Engineering to Business Data Model

The open applications group (OAGi), the organization that oversees the OAGIS, was formed in November 1994 in an effort to dramatically ease everywhere-to-everywhere integration (inside and outside of the enterprise, as well as across the supply chain). OAGi has done this by crafting standards where necessary and by recommending standards where they already exist.

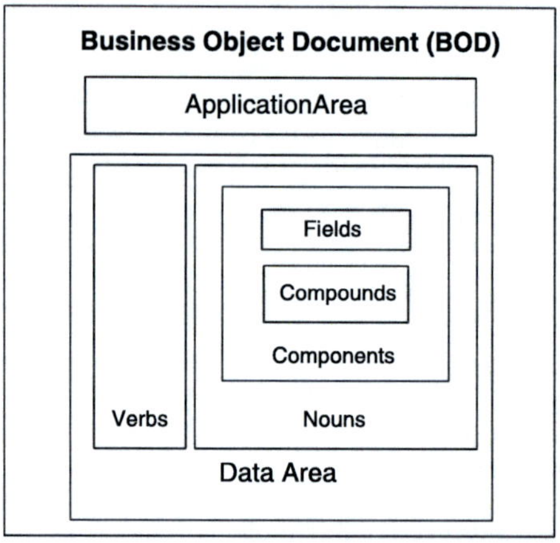

Fig. 8.21 BOD structure

The first release of OAGIS was developed in 1995 to address the need for a common business language that would enable business applications to communicate. OAGIS provides the definition of business messages in the form of business object documents (BODs) (as shown in Fig. 8.21) and example business scenarios that provide example usages of the BODs. The business scenarios identify the business applications and components being integrated and the BODs that are used. OAGIS is currently at release 9.5 and supports more than 400 business messages.

OAGi also partners with other standards bodies to provide a true canonical business language. OAGi recognizes that no one organization can be all things to all people. However, by partnering with industry vertical groups OAGIS provides the means to plug in the additional requirements and constraints that meet the specific needs of each vertical industry. Because of this long history of delivering quality usable integration standards, OAGIS has support from application vendors and implementation providers, and has been implemented by various customers in over 40 countries worldwide.

OAGIS is built as a horizontal business language, enabling it to be used in many industries worldwide. The scope of OAGIS extends the enterprise's reach across the organization, from application to application, down into the organization for enterprise application to execution systems, and outside the organization for B2B functions. The scope of OAGIS is targeted for the following types of transactions.

- eCommerce
  - e-Catalog
  - Price Lists
  - RFQ and Quote
- Order Management

- Purchasing
- Invoice
- Payments

• Manufacturing

- MES
- Shop Floor
- Plant Data Collection
- Engineering
- Warehouse Management
- Enterprise Asset Management

• Logistics

- Orders
- Shipments
- Routings

• CRM

- Opportunities
- Sales Leads
- Customer
- Sales Force Automation

• ERP

- Financials
- Human Resources
- Manufacturing
- Credit Management
- Sarbanes/Oxley and Control

The OAGIS model is said not to compete with standard electronic data interchange (EDI) business models such as ANSI X12 which was developed in 1979. However, like X12, OAGIS does target multiple business domains and does not yet have a robust quality definition although there has been significant work done for discrete manufacturing dealing with cross platform data exchange between enterprise resource planning and manufacturing execution systems.

8.5 Summary

Variation is the enemy of quality. Many types of quality improvement techniques target the reduction of variation as a key process. Quality control programs have a number of tools available including SPC, Six-Sigma, Fishbone diagrams and Pareto analysis.

Total Quality Management is concerned with improving quality across the enterprise, both small and large. Larger enterprises can afford dedicated resources and training efforts in the quality discipline but may not be agile enough to indoctrinate them. Smaller enterprises can be lean enough to adopt Quality practices but may not have sufficient resources for the effort.

Many large OEMs (including those in Automotive and Aerospace) have provided guidelines for their component suppliers. For example the QS9000 (now ISO/16949) and AS9100 standards offer manufacturers standardized approaches to quality. Data models do exist for quality as derived from the APQP and SPC guidelines from the automotive industry action group.

Some effort has been made to standardize quality models for discrete manufacturing (e.g. open applications group); however the enterprise has not fully engaged quality engineering and the shop floor.

References

1. Feo JAD, Barnard W (2004) Juran Institute's six sigma breakthrough and beyond: quality performance breakthrough methods. McGraw-Hill Professional, New York
2. Harry MJ, Schroeder R (2000) Six Sigma. Random House, UK
3. Shewhart WA (1931) Economic control of quality of manufactured product. D. Van Nostrand, New York
4. NIST (2006) Engineering statistics handbook. NIST/SEMATECH e-Handbook of statistical methods
5. Nelson LS (1984) Shewhart control chart—tests for special causes. J Qual Technol 16(4):237–239
6. Prahalad CK, Krishnan MS (1999) The new meaning of quality in the information age. Harv Bus Rev 77(5):109–118 184
7. Genichi T, Clausing D (1990) Robust quality. Harv Bus Rev January–February:65–75
8. Manoochehri GH (1988) JIT for small manufacturers. J Small Bus Manag 26(4):22–30
9. Siropolis NC (1994) Small business management: a guide to entrepreneurship. Houghton Mifflin Co, USA
10. Golhar DY, Deshpande SP (1997) HRM practices of large and small Canadian manufacturing firms. J Small Bus Manag 35(3):30–38
11. McEvoy GM (1984) Small business personnel practices. J Small Bus Manag 22(4):1–8
12. Lee GL, Oakes IK (1995) The 'pros' and 'cons' of TQM for smaller firms in manufacturing: some experiences down the supply chain. Total Qual Manag 6(4):431–444
13. Blackburn R, Rosen B (1993) Total quality and human resource management: lessons learned from baldrige award-winning companies. Acad Manag Executive 7(3):49–66
14. Lee GL (1991) The challenge of CAD/CAM: some experience of British and Canadian engineering companies. New Technol, Work Employ 1(2):100–112
15. Dawson P (1994) Total quality management. In: Storey J (ed) New wave manufacturing strategies: organizational and human resource management dimensions. P. Chapman, London
16. Bizmanualz (2006) AS 9100 Aerospace policies procedures and forms. Bizmanualz, USA
17. Barker EM (2002) Aerospace's AS9100 QMS standard, quality digest. August
18. Boeing. Advanced quality systems tools AQS D1-9000-*1*. 1998 [cited 2011 January 21]; Available from: http://www.boeingsuppliers.com/supplier/d1-9000-1.pdf
19. Drucker PF (1990) The emerging theory of manufacturing. Harv Bus Rev 68(3):94–102
20. Hutchins DC (1985) Quality circles handbook. Nichols, New York

21. Ishikawa K (1985) What is total quality control? The Japanese way. Prentice-Hall, USA
22. Tague NR (2005) The quality toolbox. ASQ Quality Press, Milwaukee
23. AIAG (1995) Advanced product quality planning and control plan (APQP) [cited 2011 January 21]; Available from: http://www.aiag.org/staticcontent/education/trainingindex.cfm?classcode=APQP

Chapter 9
Outlook for the Future of Dimensional Metrology Systems Interoperability

The development of dimensional metrology technologies follows that of manufacturing technologies. In the past 20 years, manufacturing industry has endured drastic changes and so have dimensional metrology technologies. In this information era, manufacturing industries are essential parts of digital and virtual enterprise. Information sharing and cooperation is crucial for the survival of every company. Industrial metrology is the enabling technology for information sharing between real processes, real products and digital, virtual and smart factories. It allows real people to learn and to make decisions. In this chapter, research trends of dimensional metrology in the sequence of major manufacturing technology development are discussed first. Then, the standard and technology adoption lifecycle in the industry is briefly introduced to give the reader an overview of how industry receives new standards and technologies. Finally, the two emerging international standards for dimensional metrology systems are discussed and a summary is given.

9.1 Research Trends in Dimensional Metrology Systems

In 1926, the term *mass production* was defined in an article in the Encyclopedia Britannica supplement that was written based on correspondence with Ford Motor Corporation [1]. In a mass production process, inventory has a buffer period. This process is targeted to achieve maximum efficiency; therefore, the quality requirements of the products are at acceptable level which means a product is accepted as long as the product does not break and assembles. In the past, equipment used for dimensional measurement for this type of production was mostly hand-held devices such as gages and calipers. Then, lean manufacturing techniques began to be widely employed in late 1980s. The term *lean manufacturing* was derived from the Toyota Production System (TPS) and was introduced in 1989 [2]. It is a production practice that considers the expenditure of resources for any goal other than the creation of value for the end customer to be wasteful,

Y. Zhao et al., *Information Modeling for Interoperable Dimensional Metrology*,
DOI: 10.1007/978-1-4471-2167-1_9,

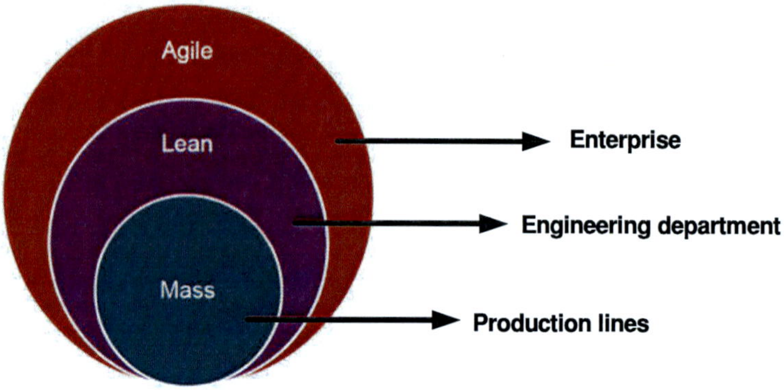

Fig. 9.1 Mass, lean and agile manufacturing

and thus a target for elimination. In lean manufacturing, inventories are kept at the minimum. Quality requirements of products and quality control during a manufacturing process become the essential parts of the process. During the transition from mass production to lean manufacturing, manufacturing technologies evolved from mostly manual operations to digital manufacturing, later on into flexible manufacturing. The technologies of dimensional measurement have also endured radical changes from manual measurement devices into flexible and computer programmable measurement equipment such as CMMs.

In the 1990s, industry leaders were trying to formulate a new paradigm for successful manufacturing enterprises in the twenty-first century; even though many manufacturing firms were still struggling to implement lean production concepts. In 1991, a group of more than 150 industry executives participated in a study. Their efforts culminated in a two-volume report titled 'Twenty-first Century Manufacturing Enterprise Strategy', which described how US industrial competitiveness would evolve during the next 15 years. As a result, the Agile Manufacturing Enterprise Forum (AMEF), affiliated with the Iacocca Institute at Lehigh University, was formed and the concept of agile manufacturing was introduced [3, 4]. For many, 'Lean manufacturing' and 'Agile manufacturing' sound similar, but they are different. Lean manufacturing is a response to competitive pressures with limited resources. Agile manufacturing, on the other hand, is a response to complexity brought about by constant change. Lean is a collection of operational techniques focused on productive use of resources. Agility is an overall strategy focused on thriving in an unpredictable environment. Focusing on the individual customer, agile competition has evolved from the unilateral producer-centered customer-responsive companies inspired by the lean manufacturing refinement of mass production to interactive producer-customer relationships [5, 6]. The development of mass production, lean manufacturing, and agile manufacturing can be seen in Fig. 9.1. Mass production mainly applies on production lines, while lean manufacturing affects the engineering department of a manufacturing firm. Agile manufacturing is a scheme that involves the entire enterprise. In order to realize

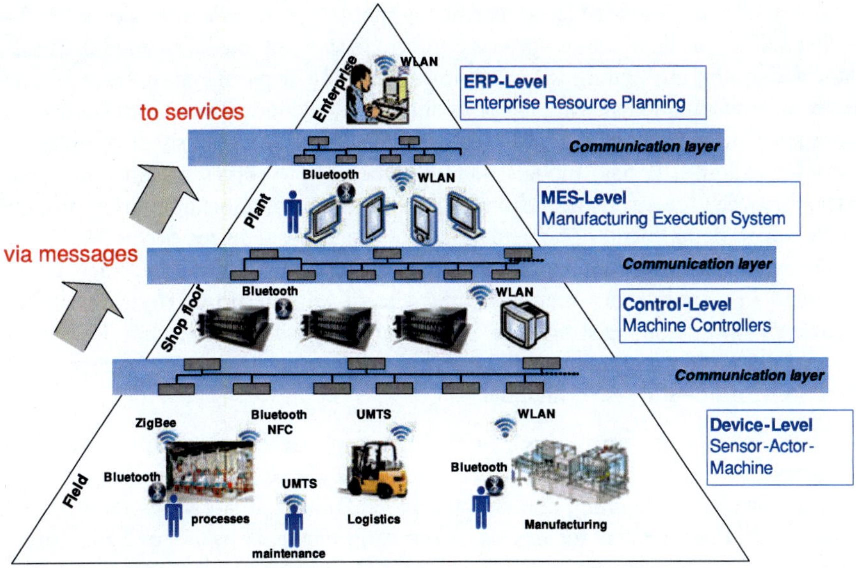

Fig. 9.2 Automation pyramid in a smart factory

agile manufacturing, every department of a manufacturing enterprise (such as supply chain, concurrent engineering department, decision support system, control system, hardware and software systems, etc.) needs to share information with each other and align their activities and effort according to fast changing customer desires. In this context, new terminologies such as smart factory, virtual factory, and cognitive factories were declared in the beginning of this century [7–10].

A typical automation pyramid in a smart factory is shown in Fig. 9.2. Among these new manufacturing research trends, how to share information seamlessly between different manufacturing departments becomes the key issue. Industrial metrology is the enabling technology for information sharing between the real process and the digital and virtual world. Dimensional measurements gather product quality information which can then be used for the improvement of product design, process efficiency, supplier selection, etc. Quality requirements from product design such as GD&T impact the choice of manufacturing equipment and processes. By associating design quality requirements and measured results of a product, certain knowledge of the chosen manufacturing process can be obtained. Therefore, dimensional metrology information is crucial not only for detecting fails and deviations in a manufacturing process, but also to understand excellence and to learn from the successes [9, 10]. For example,

- what are the process conditions when the best product results are obtained,
- what should the factory do when all manufacturing processes are perfect,
- what may be learned from successful manufacturing processes.

To attain this knowledge, dimensional metrology information needs to be connected to product value, process information, and business model. Thus, information sharing among all these manufacturing departments or elements is necessary. Information sharing is not as simple as it sounds. In order to share data seamlessly in an enterprise, certain data models must be established to represent the information. The data models must also be able to represent semantic association between the information from different sections of a manufacturing process or enterprise departments. The semantic data modeling requires expertise.

Dimensional metrology is a fundamental part of manufacturing factories. It can be found in almost every manufacturing site of an enterprise. High density of metrology information in a modern enterprise will generate information overload if the information is not managed properly. Solutions to manage dimensional metrology information may include:

- Filter the information by using metadata for information modeling. Metadata allows information inquiry to specific areas, keyword search. It also provides maintenance, cataloguing, and structuring capabilities in a data model.
- Eliminate the potential for excess information channels by interrelating information from diverse sources: calendars, CAD, metrology documents, etc.
- Automate data processing with data mining and database management technologies.
- Establish inquiry oriented methods for sharing information,
- Utilize technologies such as distributed IT, cloud computing, virtual services, etc.

The increasing complexity of manufacturing products drives dimensional metrology research in many areas, such as:

- complex mathematics for the modeling of geometry and tolerance,
- measuring methods in micro and nanoscale products,
- effects of thermal behavior of materials on dimensional measurement accuracy,
- interactions between manufacturing process and measurement process,
- knowledge management of measurement results.

It is an asset creating positive value not just a cost to prevent damage. It also provides the expertise to combine real product information with processes, customer perception, organization and business models.

9.2 Technology Adoption Lifecycle

When a new technology is developed, it takes time for industry to adopt it. The technology adoption lifecycle is a sociological model developed by Joe M. Bohlen, George M. Beal and Everett M. Rogers at Iowa State University [11]; building on earlier research conducted there about the purchase patterns of hybrid seed corn by farmers.

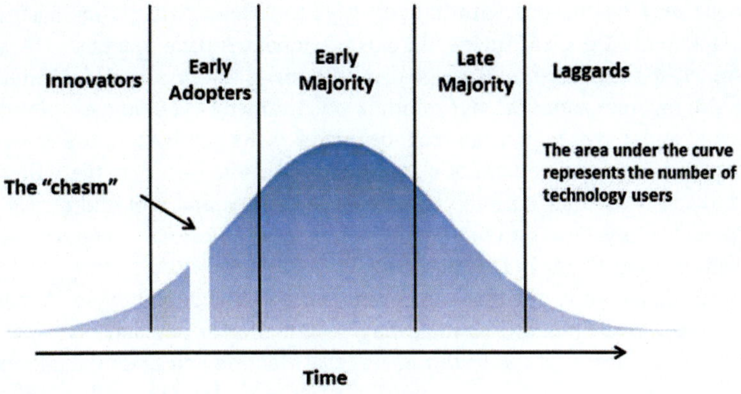

Fig. 9.3 The technology adoption lifecycle model

Beal, Rogers and Bohlen together developed a technology diffusion model [12] and later Everett Rogers generalized the use of it in his widely acclaimed book, Diffusion of Innovations [13] describing how new ideas and technologies spread in different cultures. Others have since used the model to describe how innovations spread between states in the U.S., or how new technology is proliferated in the marketplace. This model is also very useful when describing the challenges facing Interoperability in Dimensional Metrology.

The technology adoption lifecycle model (shown in Fig. 9.3) describes the adoption or acceptance of a new product or innovation, according to the characteristics of interested groups. The process of adoption over time is represented by the classical normal distribution or "bell curve." The model indicates that the first group of people to use a new product is called "innovators," followed by "early adopters." Next come the early and late majority, and the last group to eventually adopt a product are called "laggards."

This study summarized the demographics as follows:

- innovators—had larger farms, were more educated, more prosperous and more risk-oriented
- early adopters—younger, more educated, tended to be community leaders
- early majority—more conservative but open to new ideas, active in community and influence to neighbors
- late majority—older, less educated, fairly conservative and less socially active
- laggards—very conservative, had small farms and capital, oldest and least educated

Technology uptake in agriculture may not have a direct correlation to that of manufacturing but we must consider how the diffusion of innovation model applies to information modeling for interoperability in dimensional metrology.

In the real world of metrology interoperability the innovators are represented by visionaries, real people, from forward thinking organizations. Quite often these champions may be charged with finding ways to reduce costs by increasing efficiency. These are the clear intended goals of standards development.

However, it remains an elusive goal in that the lead times in development are typically long and adoption by solution practitioners even longer. Return on investment and payback periods can be many years or even decades with no guarantee of a break even condition at all. This risk often impacts the willingness of organizations to contribute the significant human capital required to standards development. Only through continuous incremental milestone successes will we see sustained investment in these works.

Another challenge is bridging the chasm in the early adopter cycle. A standard is not a standard if no one uses it. It may be a matter of demand or it may be a matter of willingness, but the truth remains that traction and practical adoption of any standard is by no means guaranteed. Just as with any new technology introduction, a critical mass must be attained for a metrology interoperability standard to gain wide acceptance. Marketing efforts are not sufficient; interoperability must be achieved through sustained vision and demand from all stakeholders. Original equipment manufacturers must strive to contribute materially to the solution, while the vendor community should drive the development of the standards that will introduce the efficiencies and cost savings that are achievable.

9.2.1 De Facto Versus De Jure Standards

There are two basic types of standards available in the industry—de facto and de jure. De facto standards may also be known as market driven. Standards that are considered market driven or de facto are those that have received wide acceptance by the industry. Some examples of de facto standards follow. Microsoft Windows, for example, is a widely accepted operating system standard with a 95% global adoption rate. Portable Document Format (PDF) and Tagged Image File Format (TIFF) are also considered de facto standards because they are widely used for transmitting documents in non-editable, non-revisable format. De facto standards result from many organizations adopting the use of them. In metrology we also have several de facto standards including VDA, HSF and JTOpen.

We also have de jure or formal standards that are developed by accredited standards organizations using rigid procedures that may periodically be audited. Formal standards development is based on openness and due process. Openness means that there are no obstacles to prevent an individual with a direct and material interest from expressing a viewpoint regardless of whether it is in agreement or disagreement with the discussion. This means that participation in the standards development activity is open to all people and organizations. The environment itself should ensure equity and fair play in the development process. These concepts of openness, due process and collaboration are found in the standards development efforts led by coalitions and consortiums. Standardization success story examples in metrology include the DMIS, DML and QMD specifications from the DMSC consortium and the AIAG group. In this realm it is sometimes difficult to find consistent harmonious implementation. We must also consider the challenge and elusiveness of standardization of semantically valid GD&T cast in the STEP AP203 edition 2 format from CAD based design models.

9.2.2 Proprietary Strategies Versus Coopetition

The traditional concept of business as a "winner takes all" contest is giving way to a realization that in the networked economy, companies must both cooperate and compete. Known as "coopetition," this emergent approach to business requires companies to create strategies that capitalize on relationships within the value chain in order to create maximum value in the marketplace.

Coopetition is a new word that has been coined to describe cooperative competition. Basic principles of this structure have been described in game theory, a scientific field that received more attention with the book Theory of Games and Economic Behavior in 1944 and the works of John Forbes Nash on Non-cooperative games.

Coopetition occurs when companies work together for parts of their business where they do not feel any distinct competitive advantage and where there is an opportunity to share common costs. For example, the arrangement between Peugeot, Citroën and Toyota to share components for a new car—simultaneously sold as the Peugeot 107, the Toyota Aygo, and the Citroën C1—qualified as coopetition. In this case, these automotive manufacturers saved on capital and shared development costs while remaining extremely competitive in other areas of the market. Other examples can be cited in constructive collaborations in the Aerospace, Electronics, Telecommunication, Pharmaceutical and many other industries whether through direct partnerships or other types of joint ventures by enabling relationships through information sharing as well as integrating and streamlining processes. In today's networked economy, coopetition is a powerful means of identifying new market opportunities and enabling the development of competitive business strategies on new technology fronts.

The quick pace of technological change in today's networked economy requires that corporate strategies and relationships must evolve over time. Metrology organizations and manufacturing firms must adapt within the changing and dynamic business environment. Companies should challenge themselves to look "outside of the box" when developing their business by initiating, leveraging and redefining relationships with other players in order to create and capture value in the dimensional metrology value chain [14].

While it is well understood how certain organizations have a well-defined "winner take all" approach with a strategy toward an end-to-end closed metrology system, these same companies should recognize the need for a broader perspective of interoperability as it relates to manufacturer desires to build best in class systems from several vendors.

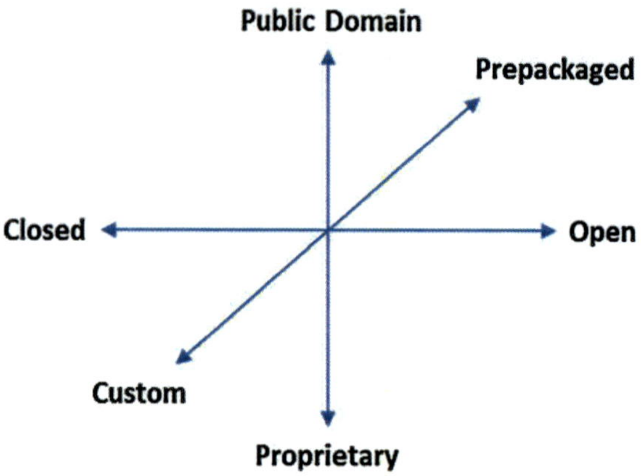

Fig. 9.4 The dimensions of software systems in manufacturing industry

9.2.3 Open Source Versus Open Standards

With software playing an increasingly vital role in manufacturing operations it is important to understand how it is developed and made available. There are three independent dimensions to be considered as shown in Fig. 9.4 [15]:

- Is the application written for general use or for a specific customer?
- Is the application developed in private or public?
- What rights do users of the application have?

9.2.3.1 Prepackaged Versus Custom Software

Software can be produced as a standard package to be used by many people and organizations without any (or substantial) changes or it can be architected and developed to suit the needs and requirements of a particular organization. These are two ends of a spectrum: off-the-shelf, shrink-wrapped software at one end and custom software at the other. Common, prepackaged software applications such as Microsoft Office, Adobe Acrobat, and Mozilla Firefox are purchased or downloaded, and can be installed easily, usually without requiring super user expertise. Custom software, on the other hand is written for a specific purpose either by in-house development staff, third party software integrators or sometimes commercial package vendors themselves. Examples include custom design systems for things like aircraft engines or special data acquisition software for automotive parts. Typically custom software requires a certain expertise to set up and use. By some estimates 90% of the money spent globally on software is for custom software.

Some software solutions are a blend of prepackaged and custom software. For example, a typical database software solution is a combination of an off-the-shelf database customized to a particular organization and combined with a set of prepackaged and custom applications.

Where it can be used prepackaged commercial software has several advantages:

- If it is in wide use, familiarity may be greater and can reduce learning curve costs.
- The development costs can be spread out over many customers.
- It already exists so you may see what you are getting through market experiences.

On the other side, the advantage of customized software is that it can meet specific needs of an organization, something that generic, prepackaged software may not be able to do. Customization comes with a price, however. There is a need to pay the development costs to write the software. This can entail major risk because a substantially large number of software projects fail to meet stakeholder expectations through functional shortcoming, late delivery and cost overrun [15].

9.2.3.2 Open Versus Closed Development

Software development involves software engineers writing source code that is compiled into binaries and executed on computer systems. For most commercial software, this source code is available only to the employees of the firm developing it; the customers purchasing the software never see the source code. For open source development, a group of developers working as individuals, or for different companies, collaborate to develop the source code, sharing both the resulting program and the source code with all interested parties. There are also intermediate approaches where the source code may be made available to a limited number of individuals or organizations, for example, consortia.

Because open source projects are typically volunteer efforts, the software is produced without firm scheduling and the only way for an organization to ensure a change will be made to the software is to do the change themselves.

For most people and organizations, having access to the source code is not important. However, the ability to examine and modify the source code gives a user with opportunities to customize the software for their specific needs in order to provide needed flexibility, scalability and maintainability [15].

9.2.3.3 Proprietary Versus Public Domain Software

The final dimension concerns who owns the software and what rights organizations have that purchase or receive it. This is all about licensing and specific business models. At one end is proprietary software that is owned and controlled by the organization that created it. Users of proprietary software may have very limited rights as to what they can do with it and the modification costs may be prohibitive. At the other end is software put into the public domain where anyone can do anything they want with it. Open source licenses tend toward the public domain end, but many of the most common ones, such as the GNU GPL, include important restrictions to make sure that the source code will always be available to all interested parties. Open source software is owned by its authors [15].

9.2.3.4 Public Good

Open source software does offer some distinct advantages to the marketplace. It is quite often of high quality, is fairly secure, and it has the ability to be customized by the software engineering community. Because the source code is readily available and is free, open source has other advantages as well, such as providing an academic backdrop that can be used to teach students learning to develop software solutions.

Because the source code is available, there is also the possibility that some software engineers in the open source community may experiment with new designs, features and implementation techniques. This has the potential to improve the overall quality of software, the state of the art, and the overall public good through experimentation and discovery. Just as the free market fuels innovation by encouraging diversity and choice, open source efforts provide a healthy competition to closed system approaches through wide availability and liberal licensing language.

The initial cost of ownership for open source software is often lower because the software is generally free, as are upgrades. However, the total cost of ownership (TCO) can be similar to other kinds of software because support, training, and maintenance are often the bulk of TCO, and open source never competes with commercial software for documentation and support based on its economic model.

Open standards also provide an essential public good. Open standards can enable anyone to contribute to shared public data structures. Open standards are at the heart of interoperability. With open standards, a person or organization can mix and match components within any given system. Without open standards—even when the system is based on open source—an organization may be unable to transform data stored in old documents or databases into the formats used by other or new applications. Open standards are a way of providing full choice to the user, and thereby efficiently and effectively lowering the total cost of ownership [16].

9.2.3.5 Internet as a Success Story for Interoperability

The Internet began in the late 1960s as an ARPA (Advanced Research Projects Agency, an agency within the Department of Defense) project that was designed to share computer resources over telephone circuits and switching nodes. This work was subsidized by the Department of Defense over a 10-year period and resulted in a set of protocols (open standards) for different computers to communicate with each other reliably. Eventually this ARPANET became the Internet which is now considered an indispensable tool in our lives.

Since its debut the Internet has facilitated open source development. In fact, many of the tools and infrastructure that support the Internet are open source. Another key to the Internet was open standards. The World Wide Web (W3C) consortium has built and maintains the foundation of the Internet protocols, including: HTTP, HTML, CSS, Ajax, XML, XSLT, SOAP, WSDL and many others.

There are some signs and significant evidence that open-source and open standards development may form a major part of the manufacturing infrastructure going forward, despite the fact that challenges to sustain an open standard and open source community remain.

One day perhaps, there may be a success story for manufacturing and private sector organizations with respect to interoperability in dimensional metrology similar to the success story of the Internet. It must all start with a fundamental business need and determining the information models to support it.

9.3 Emerging International Standards for Dimensional Metrology Systems

As new technologies emerge quickly nowadays more than ever, manufacturing and dimensional metrology companies have to adopt these technologies as fast as possible to be competitive in the market. Standards must also keep up with the technology development. In this section, two emerging standards efforts in dimensional metrology, PMI 2.0 and QIF projects, are discussed.

9.3.1 New Trends in Product Management Information Standards (PMI 2.0)

Manufacturing and dimensional metrology traditionally use information about engineering objects in the form of drawings and documents with GD&T requirements annotated on the drawing controlling geometry and dimensions. With technology development, manufactured products become more complex. This in turn results in an increasing demand for quality control of the product. The traditional geometry and dimensional tolerances are not enough to represent all the quality requirements of a product during manufacturing processes. Now, the drawings and documents have gradually faded in manufacturing. Most modern manufacturing objects are represented as 3D CAD models with GD&T and surface texture annotations. A set of new standards are being developed to represent the Product Manufacturing Information (PMI).

One of the emerging PMI standard series is the ISO/DIS 14405 standards. ISO 14405-1:2010 [16] establishes the default specification operator for linear size and

Table 9.1 Specification modifiers for linear size

Modifier	Description
(LP)	Two-point size
(LS)	Local size defined by a sphere
(GG)	Least square association criteria
(GX)	Maximum inscribed association criteria
(GN)	Minimum circumscribed association criteria
(CC)	Circumference diameter (Calculated size)
(CV)	Area diameter (Calculated size)
(CV)	Volume diameter (Calculated size)
(SX)	Maximum (rank order) size
(SN)	Minimum (rank order) size
(SA)	Average (rank order) size
(SM)	Median (rank order) size
(SD)	Mid-range (rank order) size
(SR)	Range (rank order) size

defines a number of special specification operators for linear size for feature of size types "cylinder" and "two parallel opposite planes". It also defines the specification modifiers and the drawing indications for these linear sizes. It covers the following linear sizes: local size; global size; calculated size; and rank-order size. It defines tolerances of linear sizes when there is a + and/or – limit deviation, or when there is an upper limit of size and/or lower limit of size; with an ISO size tolerance code in accordance with ISO 286-1, with or without modifiers. The new information defined in this standard includes a new series of modifiers. Table 9.1 shows the linear size modifiers defined in ISO/DIS 14405.

For CAD software vendors, they need to incorporate these symbols in their CAD systems. These new modifiers enrich the GD&T language. Nowadays, a GD&T annotation may have as many as 3 levels of compartment as shown in Fig. 9.5.

As introduced in Chap. 6, most modern measurement results analysis processes involve a measurement data fitting process that generates the measured geometry. Then, the measured geometry is compared with the design geometry. Originally, the GD&T symbols did not indicate what kind of data fitting algorithms should be used. This has changed with the publication of this new standard. In ISO/DIS 14405, a set of special elements were defined to represent the parameter and association information (shown in Table 9.2). Take symbol GQ as an example, in the drawing shown in Fig. 9.6 the tolerance means that the root-mean-square parameter of any extracted line on the upper surface, parallel to the plane of projection in which the indication is shown and measured from the total least-squares associated straight line, shall be less than or equal to 0.02 mm. We can easily notice that these special parameter and association symbols identify not only the data fitting algorithm but also the measurement method.

The new information defined in these new ISO standards is not restricted for design purposes any more. They are more considered as PMI information that

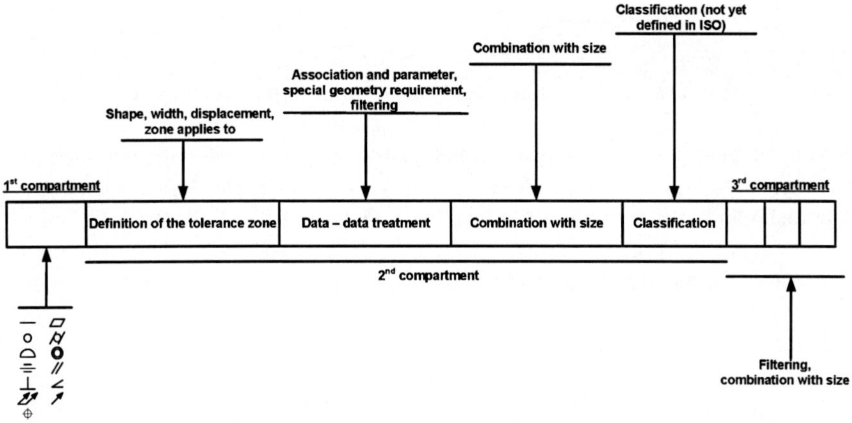

Fig. 9.5 The specification elements in the tolerance zone indicator

Table 9.2 Special parameter and association symbols

Symbol	Description
GP	Gaussian (Least squares) association reference-to-peak parameter
GV	Gaussian (Least squares) association reference-to-valley parameter
GQ	Gaussian (Least squares) association root mean square parameter
GT	Gaussian (Least squares) association peak-to-valley parameter
MI	Maximum inscribed association[a] (peak-to-valley parameter)
MC	Minimum circumscribed association[a] (peak-to-valley parameter)

[1] Only applies to roundness and cylindricity

Fig. 9.6 An example of GQ symbol

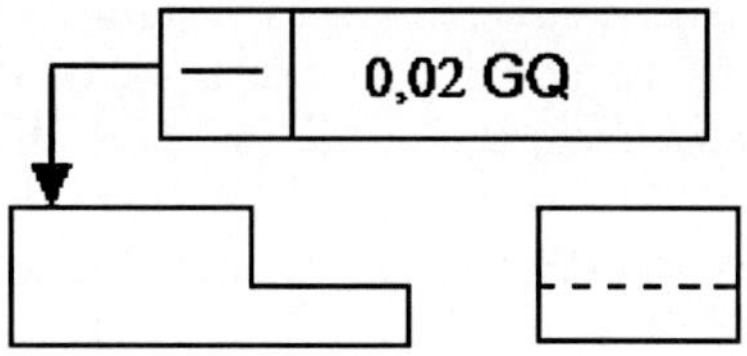

directly affects manufacturing and measurement processes. Information modeling standards introduced in this book were developed for the purpose of information exchange for dimensional metrology. They are derived directly from these base standards (i.e. ASME Y14.5, ISO 1101, ISO/DIS 14405). Any changes that occur in these base standards will influence the information exchange standards data model. Therefore, all standards working groups in interoperable dimensional metrology society should pay special attention to these emerging new standards.

9.3.2 Quality Information Framework Initiative

In early 2010, a new project named the Quality Information Framework (QIF) was initiated by the Dimensional Metrology Standards Consortium (DMSC), a consortium of businesses and experts representing the whole of the manufacturing industry including Aerospace, Automotive, National Defense, Power Generation, Nanotechnology, Electronics, Medical and Pharmaceutical, Heavy Machinery, Marine, Plastics, Consumer Goods and others. The DMSC is an ANSI accredited standards development organization and has successfully maintained, enhanced, and progressed standards such as the successful DMIS as an American and ISO standard.

The project team is comprised of domain experts from the manufacturing quality community representing a wide range of industries and quality measurement needs. Specific contributors include: Honeywell, Lockheed Martin, Chrysler, John Deere, Sandia National Laboratory, Mitutoyo America, MetroSage, Validation Technologies, NIST and others.

The purpose of QIF is to develop a new universally applicable, non-industry specific framework of interoperable standards to facilitate the exchange of manufacturing quality data between components of manufacturing quality systems. The goal of the framework is to provide a solution for the common metrology interoperability problem as identified by NIST. The solution will allow for the streamlined development, integration, and support of manufacturing quality systems and components, while maintaining the scalability necessary to adapt to an ever changing manufacturing quality landscape.

Past quality standards and specifications have been developed in isolation, each targeting a single dimension of a quality system. The QIF project is different, in that it will consist of individual standards derived from common data types and generic structures, thus ensuring interoperability between standards.

The QIF consists of a set of four standards to address the major facets of manufacturing quality systems shown in Fig. 9.7: Quality Measurement Plans (QMP), Measurement Resource Information (MRI), Quality Measurement Execution (QME), and Quality Measurement Results (QMR). Data models will be accompanied by an extensive Data Dictionary to define a common language for quality measurement enterprise. Furthermore, the framework will be capable of supporting existing database schemas currently in use by the tens of thousands of industrial customers comprising the current clientele of the DMSC membership, as well as any quality measurement enterprise, worldwide.

During the data model development in early 2011, it was found that a fifth standard was needed. This standard intends to define quality measurement statistical analysis information. However, the name of this standard has not been determined yet. It is only named as QMx for the time being.

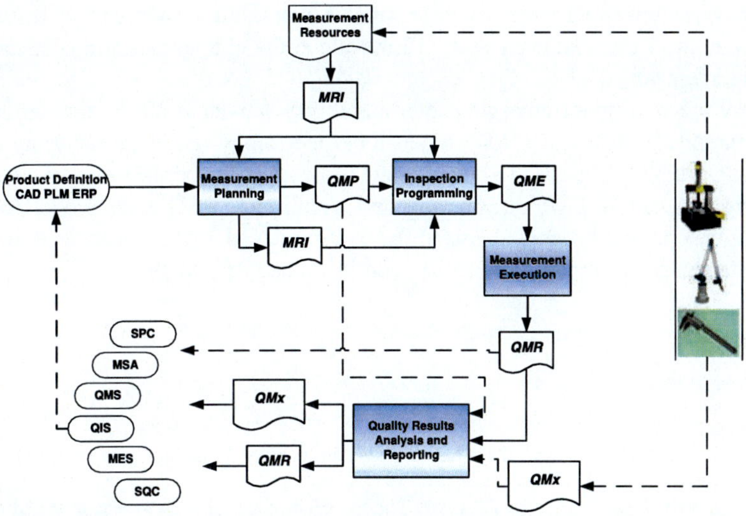

Fig. 9.7 QIF structure

9.4 Summary

The development of dimensional metrology technologies follows that of manufacturing technologies. In the past four decades, manufacturing industry has endured drastic changes from mass production in the 1960s, to lean manufacturing in the late 1980s, and later on to agile manufacturing in the late 1990s. In the new century, new manufacturing technologies to enable smart and virtual manufacturing are becoming mature. While the manufacturing industry undergoes these changes, so does the dimensional metrology industry. Dimensional measurement devices have changed from manual gages, to digital gages, and later to computer controlled CMMs and portable robotic measurement arms. The capability and accuracy of the measurement equipment also improved significantly. Dimensional measurement is now an integral part of manufacturing processes. Therefore, the information exchange among dimensional measurement systems and between dimensional measurement systems and manufacturing system becomes crucial to the competitiveness of any industrial company.

In the real world of metrology interoperability the innovators are represented by visionaries, real people, from forward thinking organizations. Quite often these champions may be charged with finding ways to reduce costs by increasing efficiency. These are the clear intended goals of standards development. However, it remains an elusive goal. Another challenge is bridging the chasm in the early adopter cycle. A standard is not a standard if no one uses it. Practical adoption of any standard is by no means guaranteed. Just as with any new technology introduction, a critical mass must be attained for a metrology interoperability standard to gain wide acceptance. Original equipment manufacturers must strive to contribute materially to the solution vendor community in order to drive the development of the standards that will introduce the efficiencies and cost savings that are achievable.

There are two major emerging interoperability standards efforts that enjoy the participations from major CAD, dimensional metrology and manufacturing organizations. These standards are being created to enable the complete representation of design GD&T + PMI information and a comprehensive data model for the entire measurement process chain. In the coming decades, the implementations of these standards are most likely to be seen in industry products.

References

1. Hounshell DA (1985) From the American system to mass production, 7th edn. JHU Press, Baltimore
2. Womack JP, Jones DT, Roos D (1990) The machine that changed the world: based on the Massachusetts Institute of Technology 5-million dollar 5-year study on the future of the automobile. Simon and Schuster, New York
3. Nagel RN, Dove R (1991) 21st century manufacturing enterprise strategy: an industry-lead view. Iacocca Institute, Bethlehem
4. Richards CW (1996) Agile manufacturing: beyond lean? Prod Inventory Manag J 37(2):60–64
5. Goldsmith SL, Nagel RN, Preiss K (1994) Agile competitors and virtual organizations: strategies for enriching the customer. Van Nostrand Reinhold, New York
6. Sanchez LM, Nagi R (2001) A review of agile manufacturing systems. Int J Prod Res 39(16):3561–3600
7. Zaeh MF et al (2009) The cognitive factory. In: ElMaraghy HA (ed) Changeable and reconfigurable manufacturing systems. Springer, London, pp 355–371
8. Zhao YF, Xu X (2010) Enabling cognitive manufacturing through automated on-machine measurement planning and feedback. Adv Eng Inform 24(3):269–284
9. Zuehlke D (2008) SmartFactory—from vision to reality in factory technologies. In: Proceedings of the 17th World Congress the International Federation of Automatic Control, Seoul, Korea
10. Zuehlke D (2010) SmartFactory-towards a factory-of-things. Annu Rev Control 34(1):129–138
11. Bohlen JM, Beal GM (1957) The diffusion process. Special Report No. 18, Agriculture Extension Service, Iowa State College
12. Beal GM, Rogers EM, Bohlen JM (1957) Validity of the concept of stages in the adoption process. Rural Sociol 22(2):166–168
13. Rogers EM (1995) Diffusion of innovations. Simon and Schuster, New York
14. Bowser J (2001) Strategic co-opetition: the value of relationships in the networked economy. IBM Global Services, New York
15. Gabriel RP, Goldman R (2011) Open standards, open source, and the public good. http://www.dreamsongs.com/OpenStandards.html. Accessed 6 July 2011
16. ISO (2010) ISO 14405-1: Geometrical product specifications (GPS)—dimensional tolerancing—Part 1: Linear sizes

Appendix A
Geometric Tolerances and the Surface They Control

Type	Characteristic	Example	Feature(s)	Datum Type
S	Spherical Diameter SØ	SØ10±0.01	Sphere	N/A
S	Diameter Ø	Ø10±0.01	Cylinder	N/A
S			Circle	N/A
S	Spherical Radius SR	SR10±0.01	Spherical Arc	N/A
S	Radius R	R10±0.01	Cylindrical Arc	N/A
S	Controlled Radius CR	CR10±0.01	Cylindrical Arc	N/A
S	Width	10±0.01	Parallel Plane	N/A
S			Parallel Points	N/A
F	Straightness —	— Ø0.1	Cylinder	N/A
F			Derived Median Line	N/A
F		— Ø0.1Ⓜ	Cylinder	N/A
F			Derived Median Line	N/A
F		— Ø0.1/10	Cylinder	N/A
F			Derived Median Line	N/A
F		— 0.1	Line	N/A
F			Parallel Planes	N/A
F			Derived Median Plane	N/A
F		— 0.1Ⓜ	Parallel Planes	N/A
F			Derived Median Plane	N/A
F		— 0.1/10	Line	N/A
F	Flatness ⏥	⏥ 0.1	Plane	N/A
F		⏥ 0.1/10X10	Plane	N/A
F	Circularity ○	○ 0.1	Sphere	N/A
F			Circle	N/A
F	Cylindricity ⌭	⌭ 0.1	Cylinder	N/A
O	Perpendicularity ⊥	⊥ Ø0.1 A	Cylinder	Cylinder
O				Plane

(continued)

Y. Zhao et al., *Information Modeling for Interoperable Dimensional Metrology*,
DOI: 10.1007/978-1-4471-2167-1, © Springer-Verlag London Limited 2011

Type	Characteristic	Example	Feature(s)	Datum Type
O				Parallel Plane
O		[⊥][⌀0.1Ⓜ][A]	Cylinder	Cylinder or (Par)plane
O		[⊥][⌀0.1Ⓛ][A]	Cylinder	Cylinder or (Par)plane
O		[⊥][⌀0.1Ⓟ10][A]	Cylinder	Cylinder or (Par)plane
O		[⊥][⌀0.1][AⓂ]	Cylinder	Cylinder
O				Parallel Plane
O		[⊥][⌀0.1][AⓁ]	Cylinder	Cylinder
O				Parallel Plane
O		[⊥][⌀0.1][A][B]	Cylinder	primary Cylinder
O		[⊥][0.1][A]	Cylinder	Cylinder
O				Plane
O				Parallel Plane
O			Plane	Cylinder
O				Plane
O				Parallel Plane
O			Parallel Plane	Cylinder
O				Plane
O				Parallel Plane
O			Line	Cylinder
O				Plane
O				Parallel Plane
O		[⊥][0.1Ⓜ][A]	Cylinder	Cylinder or (Par)plane
O			Parallel Plane	Cylinder or (Par)plane
O		[⊥][0.1Ⓛ][A]	Cylinder	Cylinder or (Par)plane
O			Parallel Plane	Cylinder or (Par)plane
O		[⊥][0.1Ⓟ10][A]	Cylinder	Cylinder or (Par)plane
O		[⊥][0.1Ⓣ][A]	Plane	Cylinder or (Par)plane
O		[⊥][0.1][AⓂ]	Cylinder, (Par)Plane, or Line	Cylinder
O				Parallel Plane
O		[⊥][0.1][AⓁ]	Cylinder, (Par)Plane, or Line	Cylinder
O				Parallel Plane
O		[⊥][0.1][A][B]	Plane	primary (Par)Plane
O			Parallel Plane	primary (Par)Plane
O			Line	primary Cylinder
O				primary (Par)Plane
O	Parallelism [//]	[//][⌀0.1][A]	Cylinder	Cylinder
O				Plane
O				Parallel Plane
O		[//][⌀0.1Ⓜ][A]	Cylinder	Cylinder or (Par)plane
O		[//][⌀0.1Ⓛ][A]	Cylinder	Cylinder or (Par)plane

Type	Characteristic	Example	Feature(s)	Datum Type
O		// Ø0.1Ⓟ10 A	Cylinder	Cylinder or (Par)plane
O		// Ø0.1 AⓂ	Cylinder	Cylinder
O				Parallel Plane
O		// Ø0.1 AⓁ	Cylinder	Cylinder
O				Parallel Plane
O		// Ø0.1 A B	Cylinder	primary (Par)Plane
O		// 0.1 A	Cylinder	Cylinder
O				Plane
O				Parallel Plane
O			Plane	Cylinder
O				Plane
O				Parallel Plane
O			Parallel Plane	Cylinder
O				Plane
O				Parallel Plane
O			Line	Cylinder
O				Plane
O				Parallel Plane
O		// 0.1Ⓜ A	Cylinder	Cylinder or (Par)plane
O			Parallel Plane	Cylinder or (Par)plane
O		// 0.1Ⓛ A	Cylinder	Cylinder or (Par)plane
O			Parallel Plane	Cylinder or (Par)plane
O		// 0.1Ⓣ A	Plane	Cylinder or (Par)plane
O		// 0.1 AⓂ	Cylinder, (Par)Plane, or Line	Cylinder
O				Parallel Plane
O		// 0.1 AⓁ	Cylinder, (Par)Plane, or Line	Cylinder
O				Parallel Plane
O		// 0.1 A B	Plane	Cylinder
O			Parallel Plane	Cylinder
O			Line	Cylinder
O				(Par)Plane
O	Angularity ∠	∠ Ø0.1 A	Cylinder	Cylinder
O				Plane
O				Parallel Plane
O		∠ Ø0.1Ⓜ A	Cylinder	Cylinder or (Par)plane
O		∠ Ø0.1Ⓛ A	Cylinder	Cylinder or (Par)plane
O		∠ Ø0.1Ⓟ10 A	Cylinder	Cylinder or (Par)plane
O		∠ Ø0.1 AⓂ	Cylinder	Cylinder
O				Parallel Plane
O		∠ Ø0.1 AⓁ	Cylinder	Cylinder
O				Parallel Plane

(continued)

Type	Characteristic	Example	Feature(s)	Datum Type
O		∠ Ø0.1 A B	Cylinder	primary Cylinder
O				primary (Par)Plane
O		∠ 0.1 A	Cylinder	Cylinder
O				Plane
O				Parallel Plane
O			Plane	Cylinder
O				Plane
O				Parallel Plane
O			Parallel Plane	Cylinder
O				Plane
O				Parallel Plane
O			Line	Cylinder
O				Plane
O				Parallel Plane
O		∠ 0.1Ⓜ A	Cylinder	Cylinder or (Par)plane
O			Parallel Plane	Cylinder or (Par)plane
O		∠ 0.1Ⓛ A	Cylinder	Cylinder or (Par)plane
O			Parallel Plane	Cylinder or (Par)plane
O		∠ 0.1Ⓣ A	Plane	Cylinder or (Par)plane
O		∠ 0.1 AⓂ	Cylinder, (Par)Plane, or Line	Cylinder
O				Parallel Plane
O		∠ 0.1 AⓁ	Cylinder, (Par)Plane, or Line	Cylinder
O				Parallel Plane
O		∠ 0.1 A B	Cylinder	primary Cylinder
O				primary (Par)Plane
O			Plane	primary Cylinder
O				primary (Par)Plane
O			Parallel Plane	primary Cylinder
O				primary (Par)Plane
O			Line	primary Cylinder
O				primary (Par)Plane
R	Circular Runout ↗	↗ 0.1 A	Circle	axis
R			Plane	axis
R	Total Runout ⌰	⌰ 0.1 A	Cylinder	axis
R			Plane	axis
R			Cone	axis
R	Profile of a Line ⌒	⌒ 0.1 A B C	Line	no DOF
R			Circle	no DOF
R			2D Curve	no DOF
R		⌒ 0.1 A B C Ⓜ	2D Curve	Sphere
R				Cylinder
R				Parallel Plane

Type	Characteristic	Example	Feature(s)	Datum Type
R		⌒ / 0.1 / A / B / C Ⓛ	2D Curve	Sphere
R				Cylinder
R				Parallel Plane
R		⌒ 0.1 / A / B / C 0.1 / A / B	Pattern of 2D Curves	
R		Simultaneous	Grp. of different 2D Curves	
R		unequal tolerance zone	2D Curve	
R		⌒ / 0.1 / A / B	Circle	
R			2D Curve	
R				
R			Pattern of 2D Curves	
R				
R		⌒ / 0.1 / A	2D Curve	
R			Pattern of 2D Curves	
P	Profile of a Surface ⌓	⌓ / 0.1 / A / B / C	Sphere	
P			Cylinder	
P			Plane	
P			3D Surface	
P		⌓ / 0.1 / A / B / C Ⓜ	3D Surface	Sphere
P				Cylinder
P				Parallel Plane
P		⌓ / 0.1 / A / B / C Ⓛ	3D Surface	Sphere
P				Cylinder
P				Parallel Plane
P		⌓ 0.1 / A / B / C 0.1 / A / B	Pattern of 3D Surfaces	
P		Simultaneous	Grp. of different 3D Surfaces	
P		Unequal tolerance zone	3D Surface	
P		⌓ / 0.1 / A / B	Sphere	
P			Cylinder	
P				
P			3D Surface	
P				
P			Pattern of 3D Surfaces	
P				
P		⌓ / 0.1 / A	3D Surface	
P				
P				
P			Pattern of 3D Surfaces	
P				
P				
P		⌓ / 0.1	3D Surface	
P			Pattern of 3D Surfaces	
L	Position ⌖	⌖ / SØ0.1 / A / B / C	Sphere	
L			Pattern of Spheres	

(continued)

Type	Characteristic	Example	Feature(s)	Datum Type
L		⌖ SØ0.1Ⓜ A B C	Sphere(s)	
L		⌖ SØ0.1Ⓛ A B C	Sphere(s)	
L		⌖ SØ0.1 A B CⓂ	Sphere(s)	Sphere
L				Cylinder
L				Parallel Plane
L		⌖ SØ0.1 A B CⓁ	Sphere(s)	Sphere
L				Cylinder
L				Parallel Plane
L		⌖ SØ0.1 A B C SØ0.1 A B	Pattern of Spheres	
L		Simultaneous	Grp. of different Spheres	
L		⌖ SØ0.1 A B	Sphere	
L				
L			Pattern of Spheres	
L				
L		⌖ SØ0.1 A	Sphere	
L				
L				
L			Pattern of Spheres	
L				
L				
L		⌖ SØ0.1	Pattern of Spheres	
L		⌖ Ø0.1 A B C	Sphere	
L			Cylinder	
L			Pattern of Spheres	
L			Pattern of Cylinders	
L		⌖ Ø0.1Ⓜ A B C	Cylinder	
L			Sphere	
L		⌖ Ø0.1Ⓛ A B C	Cylinder	
L			Sphere	
L		⌖ Ø0.1Ⓟ10 A B C	Cylinder	
L		⌖ Ø0.1 A B CⓂ	Sphere(s) or Cylinder(s)	
L				
L				
L		⌖ Ø0.1 A B CⓁ	Sphere(s) or Cylinder(s)	
L				
L				
L		⌖ Ø0.1 A B C	Pattern of Spheres	
L		Ø0.1 A B	Pattern of Cylinders	
L		Simultaneous	Grp. of different Spheres	
L			Grp. of different Cylinders	
L			Grp. of different Features	
L		⌖ Ø0.1 A B	Sphere	
L			Cylinder	
L				
L			Pattern of Spheres	
L				

Type	Characteristic	Example	Feature(s)	Datum Type
L			Pattern of Cylinders	
L				
L		⌖ Ø0.1 A	Sphere	
L				
L				
L			Cylinder	
L				
L				
L			Pattern of Spheres	
L				
L				
L			Pattern of Cylinders	
L				
L				
L		⌖ Ø0.1	Pattern of Spheres	
L			Pattern of Cylinders	
L		⌖ 0.1 A B C	Sphere	
L			Cylinder	
L			Parallel Plane	
L			Pattern of Cylinders	
L			Pattern of Parallel Planes	
L			Pattern of Spheres	
L		⌖ 0.1Ⓜ A B C	Sphere	
L			Cylinder	
L			Parallel Plane	
L		⌖ 0.1Ⓛ A B C	Sphere	
L			Cylinder	
L			Parallel Plane	
L		⌖ 0.1 A B CⓂ	Feature(s)	
L				
L				
L		⌖ 0.1 A B CⓁ	Feature(s)	
L				
L				
L		⌖ 0.1 A B C	Pattern of Spheres	
L		0.1 A B	Pattern of Cylinders	
L			Pattern of Parallel Planes	
L		Simultaneous	Grp. of different Spheres	
L			Grp. of different Cylinders	
L			Grp. of different Par. Planes	
L			Grp. of different Features	
L		⌖ 0.1 A B	Cylinder	
L			Parallel Plane	
L			Pattern of Spheres	
L				
L			Pattern of Cylinders	
L				
L			Pattern of Parallel Planes	

(continued)

Type	Characteristic	Example	Feature(s)	Datum Type
L				
L		⌖ | 0.1 | A	Pattern of Spheres	
L				
L				
L			Pattern of Cylinders	
L				
L				
L			Pattern of Parallel Planes	
L				
L				
L		⌖ | 0.1	Pattern of Spheres	
L			Pattern of Cylinders	
L			Pattern of Parallel Planes	
L	Concen- ◎	◎ | SØ0.1 | A	Sphere	Sphere
L	tricity		Median Point	Sphere
L		◎ | Ø0.1 | A	Cylinder	Cylinder
L			Median Line	Cylinder
L	Symmetry ⌯	⌯ | 0.1 | A	Parallel Plane	Cylinder
L				Parallel Plane
L			Median Plane	Cylinder
L				Parallel Plane

S Size, *F* Form, *O* Orientation, *R* Runout, *P* Profile

Appendix B
Empty Shape Representation Example File in STEP AP 203 Edition 2

Empty shape representation example file based on STEP AP 203 edition 2.

```
ISO-10303-21;
HEADER;
   /* Generated by software containing ST-Developer
    * from STEP Tools, Inc. (www.steptools.com)
    */
FILE_DESCRIPTION(
   /* description */ ('Empty Shape Representation Sample File'),
   /* implementation_level */ '2;1');
FILE_NAME(
   /* name */ 'output_file',
   /* time_stamp */ '2010-02-17T11:51:02-05:00',
   /* author */ (''),
   /* organization */ (''),
   /* preprocessor_version */ 'ST-DEVELOPER v12',
   /* originating_system */ 'Geometry Demo',
   /* authorisation */ '');
FILE_SCHEMA
(('AP203_CONFIGURATION_CONTROLLED_3D_DESIGN_OF_MECHANICAL_PARTS_AND_ASSEMBLIE
S_MIM_LF'));
ENDSEC;

DATA;
#10=AXIS2_PLACEMENT_3D('orientation',#11,$,$);
#11=CARTESIAN_POINT('',(1.1,2.2,3.3));
#12=UNCERTAINTY_MEASURE_WITH_UNIT(LENGTH_MEASURE(1.E-006),#14,
'DISTANCE_ACCURACY_VALUE','Maximum model space distance between geometric
entities at asserted connectivities');
#13=(GEOMETRIC_REPRESENTATION_CONTEXT(3)GLOBAL_UNCERTAINTY_ASSIGNED_CONTEXT(
(#12))GLOBAL_UNIT_ASSIGNED_CONTEXT((#14,#16,#19))REPRESENTATION_CONTEXT
('ID1','3D'));
#14=(LENGTH_UNIT()NAMED_UNIT(*)SI_UNIT(.MILLI.,.METRE.));
#15=DIMENSIONAL_EXPONENTS(0.,0.,0.,0.,0.,0.,0.);
#16=(CONVERSION_BASED_UNIT('degree',#18)NAMED_UNIT(#15)PLANE_ANGLE_UNIT());
#17=(NAMED_UNIT(*)PLANE_ANGLE_UNIT()SI_UNIT($,.RADIAN.));
#18=MEASURE_WITH_UNIT(PLANE_ANGLE_MEASURE(0.01745329252),#17);
#19=(NAMED_UNIT(*)SI_UNIT($,.STERADIAN.)SOLID_ANGLE_UNIT());
#20=SHAPE_DEFINITION_REPRESENTATION(#21,#22);
#21=PRODUCT_DEFINITION_SHAPE('',$,#24);
#22=SHAPE_REPRESENTATION('',(#10),#13);
#23=PRODUCT_DEFINITION_CONTEXT('3D Mechanical Parts',#27,'design');
#24=PRODUCT_DEFINITION('design','example product_definition',#25,#23);
#25=PRODUCT_DEFINITION_FORMATION('1.0','first version of our widget',#29);
```


DOI: 10.1007/978-1-4471-2167-1,

```
#26=APPLICATION_PROTOCOL_DEFINITION('technical specification',
'config_control_design',2005,#27);
#27=APPLICATION_CONTEXT('configuration controlled 3D designs of mechanical
parts and assemblies');
#28=PRODUCT_CONTEXT('3D Mechanical Parts',#27,'mechanical');
#29=PRODUCT('1234-K789','widget','a fictional product',(#28));
ENDSEC;
END-ISO-10303-21;
```

Appendix C
EXPRESS-G Diagrams of HIPP Data Model

The following figures show the EXPRESS-G diagrams of integrated data model for in-process measurement based on the HIPP data model proposed by NIST in 2007. It extends the STEP AP 238 ARM model.

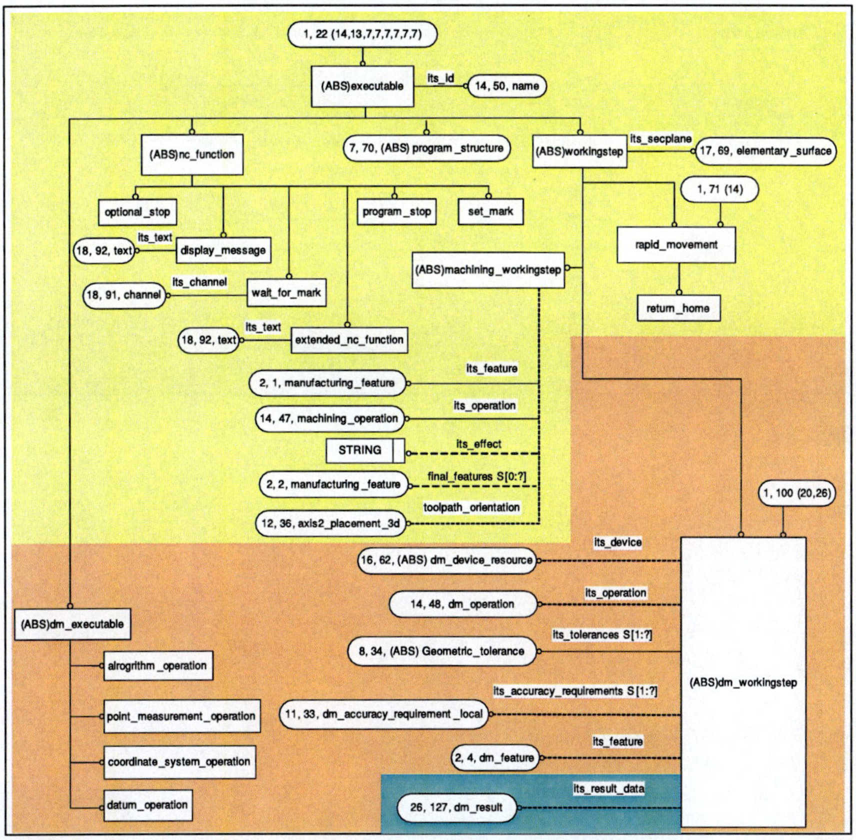

Fig. C.1

Y. Zhao et al., *Information Modeling for Interoperable Dimensional Metrology*,
DOI: 10.1007/978-1-4471-2167-1, © Springer-Verlag London Limited 2011

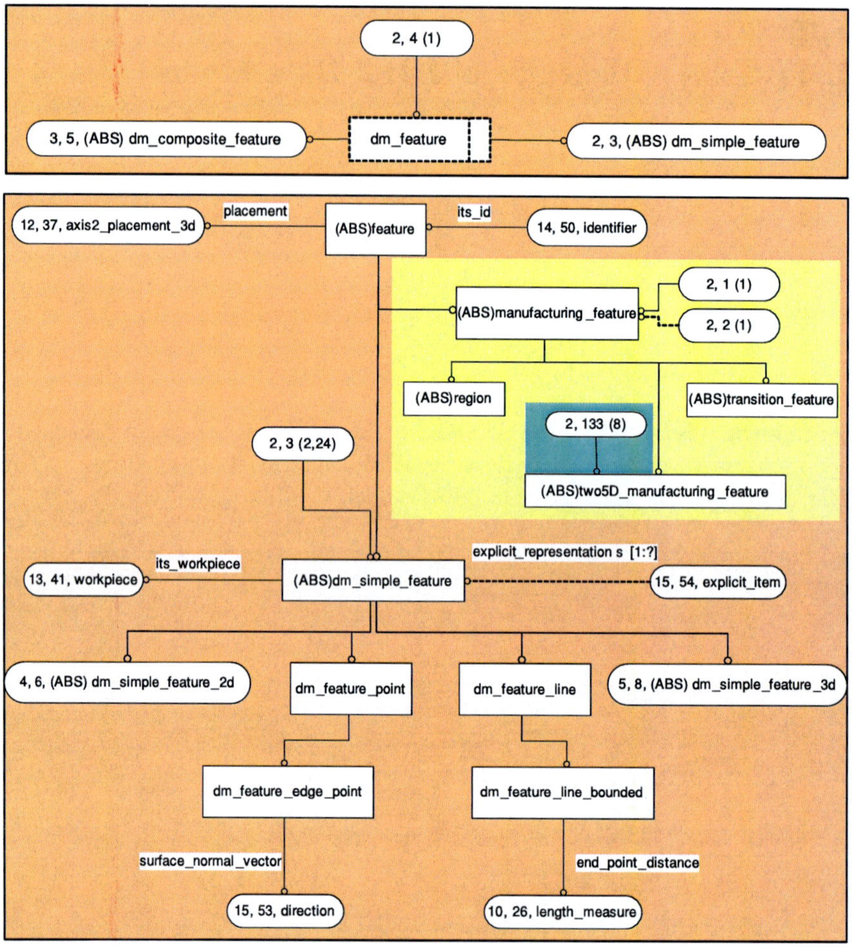
2, 4 (1)
3, 5, (ABS) dm_composite_feature
dm_feature
2, 3, (ABS) dm_simple_feature
12, 37, axis2_placement_3d
placement
(ABS)feature
its_id
14, 50, identifier
2, 1 (1)
(ABS)manufacturing_feature
2, 2 (1)
(ABS)region
(ABS)transition_feature
2, 133 (8)
2, 3 (2,24)
(ABS)two5D_manufacturing_feature
explicit_representation s [1:?]
its_workpiece
13, 41, workpiece
(ABS)dm_simple_feature
15, 54, explicit_item
4, 6, (ABS) dm_simple_feature_2d
dm_feature_point
dm_feature_line
5, 8, (ABS) dm_simple_feature_3d
dm_feature_edge_point
dm_feature_line_bounded
surface_normal_vector
end_point_distance
15, 53, direction
10, 26, length_measure

Fig. C.2

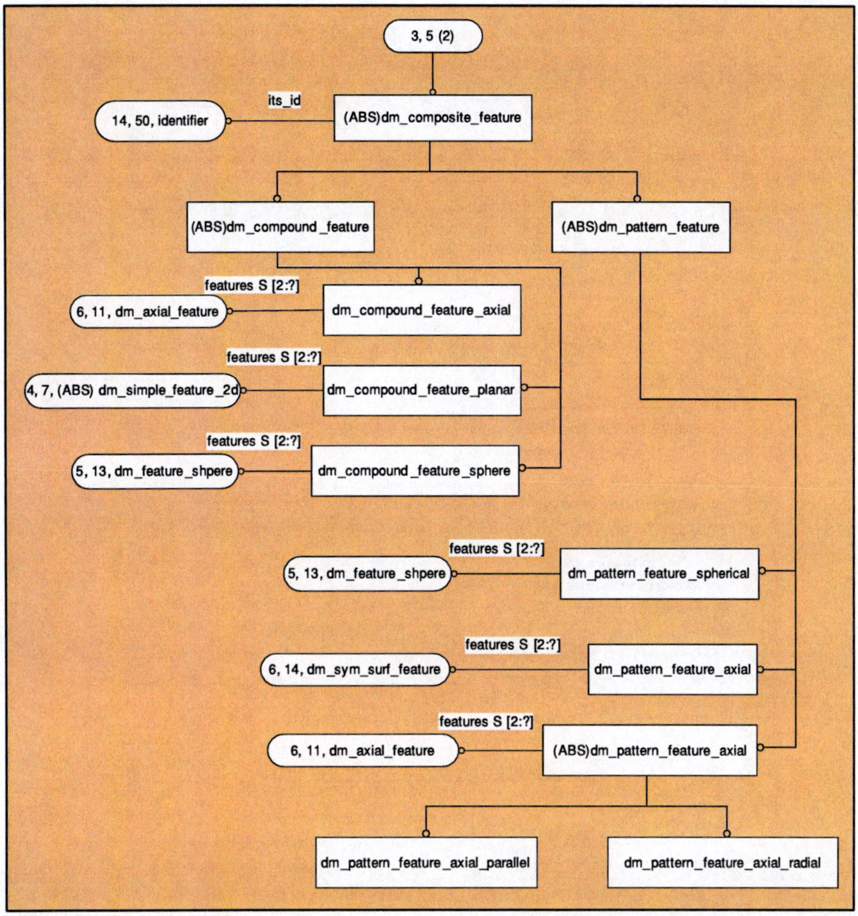
3, 5 (2)
its_id
14, 50, identifier
(ABS)dm_composite_feature
(ABS)dm_compound_feature
(ABS)dm_pattern_feature
features S [2:?]
6, 11, dm_axial_feature
dm_compound_feature_axial
features S [2:?]
4, 7, (ABS) dm_simple_feature_2d
dm_compound_feature_planar
features S [2:?]
5, 13, dm_feature_shpere
dm_compound_feature_sphere
features S [2:?]
5, 13, dm_feature_shpere
dm_pattern_feature_spherical
features S [2:?]
6, 14, dm_sym_surf_feature
dm_pattern_feature_axial
features S [2:?]
6, 11, dm_axial_feature
(ABS)dm_pattern_feature_axial
dm_pattern_feature_axial_parallel
dm_pattern_feature_axial_radial

Fig. C.3

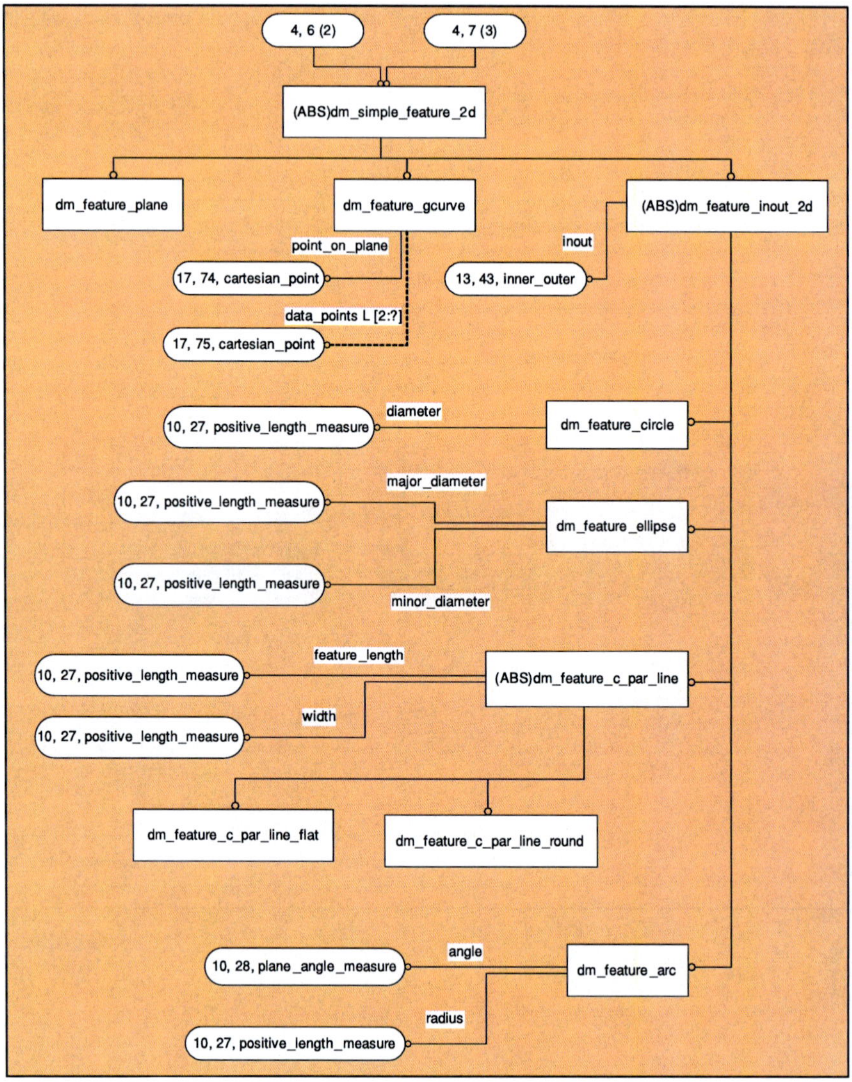

4, 6 (2)
4, 7 (3)
(ABS)dm_simple_feature_2d
dm_feature_plane
dm_feature_gcurve
(ABS)dm_feature_inout_2d
point_on_plane
17, 74, cartesian_point
inout
13, 43, inner_outer
data_points L [2:?]
17, 75, cartesian_point
diameter
10, 27, positive_length_measure
dm_feature_circle
major_diameter
10, 27, positive_length_measure
dm_feature_ellipse
10, 27, positive_length_measure
minor_diameter
feature_length
10, 27, positive_length_measure
(ABS)dm_feature_c_par_line
width
10, 27, positive_length_measure
dm_feature_c_par_line_flat
dm_feature_c_par_line_round
angle
10, 28, plane_angle_measure
dm_feature_arc
radius
10, 27, positive_length_measure

Fig. C.4

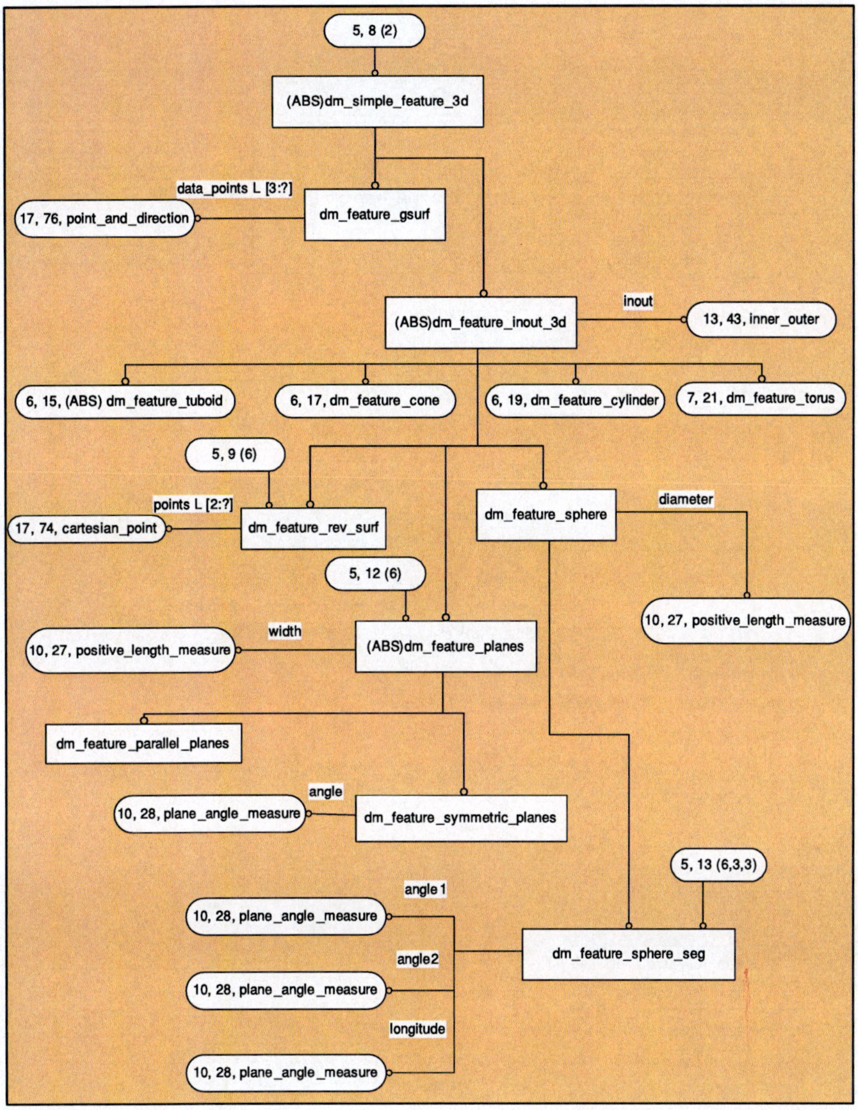
5, 8 (2)
(ABS)dm_simple_feature_3d
data_points L [3:?]
17, 76, point_and_direction
dm_feature_gsurf
(ABS)dm_feature_inout_3d
inout
13, 43, inner_outer
6, 15, (ABS) dm_feature_tuboid
6, 17, dm_feature_cone
6, 19, dm_feature_cylinder
7, 21, dm_feature_torus
5, 9 (6)
points L [2:?]
17, 74, cartesian_point
dm_feature_rev_surf
dm_feature_sphere
diameter
10, 27, positive_length_measure
5, 12 (6)
width
10, 27, positive_length_measure
(ABS)dm_feature_planes
dm_feature_parallel_planes
angle
10, 28, plane_angle_measure
dm_feature_symmetric_planes
5, 13 (6,3,3)
angle 1
10, 28, plane_angle_measure
angle2
dm_feature_sphere_seg
10, 28, plane_angle_measure
longitude
10, 28, plane_angle_measure

Fig. C.5

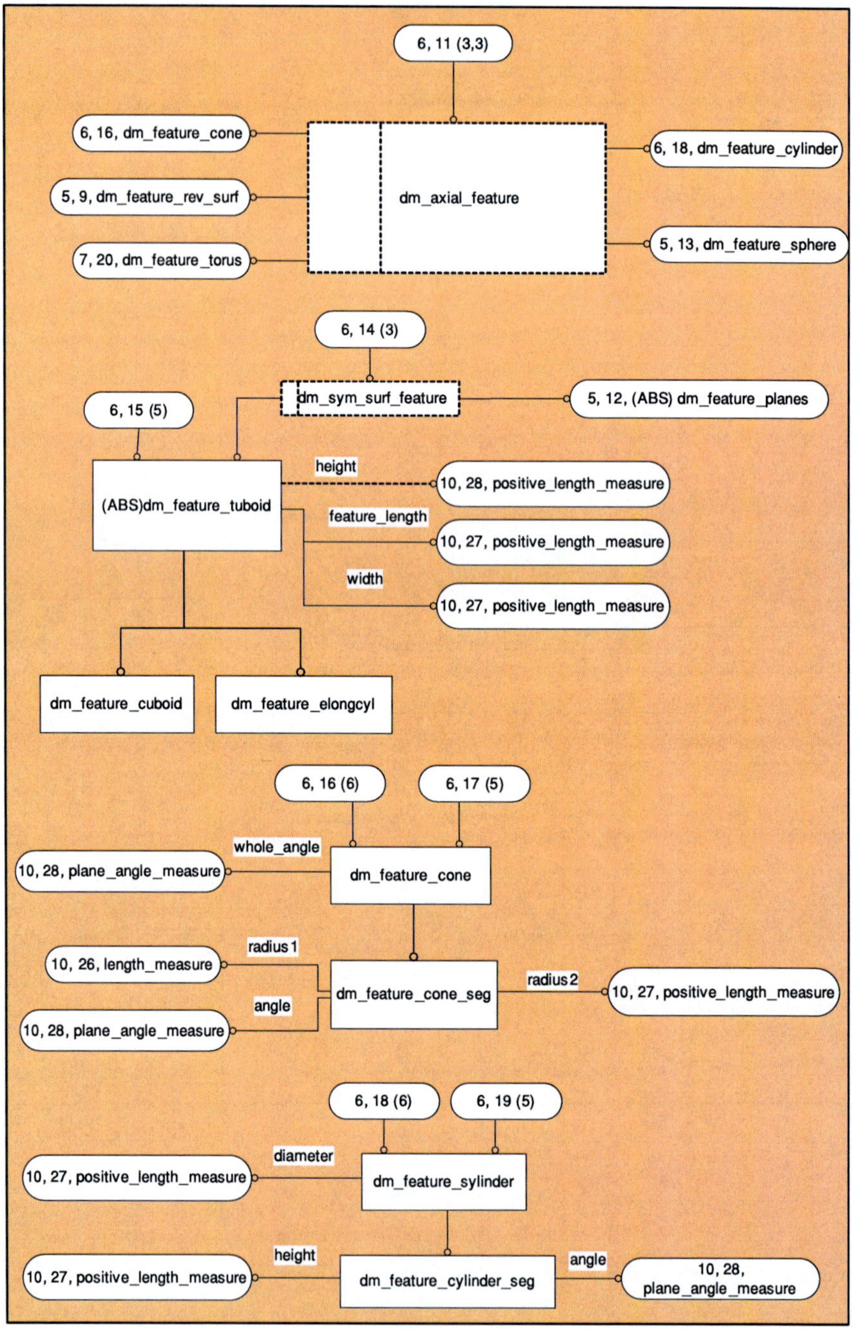
6, 11 (3,3)
6, 16, dm_feature_cone
5, 9, dm_feature_rev_surf
7, 20, dm_feature_torus
dm_axial_feature
6, 18, dm_feature_cylinder
5, 13, dm_feature_sphere
6, 14 (3)
dm_sym_surf_feature
5, 12, (ABS) dm_feature_planes
6, 15 (5)
(ABS)dm_feature_tuboid
height
10, 28, positive_length_measure
feature_length
10, 27, positive_length_measure
width
10, 27, positive_length_measure
dm_feature_cuboid
dm_feature_elongcyl
6, 16 (6)
6, 17 (5)
whole_angle
10, 28, plane_angle_measure
dm_feature_cone
radius1
10, 26, length_measure
angle
10, 28, plane_angle_measure
dm_feature_cone_seg
radius2
10, 27, positive_length_measure
6, 18 (6)
6, 19 (5)
diameter
10, 27, positive_length_measure
dm_feature_sylinder
height
10, 27, positive_length_measure
dm_feature_cylinder_seg
angle
10, 28, plane_angle_measure

Fig. C.6

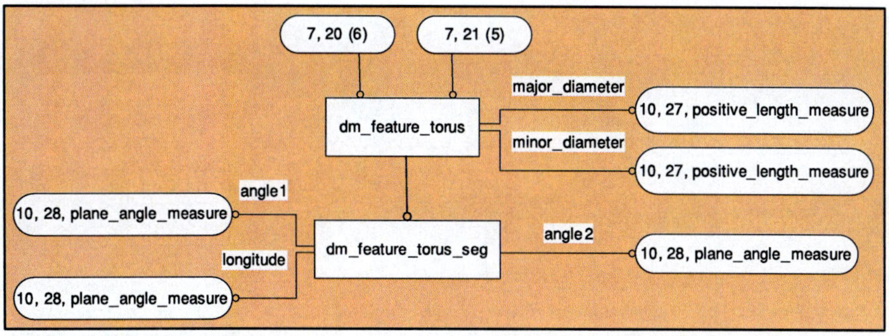
7, 20 (6)
7, 21 (5)
dm_feature_torus
major_diameter
10, 27, positive_length_measure
minor_diameter
10, 27, positive_length_measure
angle1
10, 28, plane_angle_measure
dm_feature_torus_seg
angle2
10, 28, plane_angle_measure
longitude
10, 28, plane_angle_measure

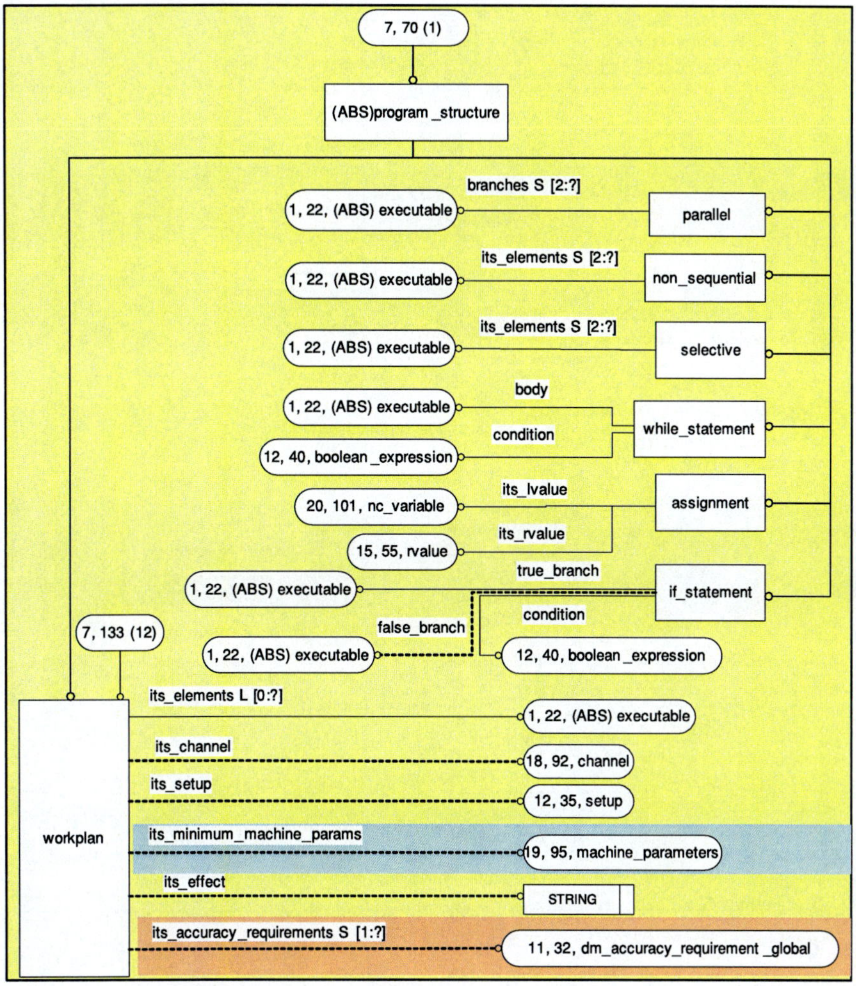
7, 70 (1)
(ABS)program_structure
branches S [2:?]
1, 22, (ABS) executable
parallel
its_elements S [2:?]
1, 22, (ABS) executable
non_sequential
its_elements S [2:?]
1, 22, (ABS) executable
selective
body
1, 22, (ABS) executable
while_statement
condition
12, 40, boolean_expression
its_lvalue
20, 101, nc_variable
assignment
its_rvalue
15, 55, rvalue
true_branch
1, 22, (ABS) executable
if_statement
condition
false_branch
1, 22, (ABS) executable
12, 40, boolean_expression
7, 133 (12)
workplan
its_elements L [0:?]
1, 22, (ABS) executable
its_channel
18, 92, channel
its_setup
12, 35, setup
its_minimum_machine_params
19, 95, machine_parameters
its_effect
STRING
its_accuracy_requirements S [1:?]
11, 32, dm_accuracy_requirement_global

Fig. C.7

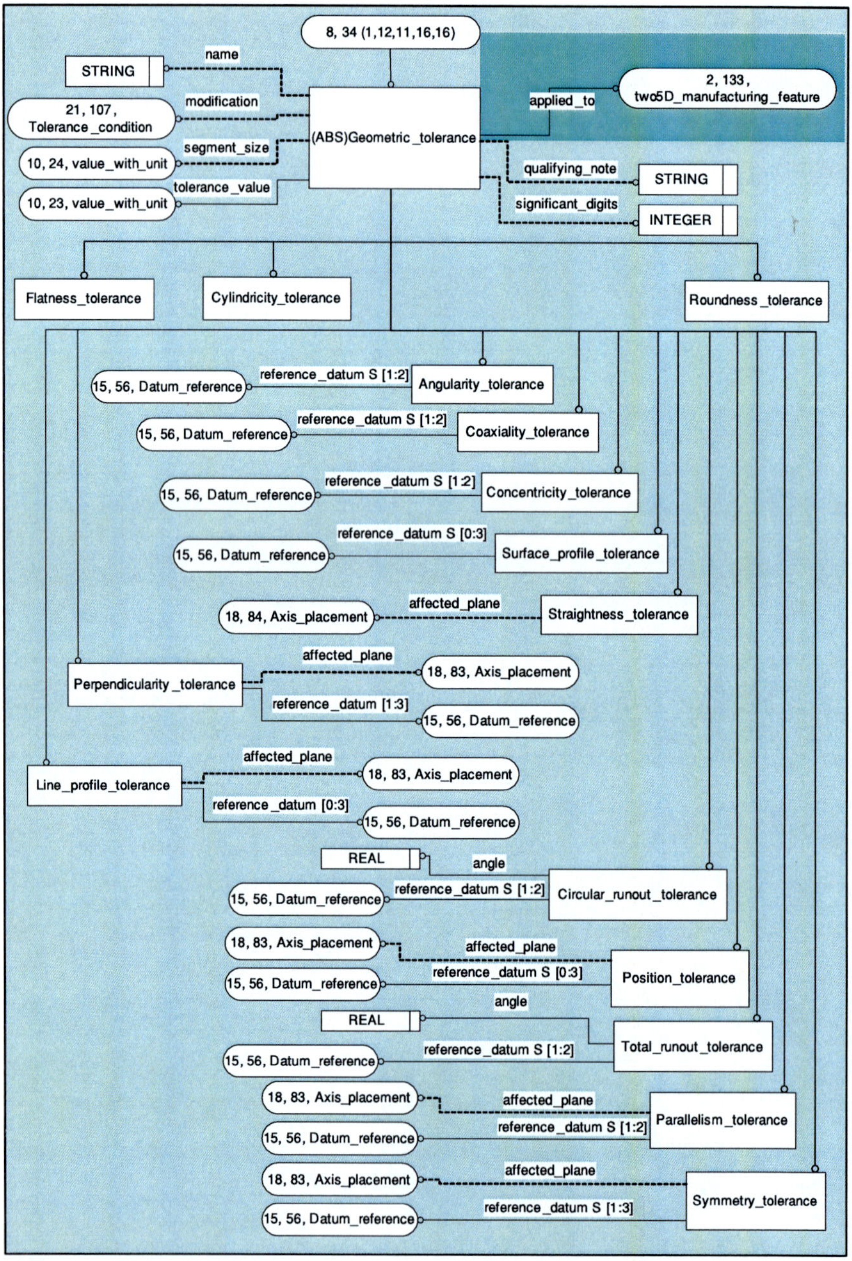
8, 34 (1,12,11,16,16)
STRING
name
21, 107, Tolerance_condition
modification
(ABS)Geometric_tolerance
applied_to
2, 133, two5D_manufacturing_feature
segment_size
10, 24, value_with_unit
tolerance_value
10, 23, value_with_unit
qualifying_note
STRING
significant_digits
INTEGER
Flatness_tolerance
Cylindricity_tolerance
Roundness_tolerance
15, 56, Datum_reference
reference_datum S [1:2]
Angularity_tolerance
15, 56, Datum_reference
reference_datum S [1:2]
Coaxiality_tolerance
15, 56, Datum_reference
reference_datum S [1:2]
Concentricity_tolerance
15, 56, Datum_reference
reference_datum S [0:3]
Surface_profile_tolerance
18, 84, Axis_placement
affected_plane
Straightness_tolerance
Perpendicularity_tolerance
affected_plane
18, 83, Axis_placement
reference_datum [1:3]
15, 56, Datum_reference
Line_profile_tolerance
affected_plane
18, 83, Axis_placement
reference_datum [0:3]
15, 56, Datum_reference
REAL
angle
15, 56, Datum_reference
reference_datum S [1:2]
Circular_runout_tolerance
18, 83, Axis_placement
affected_plane
15, 56, Datum_reference
reference_datum S [0:3]
Position_tolerance
angle
REAL
Total_runout_tolerance
15, 56, Datum_reference
reference_datum S [1:2]
18, 83, Axis_placement
affected_plane
15, 56, Datum_reference
reference_datum S [1:2]
Parallelism_tolerance
18, 83, Axis_placement
affected_plane
Symmetry_tolerance
15, 56, Datum_reference
reference_datum S [1:3]

Fig. C.8

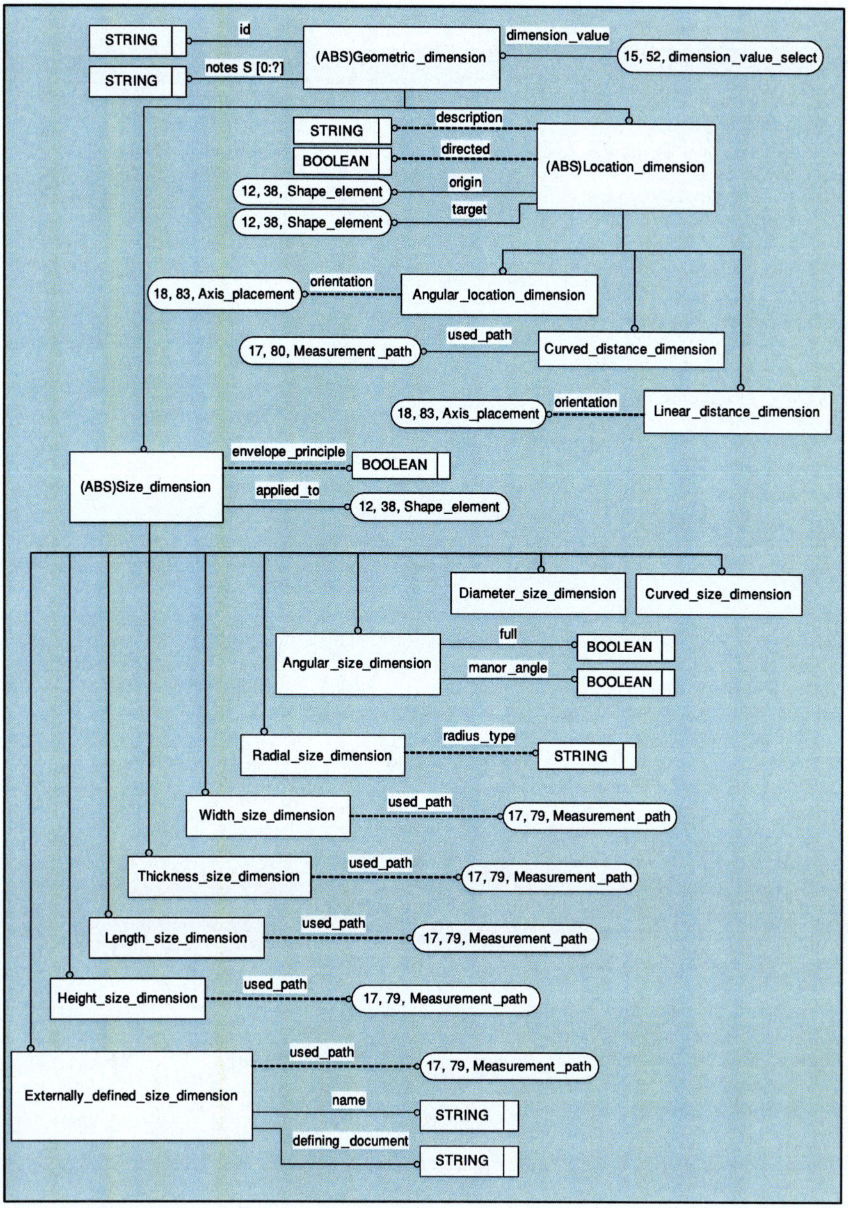
STRING
id
STRING
notes S [0:?]
(ABS)Geometric_dimension
dimension_value
15, 52, dimension_value_select
STRING
description
BOOLEAN
directed
(ABS)Location_dimension
12, 38, Shape_element
origin
12, 38, Shape_element
target
18, 83, Axis_placement
orientation
Angular_location_dimension
17, 80, Measurement_path
used_path
Curved_distance_dimension
18, 83, Axis_placement
orientation
Linear_distance_dimension
(ABS)Size_dimension
envelope_principle
BOOLEAN
applied_to
12, 38, Shape_element
Diameter_size_dimension
Curved_size_dimension
Angular_size_dimension
full
BOOLEAN
manor_angle
BOOLEAN
Radial_size_dimension
radius_type
STRING
Width_size_dimension
used_path
17, 79, Measurement_path
Thickness_size_dimension
used_path
17, 79, Measurement_path
Length_size_dimension
used_path
17, 79, Measurement_path
Height_size_dimension
used_path
17, 79, Measurement_path
Externally_defined_size_dimension
used_path
17, 79, Measurement_path
name
STRING
defining_document
STRING

Fig. C.9

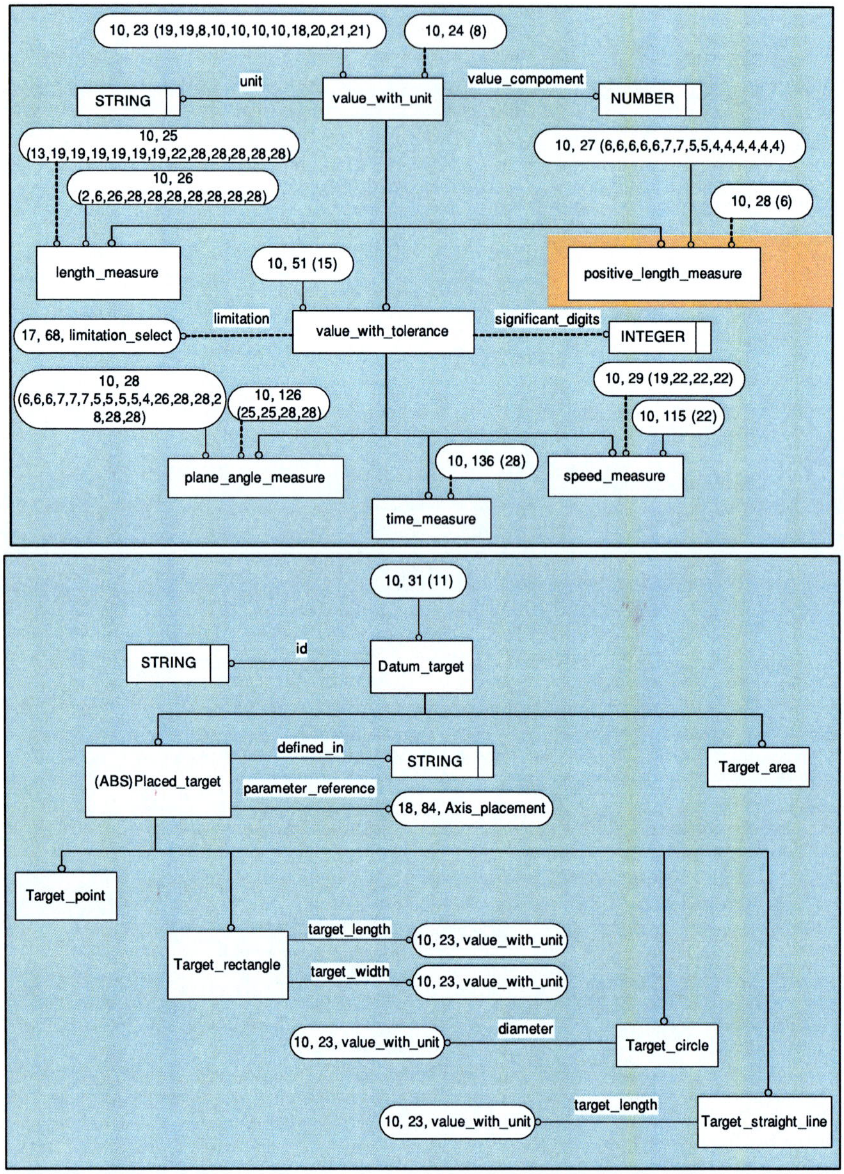
10, 23 (19,19,8,10,10,10,10,18,20,21,21)
10, 24 (8)
unit
STRING
value_with_unit
value_compoment
NUMBER
10, 25 (13,19,19,19,19,19,19,22,28,28,28,28,28)
10, 26 (2,6,26,28,28,28,28,28,28,28)
10, 27 (6,6,6,6,6,7,7,5,5,4,4,4,4,4,4)
10, 28 (6)
length_measure
10, 51 (15)
positive_length_measure
limitation
17, 68, limitation_select
value_with_tolerance
significant_digits
INTEGER
10, 28 (6,6,6,7,7,7,5,5,5,5,4,26,28,28,28,28,28)
10, 126 (25,25,28,28)
10, 29 (19,22,22,22)
10, 115 (22)
10, 136 (28)
plane_angle_measure
speed_measure
time_measure
10, 31 (11)
STRING
id
Datum_target
defined_in
STRING
(ABS)Placed_target
parameter_reference
18, 84, Axis_placement
Target_area
Target_point
target_length
10, 23, value_with_unit
Target_rectangle
target_width
10, 23, value_with_unit
10, 23, value_with_unit
diameter
Target_circle
target_length
10, 23, value_with_unit
Target_straight_line

Fig. C.10

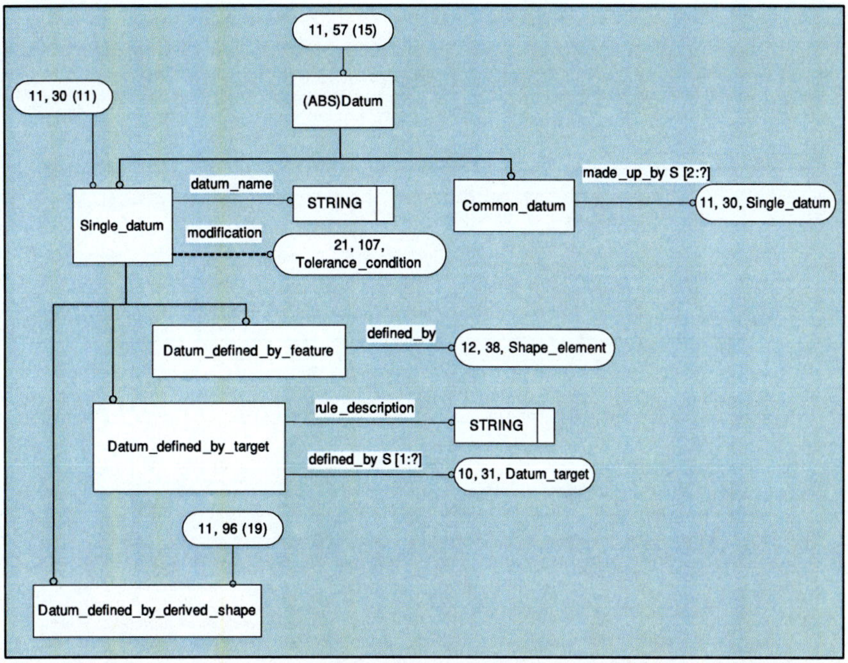
11, 57 (15)
(ABS)Datum
11, 30 (11)
Single_datum
datum_name
STRING
Common_datum
made_up_by S [2:?]
11, 30, Single_datum
modification
21, 107, Tolerance_condition
Datum_defined_by_feature
defined_by
12, 38, Shape_element
Datum_defined_by_target
rule_description
STRING
defined_by S [1:?]
10, 31, Datum_target
11, 96 (19)
Datum_defined_by_derived_shape

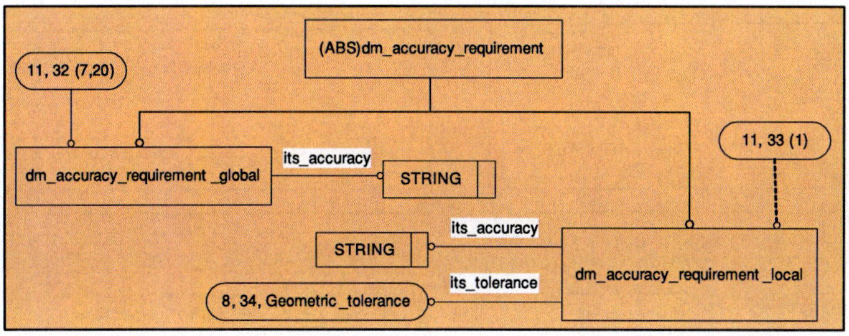
(ABS)dm_accuracy_requirement
11, 32 (7,20)
dm_accuracy_requirement _global
its_accuracy
STRING
11, 33 (1)
STRING
its_accuracy
dm_accuracy_requirement _local
its_tolerance
8, 34, Geometric_tolerance

Fig. C.11

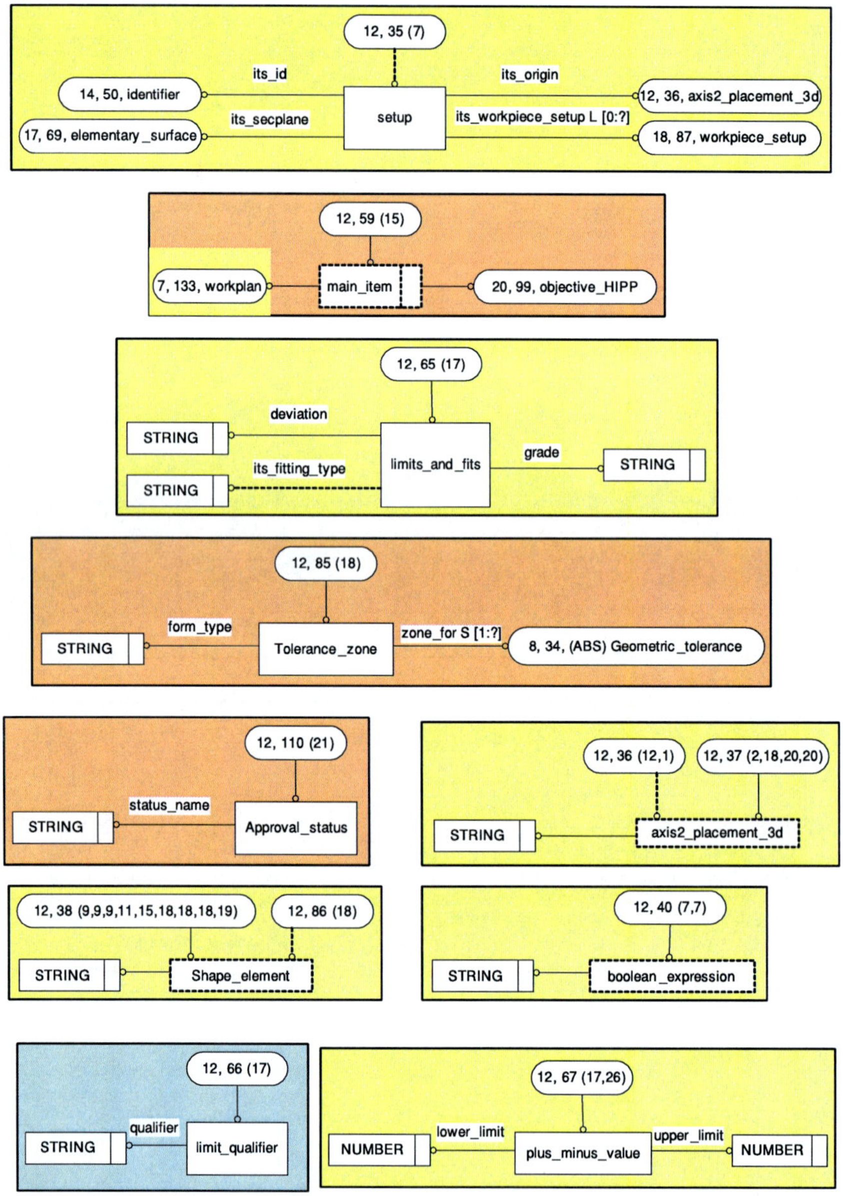
12, 35 (7)
its_id
its_origin
14, 50, identifier
12, 36, axis2_placement_3d
its_secplane
setup
its_workpiece_setup L [0:?]
17, 69, elementary_surface
18, 87, workpiece_setup
12, 59 (15)
7, 133, workplan
main_item
20, 99, objective_HIPP
12, 65 (17)
deviation
STRING
limits_and_fits
grade
STRING
its_fitting_type
STRING
12, 85 (18)
form_type
STRING
Tolerance_zone
zone_for S [1:?]
8, 34, (ABS) Geometric_tolerance
12, 110 (21)
status_name
STRING
Approval_status
12, 36 (12,1)
12, 37 (2,18,20,20)
STRING
axis2_placement_3d
12, 38 (9,9,9,11,15,18,18,18,19)
12, 86 (18)
STRING
Shape_element
12, 40 (7,7)
STRING
boolean_expression
12, 66 (17)
qualifier
STRING
limit_qualifier
12, 67 (17,26)
lower_limit
upper_limit
NUMBER
plus_minus_value
NUMBER

Fig. C.12

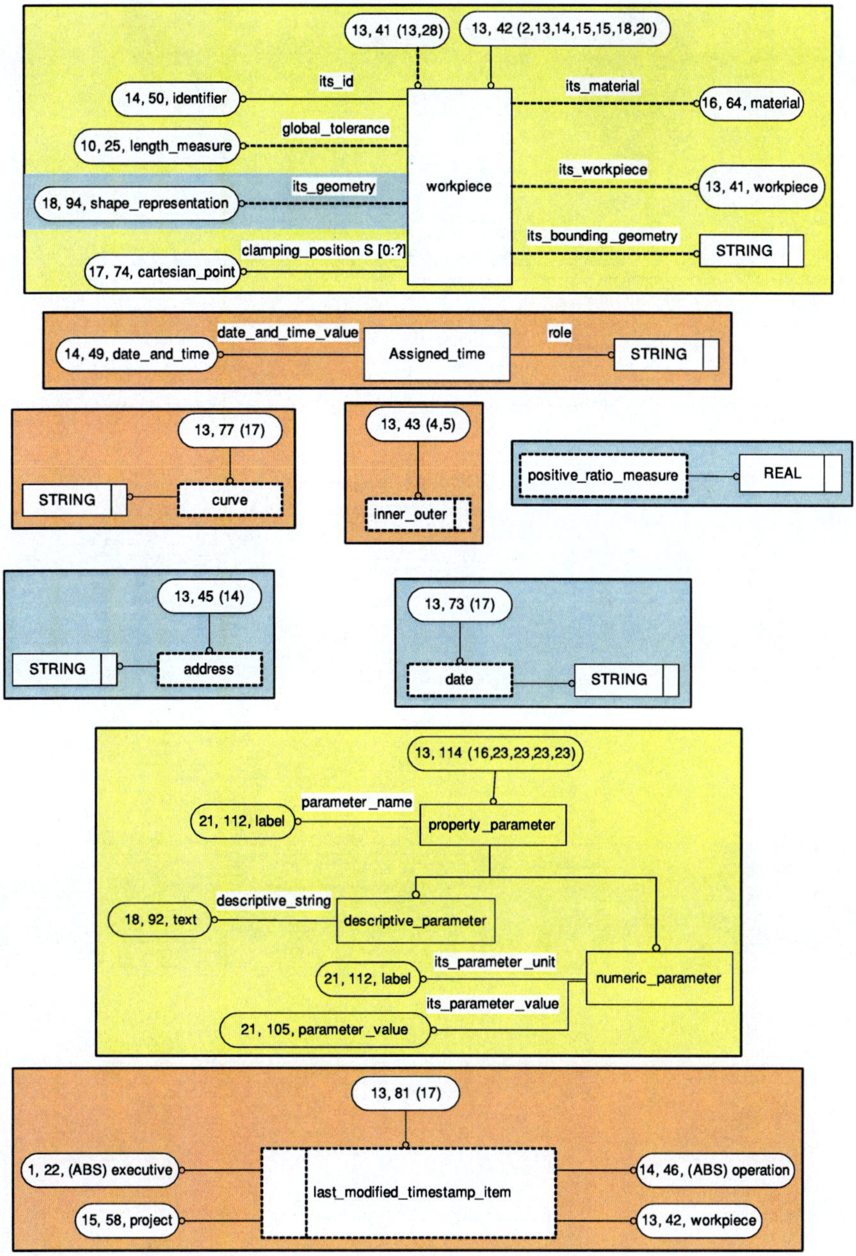
13, 41 (13,28)
13, 42 (2,13,14,15,15,18,20)
its_id
14, 50, identifier
global_tolerance
10, 25, length_measure
its_geometry
18, 94, shape_representation
clamping_position S [0:?]
17, 74, cartesian_point
workpiece
its_material
16, 64, material
its_workpiece
13, 41, workpiece
its_bounding_geometry
STRING
date_and_time_value
14, 49, date_and_time
Assigned_time
role
STRING
13, 77 (17)
STRING
curve
13, 43 (4,5)
inner_outer
positive_ratio_measure
REAL
13, 45 (14)
STRING
address
13, 73 (17)
date
STRING
13, 114 (16,23,23,23,23)
parameter_name
21, 112, label
property_parameter
descriptive_string
18, 92, text
descriptive_parameter
its_parameter_unit
21, 112, label
numeric_parameter
its_parameter_value
21, 105, parameter_value
13, 81 (17)
1, 22, (ABS) executive
15, 58, project
last_modified_timestamp_item
14, 46, (ABS) operation
13, 42, workpiece

Fig. C.13

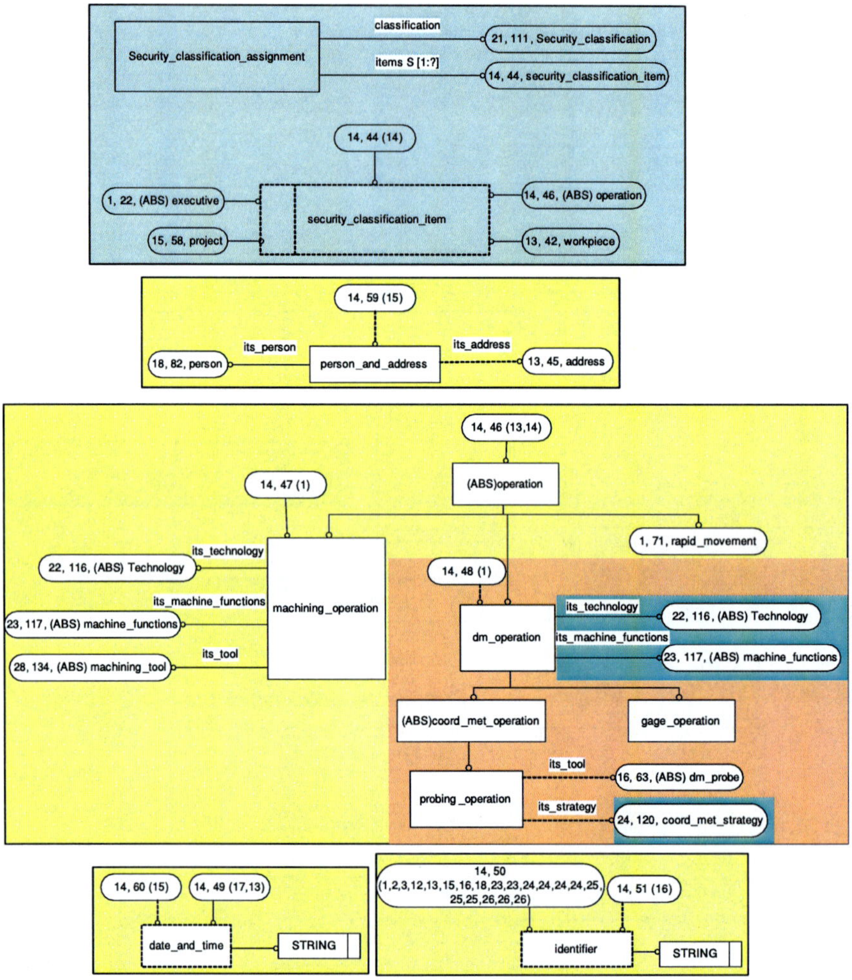

Security_classification_assignment
classification
21, 111, Security_classification
items S [1:?]
14, 44, security_classification_item
14, 44 (14)
1, 22, (ABS) executive
15, 58, project
security_classification_item
14, 46, (ABS) operation
13, 42, workpiece
14, 59 (15)
its_person
18, 82, person
person_and_address
its_address
13, 45, address
14, 46 (13,14)
(ABS)operation
14, 47 (1)
1, 71, rapid_movement
its_technology
22, 116, (ABS) Technology
its_machine_functions
23, 117, (ABS) machine_functions
its_tool
28, 134, (ABS) machining_tool
machining_operation
14, 48 (1)
dm_operation
its_technology
22, 116, (ABS) Technology
its_machine_functions
23, 117, (ABS) machine_functions
(ABS)coord_met_operation
gage_operation
probing_operation
its_tool
16, 63, (ABS) dm_probe
its_strategy
24, 120, coord_met_strategy
14, 60 (15)
14, 49 (17,13)
date_and_time
STRING
14, 50
(1,2,3,12,13,15,16,18,23,23,24,24,24,24,25,
25,25,26,26,26)
14, 51 (16)
identifier
STRING

Fig. C.14

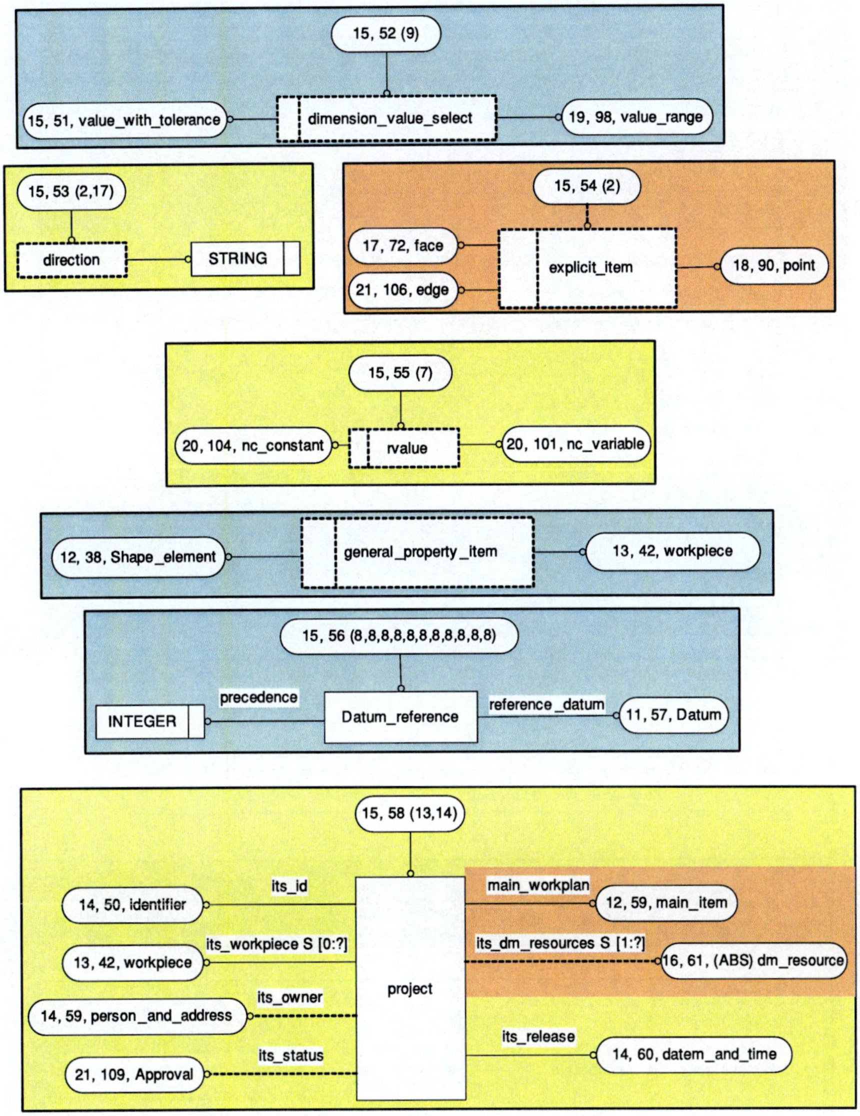

15, 52 (9)
15, 51, value_with_tolerance
dimension_value_select
19, 98, value_range
15, 53 (2,17)
direction
STRING
15, 54 (2)
17, 72, face
21, 106, edge
explicit_item
18, 90, point
15, 55 (7)
20, 104, nc_constant
rvalue
20, 101, nc_variable
12, 38, Shape_element
general_property_item
13, 42, workpiece
15, 56 (8,8,8,8,8,8,8,8,8,8,8)
INTEGER
precedence
Datum_reference
reference_datum
11, 57, Datum
15, 58 (13,14)
14, 50, identifier
its_id
13, 42, workpiece
its_workpiece S [0:?]
14, 59, person_and_address
its_owner
21, 109, Approval
its_status
project
main_workplan
12, 59, main_item
its_dm_resources S [1:?]
16, 61, (ABS) dm_resource
its_release
14, 60, datem_and_time

Fig. C.15

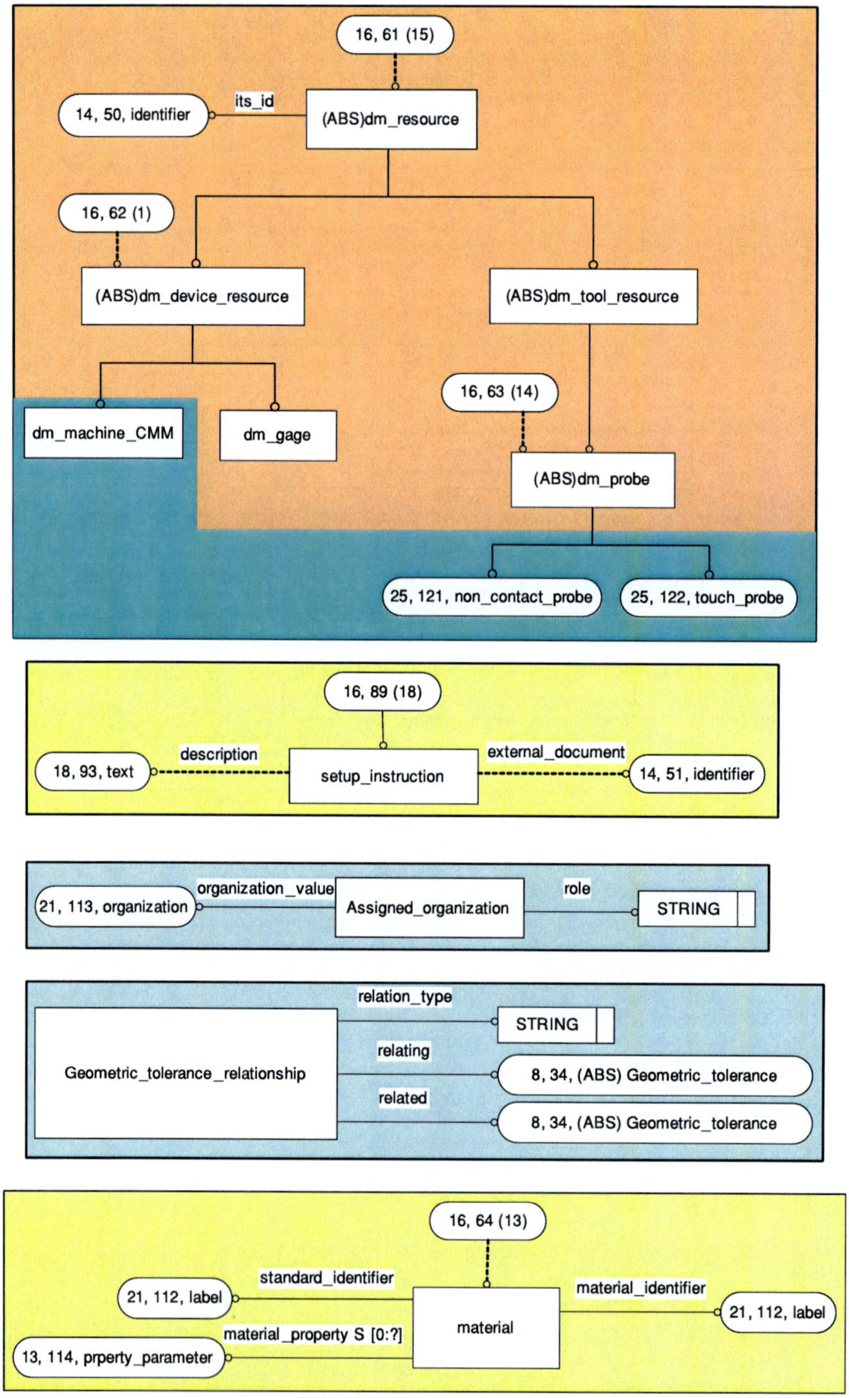

16, 61 (15)
14, 50, identifier
its_id
(ABS)dm_resource
16, 62 (1)
(ABS)dm_device_resource
(ABS)dm_tool_resource
dm_machine_CMM
dm_gage
16, 63 (14)
(ABS)dm_probe
25, 121, non_contact_probe
25, 122, touch_probe
16, 89 (18)
18, 93, text
description
setup_instruction
external_document
14, 51, identifier
21, 113, organization
organization_value
Assigned_organization
role
STRING
relation_type
STRING
Geometric_tolerance_relationship
relating
8, 34, (ABS) Geometric_tolerance
related
8, 34, (ABS) Geometric_tolerance
16, 64 (13)
21, 112, label
standard_identifier
material
material_identifier
21, 112, label
material_property S [0:?]
13, 114, prperty_parameter

Fig. C.16

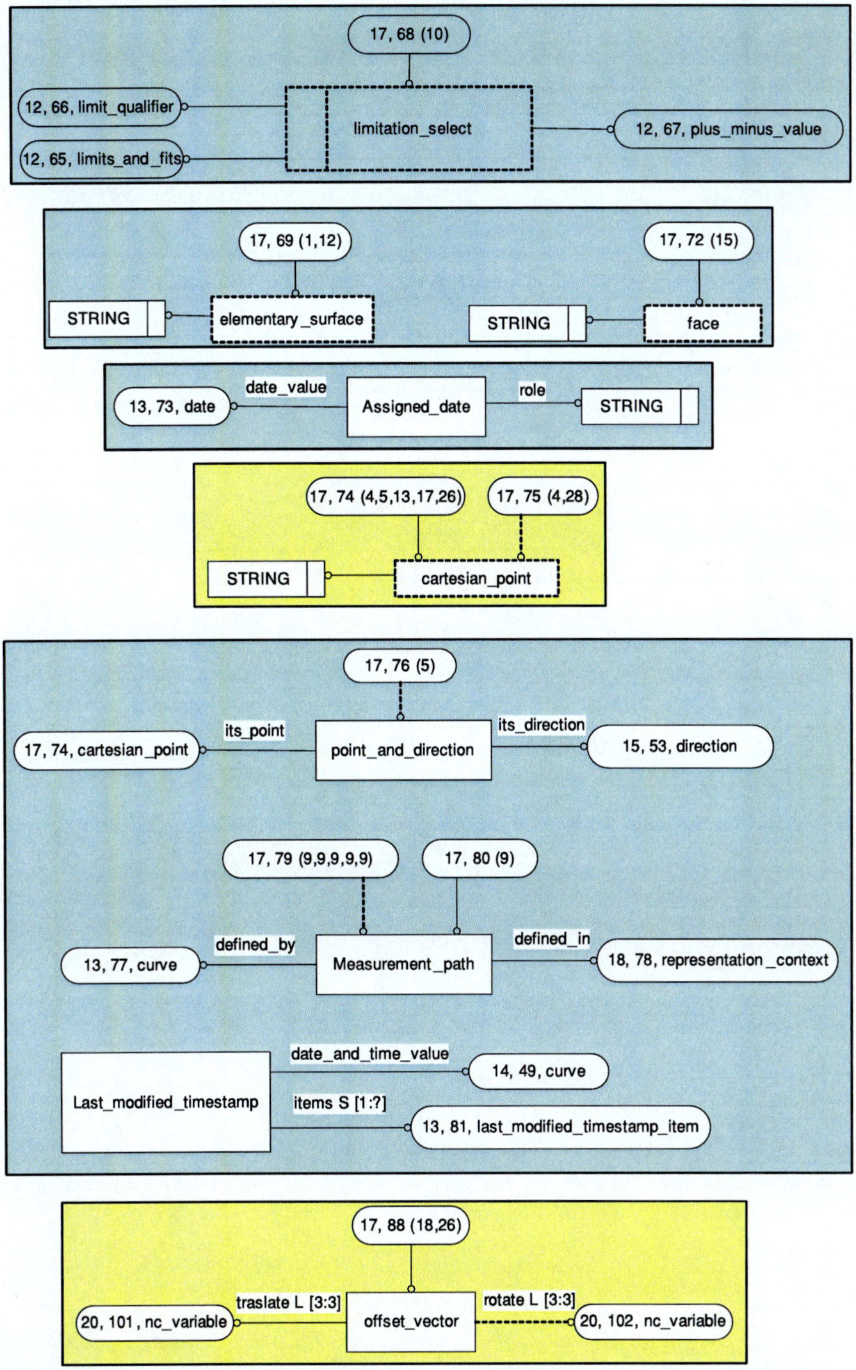
17, 68 (10)
12, 66, limit_qualifier
12, 65, limits_and_fits
limitation_select
12, 67, plus_minus_value
17, 69 (1,12)
17, 72 (15)
STRING
elementary_surface
STRING
face
date_value
role
13, 73, date
Assigned_date
STRING
17, 74 (4,5,13,17,26)
17, 75 (4,28)
STRING
cartesian_point
17, 76 (5)
its_point
its_direction
17, 74, cartesian_point
point_and_direction
15, 53, direction
17, 79 (9,9,9,9,9)
17, 80 (9)
defined_by
defined_in
13, 77, curve
Measurement_path
18, 78, representation_context
date_and_time_value
Last_modified_timestamp
14, 49, curve
items S [1:?]
13, 81, last_modified_timestamp_item
17, 88 (18,26)
traslate L [3:3]
rotate L [3:3]
20, 101, nc_variable
offset_vector
20, 102, nc_variable

Fig. C.17

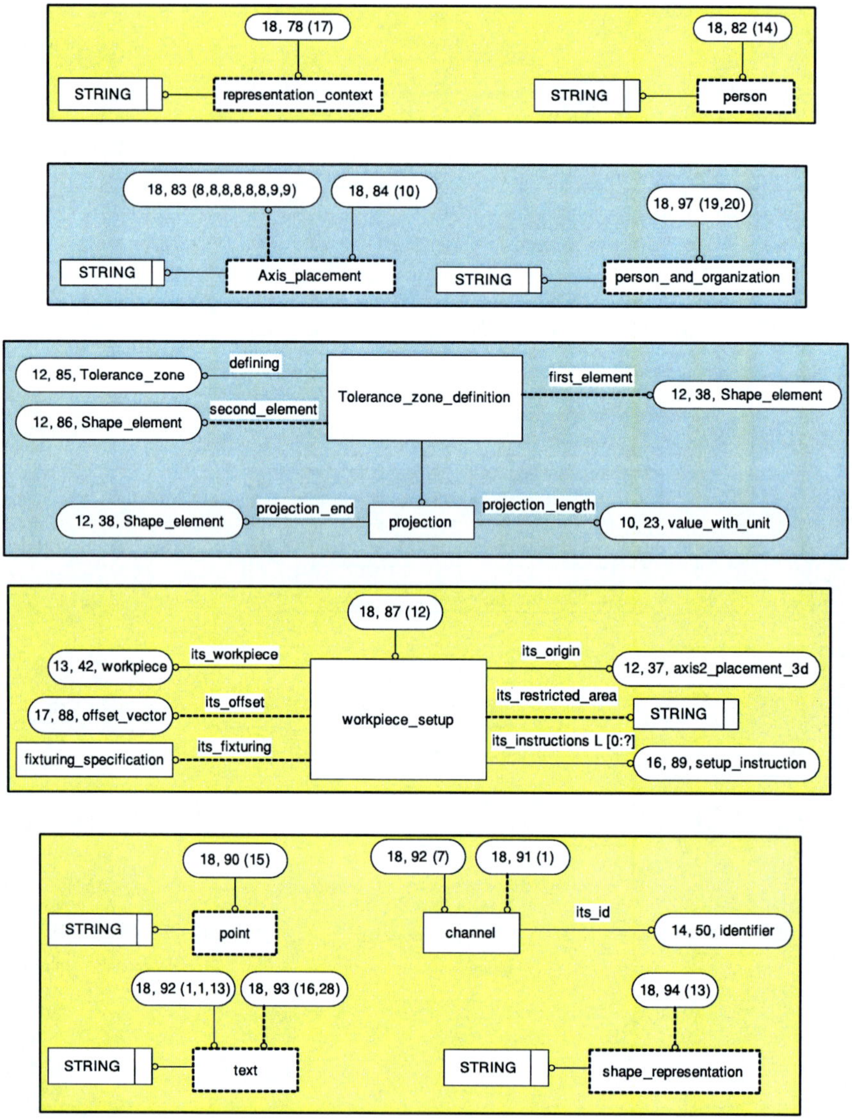
18, 78 (17)
18, 82 (14)
STRING
representation_context
STRING
person
18, 83 (8,8,8,8,8,8,9,9)
18, 84 (10)
18, 97 (19,20)
STRING
Axis_placement
STRING
person_and_organization
12, 85, Tolerance_zone
defining
Tolerance_zone_definition
first_element
12, 38, Shape_element
12, 86, Shape_element
second_element
12, 38, Shape_element
projection_end
projection
projection_length
10, 23, value_with_unit
18, 87 (12)
13, 42, workpiece
its_workpiece
its_origin
12, 37, axis2_placement_3d
its_restricted_area
17, 88, offset_vector
its_offset
workpiece_setup
STRING
its_instructions L [0:?]
fixturing_specification
its_fixturing
16, 89, setup_instruction
18, 90 (15)
18, 92 (7)
18, 91 (1)
STRING
point
channel
its_id
14, 50, identifier
18, 92 (1,1,13)
18, 93 (16,28)
18, 94 (13)
STRING
text
STRING
shape_representation

Fig. C.18

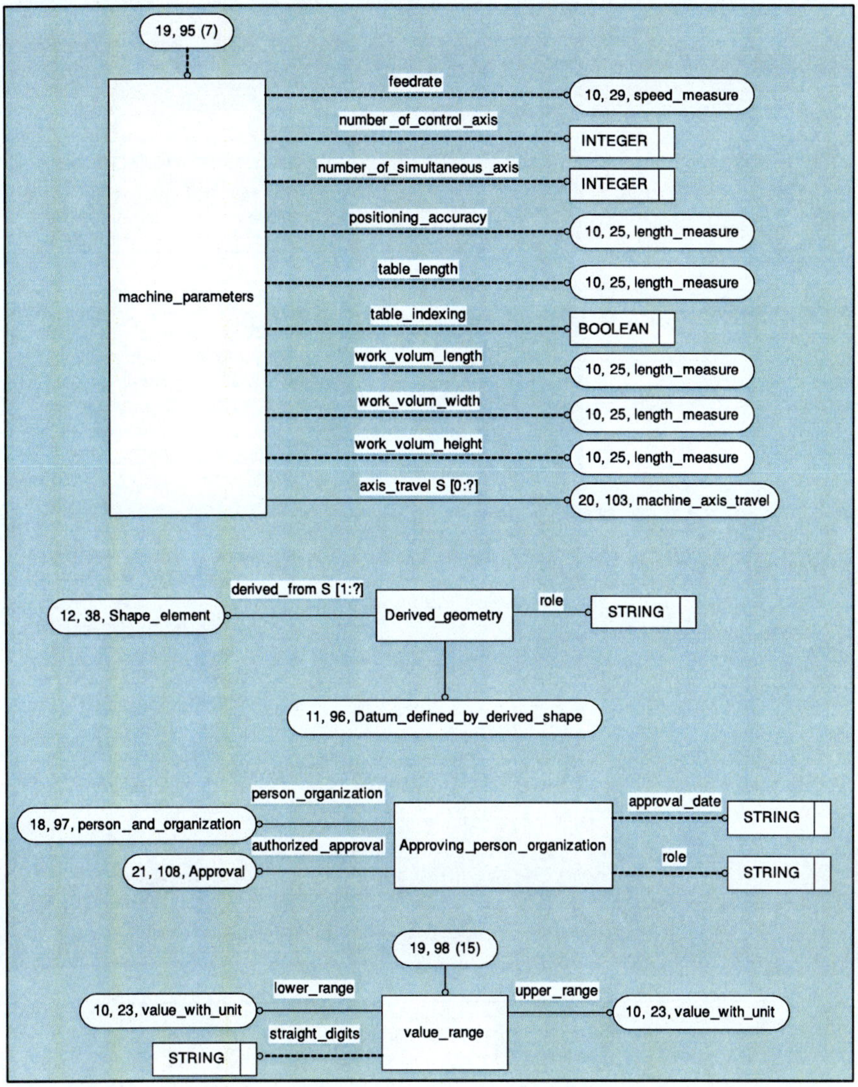
19, 95 (7)
machine_parameters
feedrate
10, 29, speed_measure
number_of_control_axis
INTEGER
number_of_simultaneous_axis
INTEGER
positioning_accuracy
10, 25, length_measure
table_length
10, 25, length_measure
table_indexing
BOOLEAN
work_volum_length
10, 25, length_measure
work_volum_width
10, 25, length_measure
work_volum_height
10, 25, length_measure
axis_travel S [0:?]
20, 103, machine_axis_travel
derived_from S [1:?]
12, 38, Shape_element
Derived_geometry
role
STRING
11, 96, Datum_defined_by_derived_shape
person_organization
18, 97, person_and_organization
authorized_approval
21, 108, Approval
Approving_person_organization
approval_date
STRING
role
STRING
19, 98 (15)
lower_range
10, 23, value_with_unit
straight_digits
STRING
value_range
upper_range
10, 23, value_with_unit

Fig. C.19

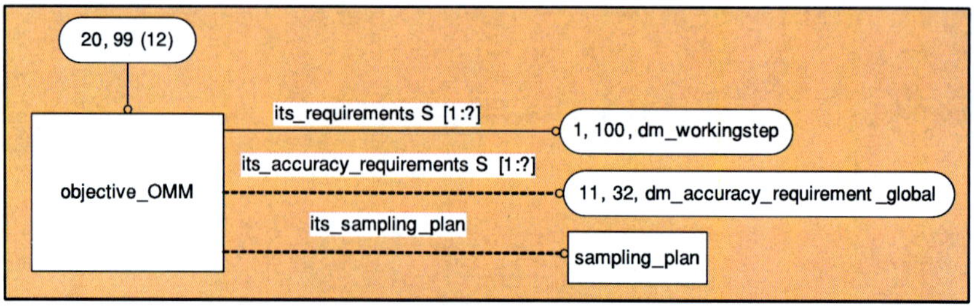
20, 99 (12)
objective_OMM
its_requirements S [1:?]
1, 100, dm_workingstep
its_accuracy_requirements S [1:?]
11, 32, dm_accuracy_requirement_global
its_sampling_plan
sampling_plan

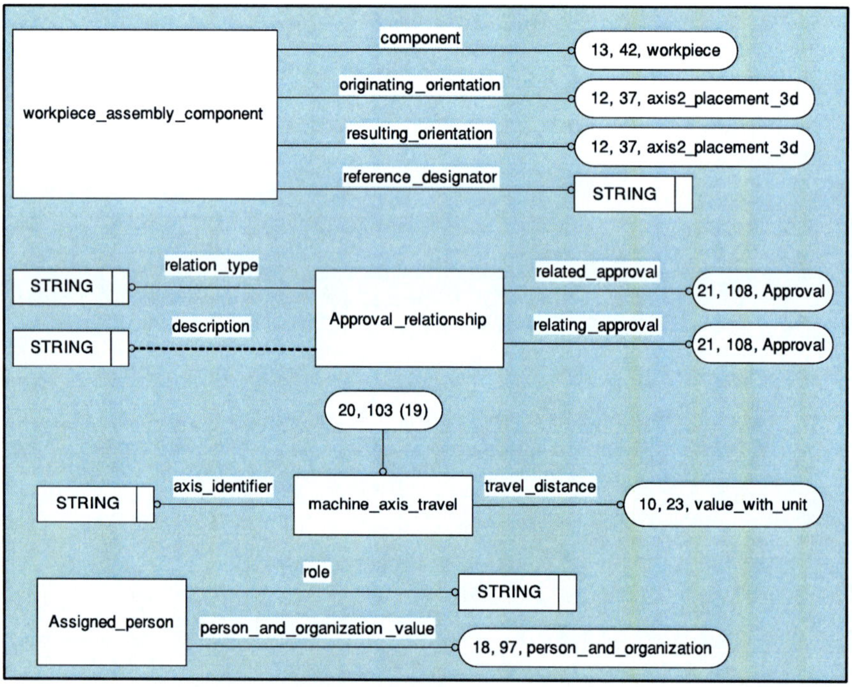
workpiece_assembly_component
component
13, 42, workpiece
originating_orientation
12, 37, axis2_placement_3d
resulting_orientation
12, 37, axis2_placement_3d
reference_designator
STRING
STRING
relation_type
Approval_relationship
related_approval
21, 108, Approval
STRING
description
relating_approval
21, 108, Approval
20, 103 (19)
STRING
axis_identifier
machine_axis_travel
travel_distance
10, 23, value_with_unit
Assigned_person
role
STRING
person_and_organization_value
18, 97, person_and_organization

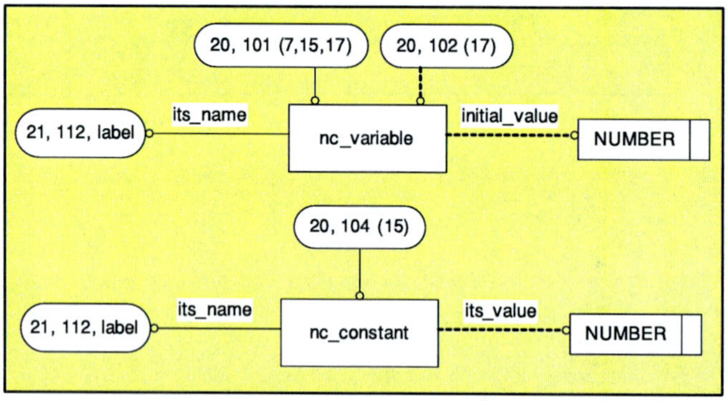
20, 101 (7,15,17)
20, 102 (17)
21, 112, label
its_name
nc_variable
initial_value
NUMBER
20, 104 (15)
21, 112, label
its_name
nc_constant
its_value
NUMBER

Fig. C.20

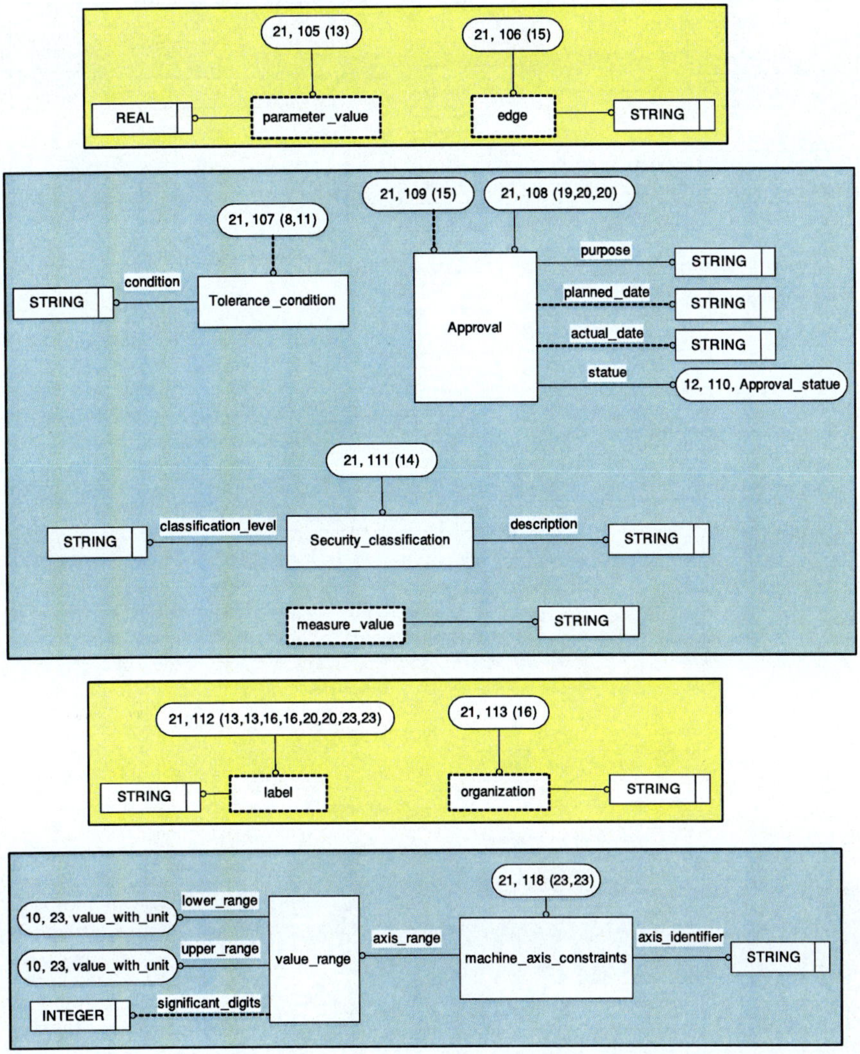
21, 105 (13)
21, 106 (15)
REAL
parameter_value
edge
STRING
21, 109 (15)
21, 108 (19,20,20)
21, 107 (8,11)
purpose
STRING
condition
STRING
Tolerance_condition
planned_date
STRING
Approval
actual_date
STRING
statue
12, 110, Approval_statue
21, 111 (14)
classification_level
STRING
Security_classification
description
STRING
measure_value
STRING
21, 112 (13,13,16,16,20,20,23,23)
21, 113 (16)
STRING
label
organization
STRING
21, 118 (23,23)
lower_range
10, 23, value_with_unit
upper_range
10, 23, value_with_unit
value_range
axis_range
machine_axis_constraints
axis_identifier
STRING
significant_digits
INTEGER

Fig. C.21

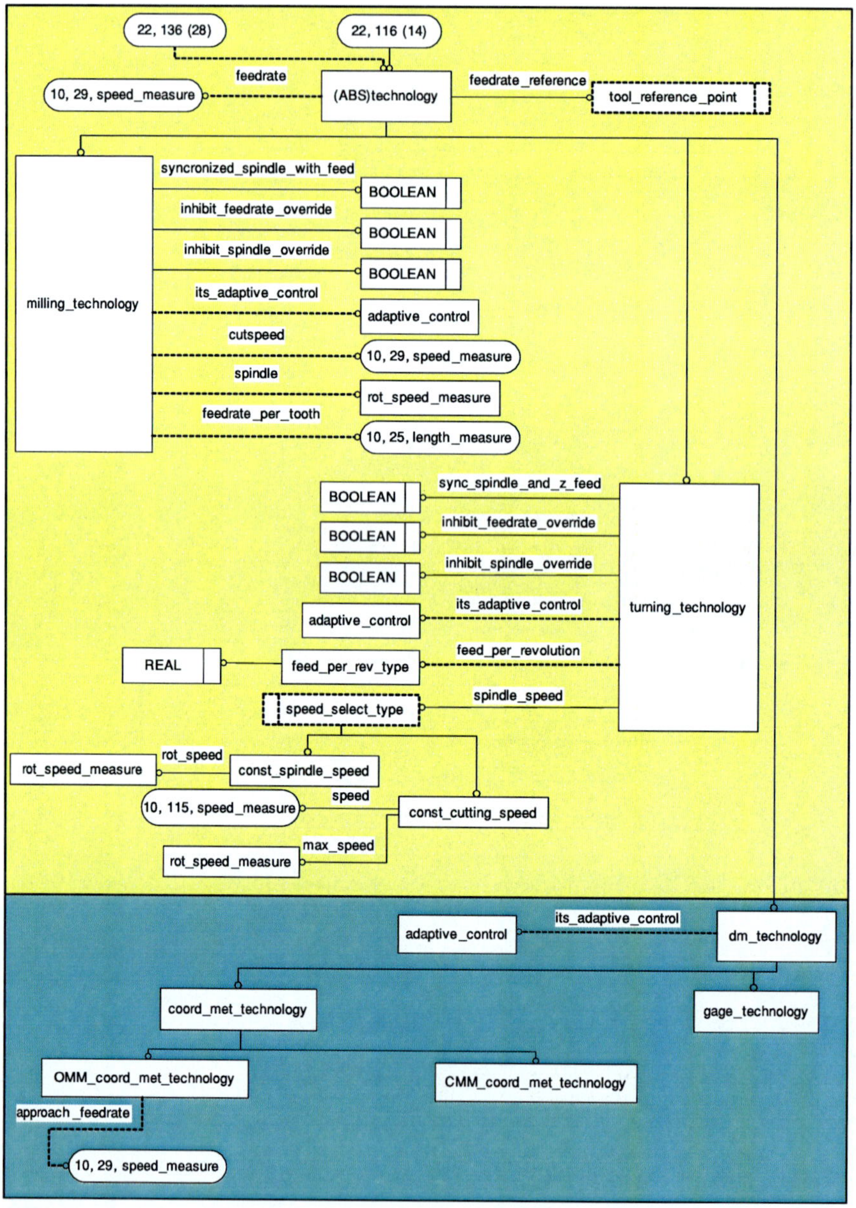
22, 136 (28)
22, 116 (14)
feedrate
10, 29, speed_measure
(ABS)technology
feedrate_reference
tool_reference_point
milling_technology
syncronized_spindle_with_feed
BOOLEAN
inhibit_feedrate_override
BOOLEAN
inhibit_spindle_override
BOOLEAN
its_adaptive_control
adaptive_control
cutspeed
10, 29, speed_measure
spindle
rot_speed_measure
feedrate_per_tooth
10, 25, length_measure
BOOLEAN
sync_spindle_and_z_feed
BOOLEAN
inhibit_feedrate_override
BOOLEAN
inhibit_spindle_override
adaptive_control
its_adaptive_control
turning_technology
REAL
feed_per_rev_type
feed_per_revolution
speed_select_type
spindle_speed
rot_speed_measure
rot_speed
const_spindle_speed
10, 115, speed_measure
speed
const_cutting_speed
rot_speed_measure
max_speed
adaptive_control
its_adaptive_control
dm_technology
coord_met_technology
gage_technology
OMM_coord_met_technology
CMM_coord_met_technology
approach_feedrate
10, 29, speed_measure

Fig. C.22

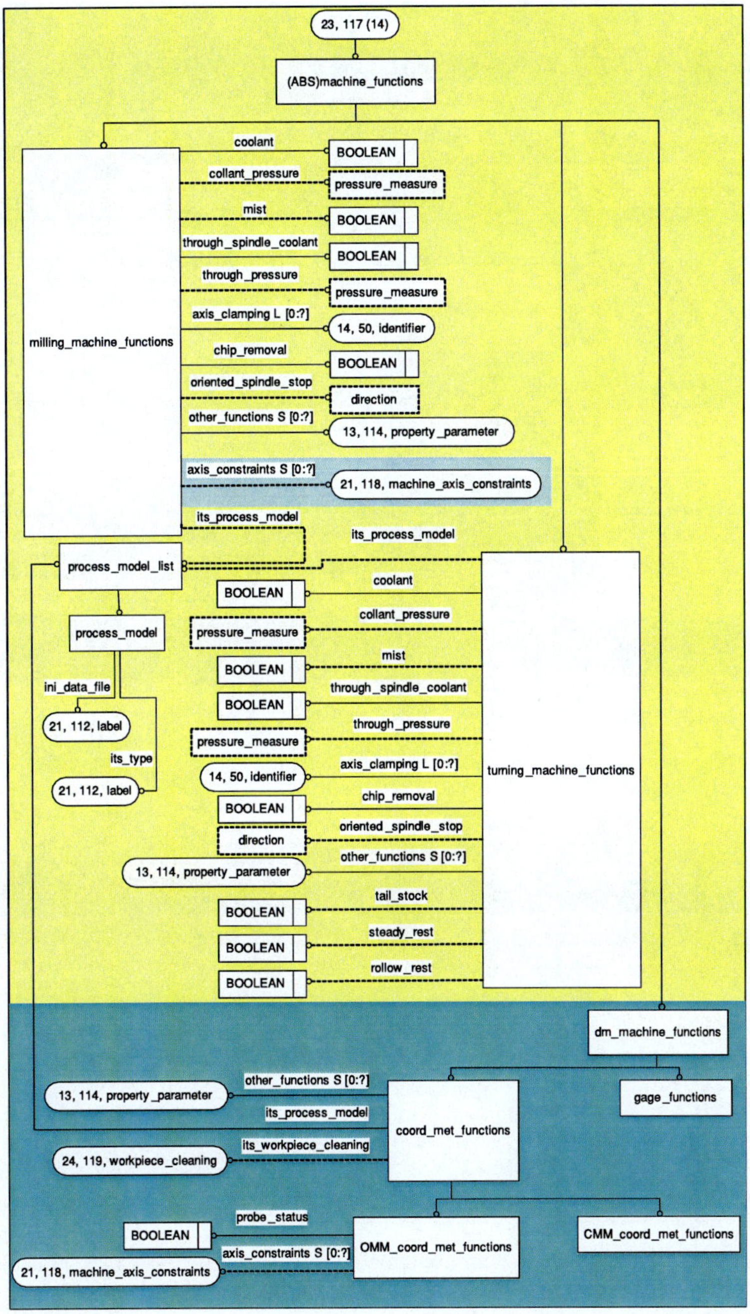
23, 117 (14)
(ABS)machine_functions
milling_machine_functions
coolant
BOOLEAN
collant_pressure
pressure_measure
mist
BOOLEAN
through_spindle_coolant
BOOLEAN
through_pressure
pressure_measure
axis_clamping L [0:?]
14, 50, identifier
chip_removal
BOOLEAN
oriented_spindle_stop
direction
other_functions S [0:?]
13, 114, property_parameter
axis_constraints S [0:?]
21, 118, machine_axis_constraints
its_process_model
process_model_list
its_process_model
process_model
ini_data_file
21, 112, label
its_type
21, 112, label
turning_machine_functions
coolant
collant_pressure
mist
through_spindle_coolant
through_pressure
axis_clamping L [0:?]
chip_removal
oriented_spindle_stop
other_functions S [0:?]
tail_stock
steady_rest
rollow_rest
BOOLEAN
pressure_measure
BOOLEAN
BOOLEAN
pressure_measure
14, 50, identifier
BOOLEAN
direction
13, 114, property_parameter
BOOLEAN
BOOLEAN
BOOLEAN
dm_machine_functions
gage_functions
coord_met_functions
other_functions S [0:?]
13, 114, property_parameter
its_process_model
its_workpiece_cleaning
24, 119, workpiece_cleaning
probe_status
BOOLEAN
OMM_coord_met_functions
axis_constraints S [0:?]
21, 118, machine_axis_constraints
CMM_coord_met_functions

Fig. C.23

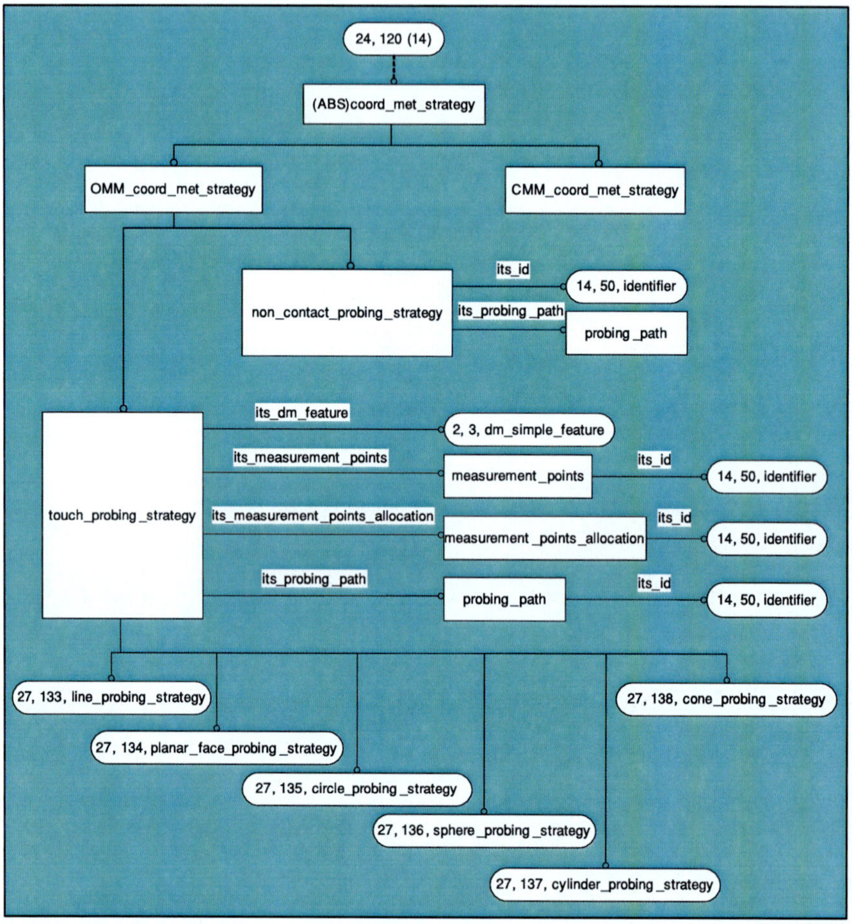
24, 120 (14)
(ABS)coord_met_strategy
OMM_coord_met_strategy
CMM_coord_met_strategy
non_contact_probing_strategy
its_id
14, 50, identifier
its_probing_path
probing_path
touch_probing_strategy
its_dm_feature
2, 3, dm_simple_feature
its_measurement_points
measurement_points
its_id
14, 50, identifier
its_measurement_points_allocation
measurement_points_allocation
its_id
14, 50, identifier
its_probing_path
probing_path
its_id
14, 50, identifier
27, 133, line_probing_strategy
27, 134, planar_face_probing_strategy
27, 135, circle_probing_strategy
27, 136, sphere_probing_strategy
27, 137, cylinder_probing_strategy
27, 138, cone_probing_strategy

Fig. C.24

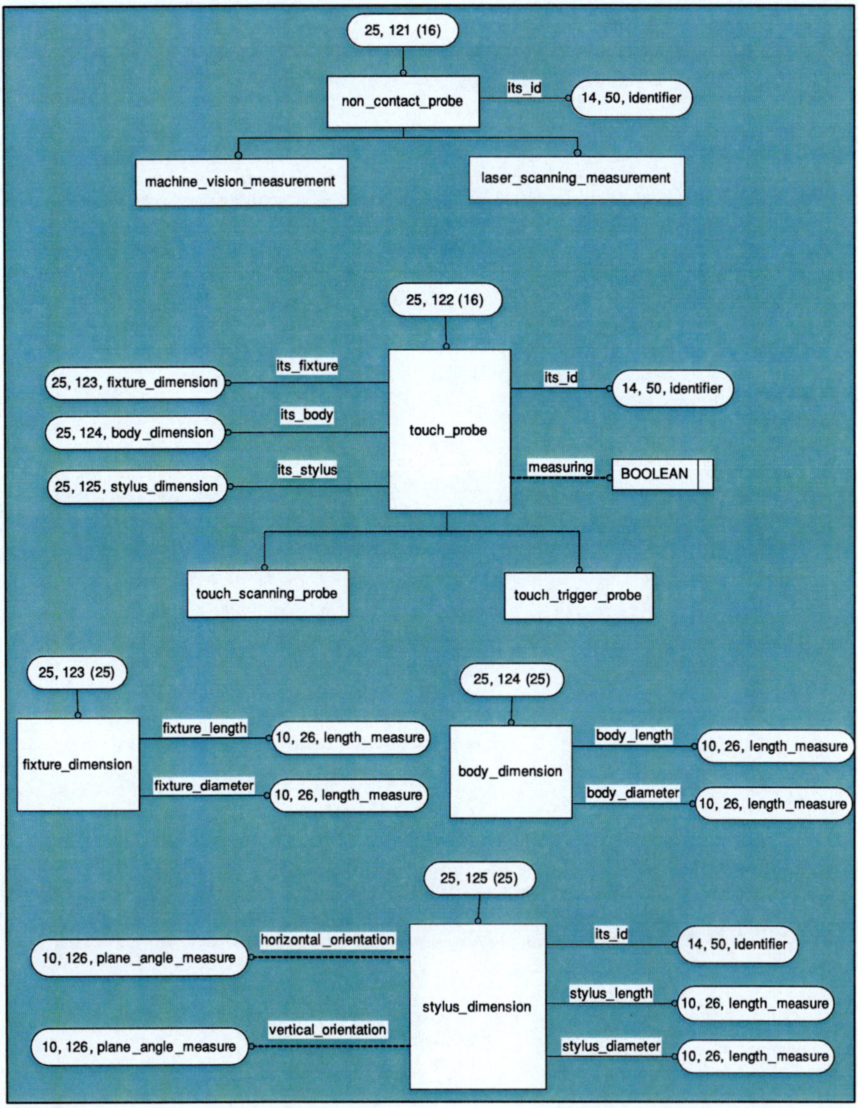
25, 121 (16)
non_contact_probe
its_id
14, 50, identifier
machine_vision_measurement
laser_scanning_measurement
25, 122 (16)
its_fixture
25, 123, fixture_dimension
its_body
25, 124, body_dimension
its_stylus
25, 125, stylus_dimension
touch_probe
its_id
14, 50, identifier
measuring
BOOLEAN
touch_scanning_probe
touch_trigger_probe
25, 123 (25)
fixture_dimension
fixture_length
10, 26, length_measure
fixture_diameter
10, 26, length_measure
25, 124 (25)
body_dimension
body_length
10, 26, length_measure
body_diameter
10, 26, length_measure
25, 125 (25)
stylus_dimension
horizontal_orientation
10, 126, plane_angle_measure
vertical_orientation
10, 126, plane_angle_measure
its_id
14, 50, identifier
stylus_length
10, 26, length_measure
stylus_diameter
10, 26, length_measure

Fig. C.25

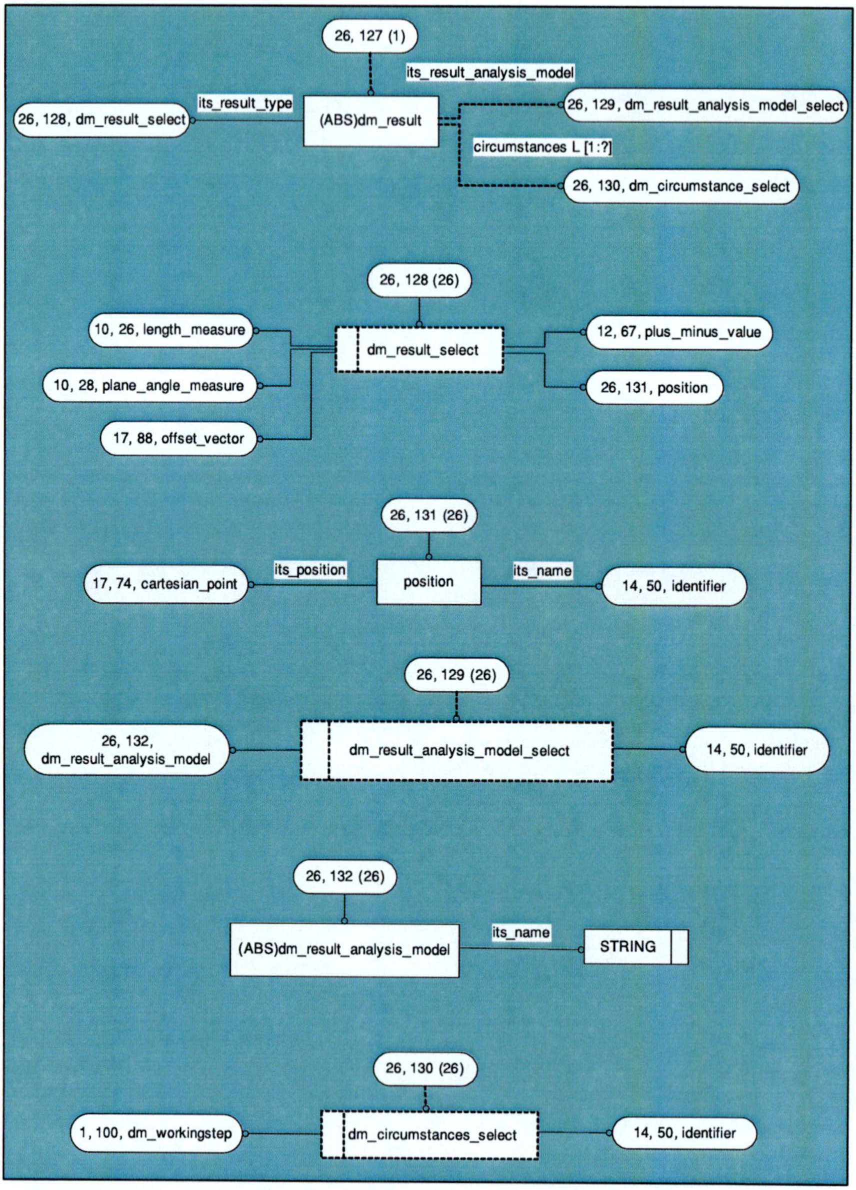

Fig. C.26

Added by AP 238	Added by HIPP	Original from ISO 14649	Added for in-process measurement

Appendix D
QMD Use Case Information

Table D.1 Inspect part or process

Use case name	Inspect part or process
Actors	Quality inspector (an individual or system that monitors quality in part or process) Quality producer (a software or system that generates an XML file from gathered quality data)
Description	The process of inspecting a part or process for quality purposes
Pre-condition	A quality requirement or criterion that has been established for a part or process in which data are to be acquired to determine conformance to some specification
Post condition	Sufficient data are collected for a specific type of quality study. The data are to be communicated to another business system
Begins when	A quality plan is initiated
Scenario (main flow)	The system allows the user to: Collect information for a variety of quality control or assurance purposes Examples include: • Collect data from various variable gauging systems to determine dimensional characteristics of part or process (variable data) • Collect data from various attribute gauging systems to determine non-conformities (attribute data) • Collect data from various go/no go gauging systems to determine non-conformances (binary data)
Scenario (sub flow)	Data collected from DME (dimensional metrology equipment) Examples include: • Digital gage (e.g., micrometer, caliper, indicator, bore gage, etc.) • CNC inspection (e.g., coordinate measuring machines, vision, laser, surface finish) • Probe sensor fixtures (e.g., LVDT gauging systems) • Attribute gage (e.g., ring gage, pin gage, thread gage, etc.) • Process sensors (e.g., temperature probes, pressure sensors, chemical concentration, etc.) Data collected through Visual Inspection. Examples include: missing holes, burrs, scratches, dents, discolorations, etc
Alternative flows	Data may be manually entered by keyboard after reading a mechanical gage
Ends when	The part or process study is concluded

Y. Zhao et al., *Information Modeling for Interoperable Dimensional Metrology*,
DOI: 10.1007/978-1-4471-2167-1, © Springer-Verlag London Limited 2011

Table D.2 Write file

Use case name	Write file
Actors	Quality producer (a software or system that generates an XML file from gathered quality data)
Description	The process of creating or producing a QMD XML data file
Pre-condition	Quality measurement data have been collected and are to be produced or transmitted in XML format
Post condition	The data are properly packaged as a QMD XML file
Begins when	A quality study is completed and is intended to be communicated
Scenario (main flow)	The system allows the user to: Generate an XML file, structure, or object in conformance with AIAG QMD XML schema The document should be well formed and carry sufficient quality information that will satisfy conformance class criteria
Scenario (sub flow)	The XML file can be manually generated by an application export function The XML file can be automatically generated by an application based on some event
Alternative flows	The XML file may be written to a message queue. In other words, a physical file is not generated, but rather the XML structure is written to an XML queue table in a relational database to be de-queued by another application or package at another time The XML file may be streamed to a Web Service for processing
Ends when	The XML file is produced

Table D.3 Publish file

Use case name	Publish file
Actors	Quality data producer (a software or system that generates an XML file from gathered quality data) Data Communicator (an individual or system that provides transport of an XML package that contains quality data)
Description	The process of publishing or transmitting an QMD XML data file
Pre-condition	A quality data XML file, structure, or object has been produced and is prepared for transmission to the consumer
Post condition	The data are properly published for consumption in another business application
Begins when	A QMD XML file is created and is intended to be communicated
Scenario (main flow)	The system allows the user to: Transmit the XML Quality Data to the consumer for processing
Scenario (sub flow)	The XML file could be emailed The XML file could be posted to a shared folder on a LAN
Alternative flows	The XML structure could be written to a database queue The XML file could be streamed to a Web service
Ends when	The XML file is transmitted or published

Table D.4 Read file

Use case name	Read file
Actors	Quality data consumer (a software or system that processes the XML package into some other business system)
Description	The process of reading or consuming a QMD XML data file
Pre-condition	A quality data XML file, structure, or object has been received by the consumer for processing
Post condition	The data are properly read or consumed in another business application
Begins when	A QMD XML file is received and is intended to be consumed
Scenario (main flow)	The system allows the user to: Parse the XML file for consumption into another business system
Scenario (sub flow)	The XML file can be manually processed by an application import function The XML file can be automatically processed by an application based on some event
Alternative flows	The XML file may be read from a message queue. In other words, a physical file has not been generated, but rather the XML structure has been written to an XML queue table in a relational database and is de-queued and processed by another application or package The XML file may be processed by a Web Service into one or more target systems The XML file may simply be transformed with a style sheet and viewed within an HTML Browser DOM viewer
Ends when	The XML file is consumed

Index

DOI: 10.1007/978-1-4471-2167-1,

In-process measurement, 10, 13–16, 18, 75, 121, 135, 155, 172, 190, 191, 227